# The Laplace Transform

David Vernon Widder

Dover Publications, Inc.
Mineola, New York

## Copyright

Copyright © 1941, 1969 by David Vernon Widder.
All rights reserved.

## Bibliographical Note

This Dover edition, first published in 2010, is an unabridged republication of the work originally published by Princeton University Press, Princeton, New Jersey, in 1941.

### Library of Congress Cataloging-in-Publication Data

Widder, D. V. (David Vernon), 1898–1990.
  The lapace transform / David Vernon Widder. — Dover ed.
    p. cm.
  Originally published: Princeton : Princeton University Press, 1941.
  ISBN-13: 978-0-486-47755-8
  ISBN-10: 0-486-47755-X
  1. Laplace transformation. I. Title.

QA432.W53 2010
515'.723—dc22

2010036235

www.doverpublications.com

## PREFACE

The material of this book is the outgrowth of a course of lectures which I have given from time to time at Harvard University on the subject of Dirichlet series and Laplace integrals. It is designed for students who have such a knowledge of analysis as might be obtained by reading the more fundamental parts of the familiar text of E. C. Titchmarsh on the theory of functions. I have taken pains to include proofs of results which such a student might not know even though they might be available elsewhere. For example, the first chapter is devoted largely to the study of the Riemann-Stieltjes integral. Although this material is in constant use by analysts it seems not to have been collected in convenient form. There are only a few instances where I have had to depart from this aim of having the book complete in itself. If he desires, the student may omit such parts without losing the fundamental ideas.

I wish to express here my gratitude to Professor G. D. Birkhoff for suggesting the writing of this book. I also wish to acknowledge my indebtedness to Professor G. H. Hardy. For it was his course of lectures delivered in Princeton University in 1928 that first aroused my interest in the type of analysis here presented. My thanks are also due to Professor Marston Morse and his committee for requesting the inclusion of the book in this series.

I gratefully acknowledge the aid provided by the Milton Fund of Harvard University for the preparation of manuscripts. Finally, I wish to extend my thanks to the John Simon Guggenheim Memorial Foundation. It was during the year in which I held a Guggenheim Fellowship that I worked out much of the material appearing in the last chapter.

D. V. W.

# CONTENTS

## Chapter I

## THE STIELTJES INTEGRAL

| SECTION | PAGE |
|---|---|
| 1. Introduction | 3 |
| 2. Stieltjes integrals | 3 |
| 3. Functions of bounded variation | 6 |
| 4. Existence of Stieltjes integrals | 7 |
| 5. Properties of Stieltjes integrals | 8 |
| 6. The Stieltjes integral as a series or a Lebesgue integral | 10 |
| 7. Further properties of Stieltjes integrals | 12 |
| 8. Normalization | 13 |
| 9. Improper Stieltjes integrals | 15 |
| 10. Laws of the mean | 16 |
| 11. Change of variable | 19 |
| 12. Variation of the indefinite integral | 20 |
| 13. Stieltjes integrals as infinite series; second method | 22 |
| 14. Further conditions of integrability | 24 |
| 15. Iterated integrals | 25 |
| 16. The selection principle | 26 |
| 17. Weak compactness | 33 |

## Chapter II

## FUNDAMENTAL FORMULAS

| | |
|---|---|
| 1. Region of convergence | 35 |
| 2. Abscissa of convergence | 38 |
| 3. Absolute convergence | 46 |
| 4. Uniform convergence | 50 |
| 5. Analytic character of the generating function | 57 |
| 6. Uniqueness of determining function | 59 |
| 7. Complex inversion formula | 63 |
| 8. Integrals of the determining function | 70 |
| 9. Summability of divergent integrals | 75 |
| 10. Inversion when the determining function belongs to $L^2$ | 80 |
| 11. Stieltjes resultant | 83 |
| 12. Classical resultant | 91 |
| 13. Order on vertical lines | 92 |
| 14. Generating function analytic at infinity | 93 |
| 15. Periodic determining function | 96 |
| 16. Relation to factorial series | 97 |

## CONTENTS

### CHAPTER III
### THE MOMENT PROBLEM

| SECTION | PAGE |
|---|---|
| 1. Statement of the problem | 100 |
| 2. Moment sequence | 101 |
| 3. An inversion operator | 107 |
| 4. Completely monotonic sequences | 108 |
| 5. Function of $L^p$ | 109 |
| 6. Bounded functions | 111 |
| 7. Hausdorff summability | 113 |
| 8. Statement of further moment problems | 125 |
| 9. The moment operator | 126 |
| 10. The Hamburger moment problem | 129 |
| 11. Positive definite sequences | 132 |
| 12. Determinant criteria | 134 |
| 13. The Stieltjes moment problem | 136 |
| 14. Moments of functions of bounded variation | 138 |
| 15. A sufficient condition for the solubility of the Stieltjes problem | 140 |
| 16. Indeterminacy of solution | 142 |

### CHAPTER IV
### ABSOLUTELY AND COMPLETELY MONOTONIC FUNCTIONS

| | |
|---|---|
| 1. Introduction | 144 |
| 2. Elementary properties of absolutely monotonic functions | 144 |
| 3. Analyticity of absolutely monotonic functions | 146 |
| 4. Bernstein's second definition | 147 |
| 5. Existence of one-sided derivatives | 149 |
| 6. Higher differences of absolutely monotonic functions | 150 |
| 7. Equivalence of Bernstein's two definitions | 151 |
| 8. Bernstein polynomials | 152 |
| 9. Definition of Grüss | 154 |
| 10. Equivalence of Bernstein and Grüss definitions | 155 |
| 11. Additional properties of absolutely monotonic functions | 156 |
| 12. Bernstein's theorem | 160 |
| 13. Alternative proof of Bernstein's theorem | 162 |
| 14. Interpolation by completely monotonic functions | 163 |
| 15. Absolutely monotonic functions with prescribed derivatives at a point | 165 |
| 16. Hankel determinants whose elements are the derivatives of an absolutely monotonic function | 167 |
| 17. Laguerre polynomials | 168 |
| 18. A linear functional | 171 |
| 19. Bernstein's theorem | 175 |
| 20. Completely convex functions | 177 |

### CHAPTER V
### TAUBERIAN THEOREMS

| | |
|---|---|
| 1. Abelian theorems for the Laplace transform | 180 |
| 2. Abelian theorems for the Stieltjes transform | 183 |

| SECTION | PAGE |
|---|---|
| 3. Tauberian theorems | 185 |
| 4. Karamata's theorem | 189 |
| 5. Tauberian theorems for the Stieltjes transform | 198 |
| 6. Fourier transforms | 202 |
| 7. Fourier transforms of functions of $L$ | 204 |
| 8. The quotient of Fourier transforms | 207 |
| 9. A special Tauberian theorem | 209 |
| 10. Pitt's form of Wiener's theorem | 210 |
| 11. Wiener's general Tauberian theorem | 212 |
| 12. Tauberian theorem for the Stieltjes integral | 213 |
| 13. One-sided Tauberian condition | 215 |
| 14. Application of Wiener's theorem to the Laplace transform | 221 |
| 15. Another application | 222 |
| 16. The prime-number theorem | 224 |
| 17. Ikehara's theorem | 233 |

## Chapter VI

## THE BILATERAL LAPLACE TRANSFORM

| | |
|---|---|
| 1. Introduction | 237 |
| 2. Region of convergence | 238 |
| 3. Integration by parts | 239 |
| 4. Abscissae of convergence | 240 |
| 5. Inversion formulas | 241 |
| 6. Uniqueness | 243 |
| 7. Summability | 244 |
| 8. Determining function belonging to $L^2$ | 245 |
| 9. The Mellin transform | 246 |
| 10. Stieltjes resultant | 248 |
| 11. Stieltjes resultant at infinity | 249 |
| 12. Stieltjes resultant completely defined | 250 |
| 13. Preliminary results | 251 |
| 14. The product of Fourier-Stieltjes transforms | 252 |
| 15. Stieltjes resultant of indefinite integrals | 256 |
| 16. Product of bilateral Laplace integrals | 257 |
| 17. Resultants in a special case | 259 |
| 18. Iterates of the Stieltjes kernel | 262 |
| 19. Representation of functions | 265 |
| 20. Kernels of positive type | 270 |
| 21. Necessary and sufficient conditions for representation | 272 |

## Chapter VII

## INVERSION AND REPRESENTATION PROBLEMS FOR THE LAPLACE TRANSFORM

| | |
|---|---|
| 1. Introduction | 276 |
| 2. Laplace's asymptotic evaluation of an integral | 277 |
| 3. Application of the Laplace method | 280 |
| 4. Uniform convergence | 283 |

| SECTION | PAGE |
|---|---|
| 5. Uniform convergence; continuation | 285 |
| 6. The inversion operator for the Laplace-Lebesgue integral | 288 |
| 7. The inversion operator for the Laplace-Stieltjes integral | 290 |
| 8. Laplace method for a new integral | 296 |
| 9. The jump operator | 298 |
| 10. The variation of the determining function | 299 |
| 11. A general representation theorem | 302 |
| 12. Determining function of bounded variation | 306 |
| 13. Modified conditions for determining functions of bounded variation | 308 |
| 14. Determining function non-decreasing | 310 |
| 15. The class $L^p$, $p > 1$ | 312 |
| 16. Determining function the integral of a bounded function | 315 |
| 17. The class $L$ | 317 |
| 18. The general Laplace-Stieltjes integral | 320 |

## Chapter VIII

## THE STIELTJES TRANSFORM

| | |
|---|---|
| 1. Introduction | 325 |
| 2. Elementary properties of the transform | 325 |
| 3. Asymptotic properties of Stieltjes transforms | 329 |
| 4. Relation to the Laplace transform | 334 |
| 5. Uniqueness | 336 |
| 6. The Stieltjes transform singular at the origin | 336 |
| 7. Complex inversion formula | 338 |
| 8. A singular integral | 341 |
| 9. The inversion operator for the Stieltjes transform with $\alpha(t)$ an integral | 345 |
| 10. The inversion operator for the Stieltjes transform in the general case | 347 |
| 11. The jump operator | 351 |
| 12. The variation of $\alpha(t)$ | 353 |
| 13. A general representation theorem | 355 |
| 14. Order conditions | 357 |
| 15. General representation theorems | 360 |
| 16. The function $\alpha(t)$ of bounded variation | 361 |
| 17. The function $\alpha(t)$ non-decreasing and bounded | 363 |
| 18. The function $\alpha(t)$ non-decreasing and unbounded | 365 |
| 19. The class $L^p$, $p > 1$ | 368 |
| 20. The function $\varphi(t)$ bounded | 372 |
| 21. The class $L$ | 374 |
| 22. The function $\alpha(t)$ of bounded variation in every finite interval | 377 |
| 23. Operational considerations | 381 |
| 24. The iterated Stieltjes transform | 383 |
| 25. Application to the Laplace transform | 384 |
| 26. Solution of an integral equation | 387 |
| 27. A related integral equation | 390 |

# The Laplace Transform

# CHAPTER I

# THE STIELTJES INTEGRAL

## 1. Introduction

In this chapter we shall collect certain results of a general nature which we shall need later for the study of the Laplace transform. We shall give proofs of the more fundamental results, but for the proofs of a few theorems, rarely used in the text, we shall merely refer the reader to a source. We begin with a definition of a Stieltjes integral.

The importance of the Stieltjes integral for our purposes is that it includes finite or infinite series as well as the Lebesgue integral. For example, the Laplace integral in classical form is

$$(1) \qquad f(s) = \int_0^\infty e^{-st} \varphi(t) \, dt.$$

A Dirichlet series has the form

$$(2) \qquad f(s) = \sum_{n=1}^\infty a_n e^{-\lambda_n s}.$$

The analogy between the two is apparent. The discrete set of exponents $\lambda_n$ in the series becomes the continuous variable $t$ of the integral; the sequence of coefficients $a_n$ corresponds to the function $\varphi(t)$. As we shall see, both series and integral are included in the Laplace-Stieltjes **integral**

$$(3) \qquad f(s) = \int_0^\infty e^{-st} \, d\alpha(t).$$

If $\alpha(t)$ is absolutely continuous, this becomes the Laplace integral (1); if $\alpha(t)$ is a step-function, it becomes a Dirichlet series (2). If $\alpha(t)$ is an increasing continuous function which is not absolutely continuous, the integral (3) defines a class of functions $f(s)$ which cannot be expressed either in the form (1) or (2).

## 2. Stieltjes Integrals

We now define the Stieltjes integral.* Let $\alpha(x)$ and $f(x)$ be **real** functions of the real variable $x$ defined for $a \leq x \leq b$. Denote by $\Delta$ a *subdivision* of the interval $(a, b)$ by the points $x_0, x_1, \cdots, x_n$, where

$$a = x_0 < x_1 < \cdots < x_n = b.$$

---

* For a list of references to works on the Stieltjes integral see T. H. Hildebrandt [1938].

By the *norm* $\delta$ of $\Delta$ we mean the largest of the numbers
$$x_{i+1} - x_i \quad (i = 0, 1, \cdots, n-1)$$

**DEFINITION 2.** *If the limit*
$$\lim_{\delta \to 0} \sum_{i=0}^{n-1} f(\xi_i)[\alpha(x_{i+1}) - \alpha(x_i)],$$

*where*
$$x_i \leq \xi_i \leq x_{i+1} \quad (i = 0, 1, \cdots, n-1),$$

*exists independently of the manner of subdivision and of the choice of the numbers $\xi_i$, then the limit is called the Stieltjes integral of $f(x)$ with respect to $\alpha(x)$ from $a$ to $b$ and is denoted by*

(1) $$\int_a^b f(x) \, d\alpha(x)$$

The definition is easily extended to include complex functions. Thus, if
$$f(x) = f_1(x) + if_2(x)$$
$$\alpha(x) = \alpha_1(x) + i\alpha_2(x),$$

where $f_1(x)$, $f_2(x)$, $\alpha_1(x)$, $\alpha_2(x)$ are real, we define the integral (1) by the equation
$$\int_a^b f(x) \, d\alpha(x) = \int_a^b f_1(x) \, d\alpha_1(x) - \int_a^b f_2(x) \, d\alpha_2(x)$$
$$+ i \int_a^b f_2(x) \, d\alpha_1(x) + i \int_a^b f_1(x) \, d\alpha_2(x),$$

provided all of the integrals on the right exist.

The integral (1) reduces to the Riemann integral if $\alpha(x) = x$. For this reason it is sometimes called the Riemann-Stieltjes integral in contradistinction to the Lebesgue-Stieltjes integral which we shall mention later.

Introduce the following notation:*

$$M_k = \operatorname*{u.b.}_{x_k \leq x \leq x_{k+1}} f(x)$$

$$m_k = \operatorname*{l.b.}_{x_k \leq x \leq x_{k+1}} f(x)$$

(2) $$S_\Delta = \sum_{k=0}^{n-1} M_k[\alpha(x_{k+1}) - \alpha(x_k)]$$

$$s_\Delta = \sum_{k=0}^{n-1} m_k[\alpha(x_{k+1}) - \alpha(x_k)].$$

\* We abbreviate *least upper bound* by u.b. and *greatest lower bound* by l.b.

**THEOREM 2.** *If $f(x)$ and $\alpha(x)$ are real bounded functions in $a \leq x \leq b$ and $\alpha(x)$ is in addition non-decreasing, then a necessary and sufficient condition that* (1) *should exist is that*

(3) $$\lim_{\delta \to 0} (S_\Delta - s_\Delta) = 0,$$

*independently of the manner of subdivision.*

The condition is necessary. For suppose (1) exists and has the value $I$. Let $\epsilon$ be an arbitrary positive number. Then it is possible to determine $n$ and the $\xi_k$ so that

$$0 \leq M_k - f(\xi_k) < \epsilon \qquad (k = 0, 1, 2, \cdots, n-1).$$

Hence

$$\sum_{k=0}^{n-1} f(\xi_k)[\alpha(x_{k+1}) - \alpha(x_k)] \leq S_\Delta$$

$$\leq \sum_{k=0}^{n-1} f(\xi_k)[\alpha(x_{k+1}) - \alpha(x_k)] + \epsilon[\alpha(b) - \alpha(a)],$$

so that

$$I \leq \varliminf_{\delta \to 0} S_\Delta \leq \varlimsup_{\delta \to 0} S_\Delta \leq I + \epsilon[\alpha(b) - \alpha(a)],$$

$$\lim_{\delta \to 0} S_\Delta = I.$$

In a similar way

$$\lim_{\delta \to 0} s_\Delta = I,$$

so that (3) is necessary.

Conversely, let the condition (3) be satisfied. Set

$$\sigma_\Delta = \sum_{k=0}^{n-1} f(\xi_k)[\alpha(x_{k+1}) - \alpha(x_k)].$$

Then

$$s_\Delta \leq \sigma_\Delta \leq S_\Delta$$

To show that $\sigma_\Delta$ approaches a limit we need only show, by virtue of (3), that $S_\Delta$ approaches a limit. Given $\epsilon > 0$, we wish to show that there corresponds a number $\delta_0$ such that if $S_{\Delta_1}$ and $S_{\Delta_2}$ are any two sums (2) for which the norms $\delta_1$ and $\delta_2$ are both less than $\delta_0$ then

$$|S_{\Delta_1} - S_{\Delta_2}| < \epsilon$$

Let $\Delta_3$ be a subdivision of $(a, b)$ by all the points $x_j$ involved in the subdivision $\Delta_1$ and $\Delta_2$. Then

$$s_{\Delta_1} \leq S_{\Delta_3} \leq S_{\Delta_1}$$

$$s_{\Delta_2} \leq S_{\Delta_3} \leq S_{\Delta_2}.$$

By (3) we can determine $\delta_0$ so that

$$S_\Delta - s_\Delta < \frac{\epsilon}{2} \qquad (\delta < \delta_0).$$

Hence for $\delta_1$ and $\delta_2$ less than $\delta_0$

$$0 \leq S_{\Delta_1} - S_{\Delta_3} < \frac{\epsilon}{2}$$

$$0 \leq S_{\Delta_2} - S_{\Delta_3} < \frac{\epsilon}{2}$$

$$|S_{\Delta_1} - S_{\Delta_2}| < \epsilon.$$

This completes the proof of the theorem.

### 3. Functions of Bounded Variation

Let us consider certain properties of non-decreasing functions and of functions of bounded variation.

DEFINITION 3a. *Let $\alpha(x)$ be non-decreasing in $a \leq x \leq b$. A point $x_1$ of this interval is a point of invariability of $\alpha(x)$ if there exists a two-sided (one-sided if $x$ is $a$ or $b$) neighborhood of $x_1$ in which $\alpha(x)$ is constant.*

DEFINITION 3b. *A point of increase of a non-decreasing function is one which is not a point of invariability.*

THEOREM 3a. *If a non-decreasing function has a finite number of points of increase it is discontinuous at these points and constant between them. It is then a step-function.*

The proof is simple and is omitted.

We shall have frequent use for functions of bounded variation. A complex function is of bounded variation if and only if its real and imaginary parts are of bounded variation. If $\alpha(x)$ is of bounded variation in $a \leq x \leq b$ we denote the total variation of $\alpha(x)$ in the interval $a \leq x \leq t$, where $t \leq b$, by $V_\alpha(t)$ or by $V[\alpha]_a^t$. The following familiar results* frequently will be useful.

THEOREM 3b. *If $\alpha(x)$ is a real function of bounded variation in $(a, b)$, then there exist two functions $\alpha_1(x)$, $\alpha_2(x)$ which are non-negative, non-*

---

\* Compare E. C. Titchmarsh [1932] p. 355.

*decreasing and bounded in $(a, b)$, which vanish at $x = a$, have the same points of continuity and discontinuity as $\alpha(x)$, and are such that*

$$\alpha(x) - \alpha(a) = \alpha_1(x) - \alpha_2(x)$$
$$V_\alpha(x) = \alpha_1(x) + \alpha_2(x) \qquad (a \leqq x \leqq b).$$

THEOREM 3c.  *If $\alpha_1(x)$ and $\alpha_2(x)$ are monotonic and bounded in $(a, b)$, and if $\alpha(x) = \alpha_1(x) + \alpha_2(x)$, then*

$$V_\alpha(x) \leqq |\alpha_1(x) - \alpha_1(a)| + |\alpha_2(x) - \alpha_2(a)| \qquad (a \leqq x \leqq b).$$

## 4. Existence of Stieltjes Integrals

We now consider an important case in which the Stieltjes integral exists.

THEOREM 4a.  *If $f(x)$ is continuous and $\alpha(x)$ is of bounded variation in $(a, b)$, then the Stieltjes integral of $f(x)$ with respect to $\alpha(x)$ from $a$ to $b$ exists.*

It is clearly no restriction to assume that $f(x)$ and $\alpha(x)$ are real. By Theorem 3b we may also assume that $\alpha(x)$ is non-decreasing and bounded. We may now apply Theorem 2. Since $f(x)$ is uniformly continuous in $a \leqq x \leqq b$, to an arbitarary positive $\epsilon$ there corresponds a $\delta_0$ such that

$$M_k - m_k < \epsilon \qquad (k = 0, 1, \cdots, n - 1)$$

if the norm $\delta$ is less than $\delta_0$. Hence

$$0 \leqq S_\Delta - s_\Delta \leqq \epsilon[\alpha(b) - \alpha(a)]$$

for such a subdivision. It thus becomes clear that the condition §2 (3) is satisfied, so that our result is established.

THEOREM 4b.  *If $f(x)$ is of bounded variation and $\alpha(x)$ is continuous in $(a, b)$, then the Stieltjes integral of $f(x)$ with respect to $\alpha(x)$ from $a$ to $b$ exists and*

$$(1) \qquad \int_a^b f(x)\, d\alpha(x) = f(b)\alpha(b) - f(a)\alpha(a) - \int_a^b \alpha(x)\, df(x).$$

To prove this choose an arbitrary subdivision $\Delta$ of $(a, b)$ and form the sum

$$\sigma_\Delta = \sum_{k=0}^{n-1} f(\xi_k)[\alpha(x_{k+1}) - \alpha(x_k)] \qquad (x_k \leqq \xi_k \leqq x_{k+1}).$$

By use of partial summation we have

$$\sigma_\Delta = \sum_{k=1}^{n-1} \alpha(x_k)[f(\xi_k) - f(\xi_{k-1})] - \alpha(a)[f(\xi_0) - f(a)]$$
$$- \alpha(b)[f(b) - f(\xi_{n-1})] + f(b)\alpha(b) - f(a)\alpha(a).$$

As the maximum of the differences $x_{k+1} - x_k$ ($k = 0, 1, \cdots, n - 1$) approaches zero so too will the maximum of the differences $\xi_0 - a$, $\xi_1 - \xi_0, \cdots, \xi_{n-1} - \xi_{n-2}, b - \xi_{n-1}$. Hence by Theorem 4a $\sigma_\Delta$ will approach

$$-\int_a^b \alpha(x)\, df(x) + f(b)\alpha(b) - f(a)\alpha(a)$$

as the norm of $\Delta$ approaches zero. This proves the theorem.

Equation (1) is called the formula for integration by parts for the Stieltjes integral.

### 5. Properties of Stieltjes Integrals

We now collect certain useful properties of the Stieltjes integral. The functions $f(x)$ and $\alpha(x)$ in the following theorems are complex unless otherwise stated. It is obvious that

$$\int_a^b d\alpha(x) = \alpha(b) - \alpha(a),$$

and that

$$\int_a^b f(x)\, d\alpha(x) = 0$$

if $\alpha(x)$ is constant.

**THEOREM 5a.** *If $f_1(x)$ and $f_2(x)$ are continuous, if $\alpha_1(x)$ and $\alpha_2(x)$ are of bounded variation in $a \leqq x \leqq b$, and if $k_1, k_2, l_1, l_2$ are any constants, then*

(1)
$$\int_a^b [k_1 f_1(x) + k_2 f_2(x)]\, d[l_1 \alpha_1(x) + l_2 \alpha_2(x)]$$
$$= \sum_{i=1}^2 \sum_{j=1}^2 k_i l_j \int_a^b f_i(x)\, d\alpha_j(x)$$

**THEOREM 5b.** *If $f(x)$ is continuous and $\alpha(x)$ is of bounded variation in $a \leqq x \leqq b$, then for any $c$ in $a < x < b$*

(2) $$\int_a^b f(x)\, d\alpha(x) = \int_a^c f(x)\, d\alpha(x) + \int_c^b f(x)\, d\alpha(x)$$

(3) $$\left| \int_a^b f(x)\, d\alpha(x) \right| \leqq \int_a^b |f(x)|\, dV_\alpha(x) \leqq \underset{a \leqq x \leqq b}{\text{Max}} |f(x)|\, V_\alpha(b)$$

§5]  PROPERTIES OF STIELTJES INTEGRALS  9

To prove (3) one needs the more obvious property that if $f(x)$ and $g(x)$ are real and continuous with $f(x) \leq g(x)$, and if $\alpha(x)$ is non-decreasing, then

$$\int_a^b f(x)\, d\alpha(x) \leq \int_a^b g(x)\, d\alpha(x).$$

The first inequality (3) is sometimes written in the form

$$\left| \int_a^b f(x)\, d\alpha(x) \right| \leq \int_a^b |f(x)|\, |d\alpha(x)|$$

In connection with (2) we observe that the existence of the two integrals on the right does not imply the existence of the one on the left. For example, if

$$f(x) = 1 \quad (0 \leq x \leq 1) \qquad \alpha(x) = 1 \quad (0 \leq x < 1)$$
$$= 0 \quad (1 < x \leq 2) \qquad\qquad\quad = 0 \quad (1 \leq x \leq 2),$$

we have

$$1 + \int_0^1 f(x)\, d\alpha(x) = \int_1^2 f(x)\, d\alpha(x) = 0;$$

but

$$\int_0^2 f(x)\, d\alpha(x)$$

cannot exist since $f(x)$ and $\alpha(x)$ have a common point of discontinuity at $x = 1$. The defining limit may be made to have several different values by different choices of the points $\xi_i$.

THEOREM 5c. *If $f(x)$ is continuous and $\alpha(x)$ is of bounded variation in $a \leq x \leq b$, then*

(4) $$F(x) = \int_a^x f(t)\, d\alpha(t) \qquad (a \leq x \leq b)$$

*is also of bounded variation in $a \leq x \leq b$ and*

$$V_F(x) \leq \int_a^x |f(t)|\, |d\alpha(t)|$$

*Moreover*

(5) $$F(x+) - F(x) = f(x)[\alpha(x+) - \alpha(x)] \qquad (a \leq x < b)$$
(6) $$F(x) - F(x-) = f(x)[\alpha(x) - \alpha(x-)] \qquad (a < x \leq b).$$

In the definition (4) of $F(x)$ it is of course assumed that

$$F(a) = \int_a^a f(t)\, d\alpha(t) = 0.$$

It must not be supposed, however, that $F(a+)$ is also zero as in the case of the Lebesgue integral. The example $f(t) = 1$ as well as equation (5) shows that this is not always the case. We shall use the following notation:

$$F(b) - F(a+) = \int_{a+}^b f(x)\, d\alpha(x)$$

$$F(b-) = \int_a^{b-} f(x)\, d\alpha(x).$$

THEOREM 5d. *If the functions*

$$f_n(x) \qquad (n = 0, 1, 2, \cdots)$$

*are continuous, $\alpha(x)$ of bounded variation in $a \leq x \leq b$, and if the series*

$$\sum_{n=0}^{\infty} f_n(x)$$

*converges uniformly to $f(x)$ on that interval, then*

$$\int_a^b f(x)\, d\alpha(x) = \sum_{n=0}^{\infty} \int_a^b f_n(x)\, d\alpha(x).$$

The proof is similar to that of the corresponding theorem for Riemann integrals.

## 6. The Stieltjes Integral as a Series or a Lebesgue Integral

We consider in this section certain useful special cases of the Stieltjes integral. It is clear from the definition of the integral that if $\alpha(x)$ is a function with a continuous first derivative and $f(x)$ is continuous in $a \leq x \leq b$, then

$$\int_a^b f(x)\, d\alpha(x) = \int_a^b f(x) \alpha'(x)\, dx.$$

On the other hand, if $\alpha(x)$ is a step-function

$$\alpha(x) = 0 \qquad\qquad\qquad\qquad (a \leq x \leq x_0)$$
$$\phantom{\alpha(x)} = a_0 + a_1 + \cdots + a_{k-1} \quad (x_{k-1} < x < x_k\,;\, k = 1, 2, \cdots, n)$$
$$\alpha(x) = a_0 + a_1 + \cdots + a_n \qquad\qquad (x_n \leq x \leq b)$$
$$a \leq x_0 < x_1 < x_2 < \cdots < x_n \leq b,$$

then

(1) $$\int_a^b f(x)\,d\alpha(x) = \sum_{k=0}^{n} a_k f(x_k).$$

Note that we have not defined $\alpha(x)$ at those points $x_k$ which cannot coincide with $a$ or $b$. The definition of $\alpha(x)$ at such points may be made in an arbitrary fashion without affecting (1).

It was stated earlier that the integral of a continuous function with respect to an absolutely continuous function is a Lebesgue integral. Let us prove this result.

THEOREM 6a. *If $f(x)$ is continuous and $\varphi(x)$ belongs\* to $L$ in $a \leqq x \leqq b$, and if*

$$\alpha(x) = \int_c^x \varphi(t)\,dt \quad (a \leqq c \leqq b; a \leqq x \leqq b),$$

*then*

(1) $$\int_a^b f(x)\,d\alpha(x) = \int_a^b f(x)\varphi(x)\,dx = \int_a^b f(x)\alpha'(x)\,dx.$$

Here the Riemann-Stieltjes integral (1) clearly exists since $\alpha(x)$ is of bounded variation on $(a, b)$. The integral on the right of (1) exists as a Lebesgue integral. For an arbitrary subdivision $\Delta$ of norm $\delta$ we have

$$\sigma_\Delta = \sum_{k=0}^{n-1} f(x_k) \int_{x_k}^{x_{k+1}} \varphi(t)\,dt$$

$$\sigma_\Delta - \int_a^b f(t)\varphi(t)\,dt = \sum_{k=0}^{n-1} \int_{x_k}^{x_{k+1}} [f(x_k) - f(t)]\varphi(t)\,dt$$

$$\left|\sigma_\Delta - \int_a^b f(t)\varphi(t)\,dt\right| \leqq \sum_{k=0}^{n-1} \max_{x_k \leqq t \leqq x_{k+1}} |f(x_k) - f(t)| \int_{x_k}^{x_{k+1}} |\varphi(t)|\,dt$$

$$\leqq M_\Delta \int_a^b |\varphi(t)|\,dt,$$

where $M_\Delta$ is the largest of the numbers

$$\max_{x_k \leqq t \leqq x_{k+1}} |f(x_k) - f(t)| \qquad (k = 0, 1, \cdots, n-1).$$

---

\* A function $\varphi(x)$ belongs to $L^p$ $(p > 0)$ in $a \leqq x \leqq b$ if measurable there and if

$$\int_a^b |\varphi(x)|^p\,dx < \infty.$$

*Without explicit statement we assume that all functions employed are measurable.*

By the uniform continuity of $f(x)$ on $a \leq x \leq b$ it follows that $M_\Delta$ tends to zero with $\delta$. Since $\sigma_\Delta$ tends to the left-hand side of (1) as $\delta$ approaches zero, our theorem is proved.

An important result of a similar nature follows.

**Theorem 6b.** *If $f(x)$ and $\varphi(x)$ are continuous and $\alpha(x)$ is of bounded variation in $a \leq x \leq b$, and if*

$$\beta(x) = \int_c^x \varphi(t)\,d\alpha(t) \quad (a \leq x \leq b; a \leq c \leq b),$$

*then*

$$\int_a^b f(x)\,d\beta(x) = \int_a^b f(x)\varphi(x)\,d\alpha(x)$$

It is clear that both integrals exist by Theorem 4a and Theorem 5c. Then if

$$\sigma_\Delta = \sum_{k=0}^{n-1} f(x_k) \int_{x_k}^{x_{k+1}} \varphi(t)\,d\alpha(t),$$

we have as in the previous proof

$$\left| \sigma_\Delta - \int_a^b f(t)\varphi(t)\,d\alpha(t) \right| \leq M_\Delta \int_a^b |\varphi(t)|\,|d\alpha(t)|,$$

from which the conclusion is evident.

## 7. Further Properties of Stieltjes Integrals

Let us investigate to what extent the definition of $f(x)$ or of $\alpha(x)$ can be changed without affecting the value of

(1) $$\int_a^b f(x)\,d\alpha(x)$$

**Theorem 7a.** *If $f(x)$ is continuous and $\alpha(x)$ is of bounded variation in $a \leq x \leq b$, and if $\alpha(x)$ takes on a constant value $c$ at a set of points $E$ which is dense in $(a, b)$ and which includes the points $a$ and $b$, then*

$$\int_a^b f(x)\,d\alpha(x) = 0.$$

This follows at once from the definition of the integral. The limit of Definition 2 exists independently of the manner of subdivision. Hence we may choose the $x_k$ in $E$, so that the limit is surely zero.

As a result of this theorem we see that the value of (1) is unchanged if the definition of $\alpha(x)$ is altered at a countable set of points of the interval $a < x < b$ provided $\alpha(x)$ remains of bounded variation.

THEOREM 7b. *If $\alpha_1(x)$ and $\alpha_2(x)$ are of bounded variation in $a \leq x \leq b$ and differ by a constant $c$ in the set $E$ of Theorem 7a, and if $f(x)$ is continuous in $a \leq x \leq b$, then*

$$\int_a^b f(x)\, d\alpha_1(x) = \int_a^b f(x)\, d\alpha_2(x).$$

The proof is obvious.

THEOREM 7c. *If $\alpha(x)$ is non-decreasing with at least $n + 1$ points of increase, if $f(x)$ is continuous and non-negative with at most $n$ zeros in $a \leq x \leq b$, then*

$$\int_a^b f(x)\, d\alpha(x) > 0.$$

There is at least one point of increase $x_0$ of $\alpha(x)$ at which $f(x)$ is not zero. Suppose $a < x_0 < b$. Then for all positive numbers $\delta$ sufficiently small

$$\alpha(x_0 + \delta) - \alpha(x_0 - \delta) > 0 \qquad (a < x_0 - \delta < x_0 + \delta < b).$$

Hence

$$\int_a^b f(x)\, d\alpha(x) \geq \int_{x_0 - \delta}^{x_0 + \delta} f(x)\, d\alpha(x)$$

$$\geq \min_{x_0 - \delta \leq x \leq x_0 + \delta} f(x)\, [\alpha(x_0 + \delta) - \alpha(x_0 - \delta)].$$

By the continuity of $f(x)$ the right-hand side of this inequality can certainly be made positive by choice of $\delta$. An obvious modification of the argument gives the result when $x_0$ is $a$ or $b$.

The application of this theorem usually made is to the case in which $\alpha(x)$ has infinitely many points of increases and $f(x)$ has but a finite number of zeros.

## 8. Normalization

We now introduce the notion of a normalized function of bounded variation.

DEFINITION 8. *If $\alpha(x)$ is of bounded variation in $a \leq x \leq b$, it is said to be normalized there if*

$$\alpha(a) = 0$$

$$\alpha(x) = \frac{\alpha(x+) + \alpha(x-)}{2} \qquad (a < x < b).$$

We observe that *normalization* of a function $\alpha(x)$ does not change the value of the integral of a continuous function with respect to $\alpha(x)$. More precisely, we have:

**Theorem 8a.** *If $f(x)$ is continuous and $\alpha(x)$ is of bounded variation in $a \leq x \leq b$, then there exists a normalized function $\alpha^*(x)$ of bounded variation in $a \leq x \leq b$ such that*

$$\int_a^b f(x)\,d\alpha(x) = \int_a^b f(x)\,d\alpha^*(x).$$

In fact we may define $\alpha^*(x)$ as follows:

$$\alpha^*(a) = 0$$

$$\alpha^*(x) = \frac{\alpha(x+) + \alpha(x-)}{2} - \alpha(a) \qquad (a < x < b)$$

$$\alpha^*(b) = \alpha(b) - \alpha(a).$$

Then it is clear that $\alpha^*(x)$ is normalized and by Theorem 7a that

$$\int_a^b f(x)\,d[\alpha(x) - \alpha(a) - \alpha^*(x)] = 0,$$

from which our result follows.

**Theorem 8b.** *If $f(x)$ is continuous and $\alpha(x)$ is a normalized function of bounded variation in $a \leq x \leq b$, then the function*

$$F(x) = \int_a^x f(t)\,d\alpha(t) \qquad (a \leq x \leq b)$$

*is also normalized.*

For, by subtracting equation §5 (6) from §5 (5) we have

$$\frac{F(x+) + F(x-)}{2} - F(x) = f(x)\left[\frac{\alpha(x+) + \alpha(x-)}{2} - \alpha(x)\right].$$

**Theorem 8c.** *If $\alpha(x)$ is a normalized function of bounded variation in $a \leq x \leq R$ for every $R > a$, and if*

$$\lim_{x \to \infty} \alpha(x) = A. \qquad (x \,\varepsilon\, E),$$

*$x$ becoming infinite through the set $E$ of points of continuity of $\alpha(x)$, then*

$$\lim_{x \to \infty} \alpha(x) = A.$$

For if $\epsilon > 0$, we can find an $x_0$ such that

$$|\alpha(x) - A| < \epsilon \qquad (x \,\varepsilon\, E,\ x > x_0),$$

from which it is clear that for any $x > x_0$

(1) $\qquad |\alpha(x+) - A| \leqq \epsilon$

(2) $\qquad |\alpha(x-) - A| \leqq \epsilon$

(3) $\qquad |\alpha(x) - A| \leqq \epsilon.$

Inequality (3) follows from (1) and (2) since $\alpha(x)$ is normalized. This establishes our result.

## 9. Improper Stieltjes Integrals

Let $f(x)$ be continuous in $a \leqq x < \infty$ and let $\alpha(x)$ be of bounded variation and normalized in $a \leqq x \leqq R$ for every $R > 0$. Then we define the improper integral of $f(x)$ with respect to $\alpha(x)$ on the infinite interval $(a, \infty)$ by the equation

(1) $$\int_a^\infty f(x)\,d\alpha(x) = \lim_{R \to \infty} \int_a^R f(x)\,d\alpha(x).$$

When the limit (1) exists, the integral (1) *converges*; otherwise it *diverges*. In a similar way we define improper integrals over $(-\infty, a)$ and $(-\infty, \infty)$ by the equations

$$\int_{-\infty}^a f(x)\,d\alpha(x) = \lim_{R \to \infty} \int_{-R}^a f(x)\,d\alpha(x)$$

$$\int_{-\infty}^\infty f(x)\,d\alpha(x) = \int_0^\infty f(x)\,d\alpha(x) + \int_{-\infty}^0 f(x)\,d\alpha(x).$$

Thus an integral with an infinite limit of integration is to be regarded as a "Cauchy-value," as defined above.

In particular if $\alpha(x)$ is a step-function, (1) may become an infinite series

(2) $$\int_a^\infty f(x)\,d\alpha(x) = \sum_{k=0}^\infty a_k f(x_k).$$

Note that the meaning of the integral (1) has been chosen in such a way that the series and integral converge or diverge together.

We say that (1) converges absolutely if and only if

$$\int_a^\infty |f(x)|\,dV_\alpha(x) < \infty.$$

Again the series (2) converges absolutely if and only if the integral (2) does.

THEOREM 9. *If $f(x)$ is continuous in $a \leqq x < \infty$, if $\alpha(x)$ is of bounded* variation in $a \leqq x < R$ for every $R > a$, and if $\alpha^*(x)$ is the normalized function of $\alpha(x)$, then

(3) $$\int_a^\infty f(x)\,d\alpha(x) = \int_a^\infty f(x)\,d\alpha^*(x)$$

provided the first integral converges.

First let $R$ belong to $E$, the set of points of continuity of $\alpha(x)$ (and of $\alpha^*(x)$). Then by Theorem 8a

(4) $$\int_a^R f(x)\,d\alpha(x) = \int_a^R f(x)\,d\alpha^*(x).$$

By Theorem 8b the integral on the right of (4) considered as a function of $R$ is normalized. But

$$\lim_{\substack{R \to \infty \\ R \in E}} \int_a^R f(x)\,d\alpha^*(x) = \int_a^\infty f(x)\,d\alpha(x)$$

by (4). Hence by Theorem 8c

$$\lim_{R \to \infty} \int_a^R f(x)\,d\alpha^*(x) = \int_a^\infty f(x)\,d\alpha(x),$$

and the result is established.

Note that the convergence of the second integral of (3) does not imply that of the first, as one may easily see by taking $f(x) = 1$, and

$$\alpha(x) = 0 \qquad x \neq 1, 2, 3, \cdots$$
$$\alpha(x) = 1 \qquad x = 1, 2, 3, \cdots.$$

## 10. Laws of the Mean

We discuss here the laws of the mean for Stieltjes integrals.

**THEOREM 10a.** If $\alpha(x)$ is non-decreasing (or non-increasing) and $f(x)$ is a real continuous function in $a \leqq x \leqq b$, then

(1) $$\int_a^b f(x)\,d\alpha(x) = f(\xi)[\alpha(b) - \alpha(a)] \qquad (a \leqq \xi \leqq b).$$

The proof is similar to the classic proof of the first law of the mean for Riemann integrals. If

(2) $$\alpha(x) = \int_a^x \varphi(t)\,dt \qquad (a \leqq x \leqq b),$$

(3) $$\varphi(x) \, \varepsilon \, L \qquad (a \leqq x \leqq b).$$

where $\varphi(x)$ is non-negative, equation (1) becomes

$$\int_a^b f(x)\varphi(x)\,dx = f(\xi)\int_a^b \varphi(x)\,dx.$$

THEOREM 10b. *If $\alpha(x)$ is real and continuous and $f(x)$ is non-decreasing (or non-increasing) in $a \leqq x \leqq b$, then*

(4) $$\int_a^b f(x)\,d\alpha(x) = f(a)\int_a^\xi d\alpha(x) + f(b)\int_\xi^b d\alpha(x) \qquad (a \leqq \xi \leqq b).$$

For,

$$\int_a^b f(x)\,d\alpha(x) = f(b)\alpha(b) - f(a)\alpha(a) - \int_a^b \alpha(x)\,df(x)$$

$$= f(b)\alpha(b) - f(a)\alpha(a) - \alpha(\xi)\int_a^b df(x),$$

by Theorem 10a. This is precisely equation (4).

In particular if $\alpha(x)$ is defined by (2) and (3), equation (4) becomes*

$$\int_a^b f(x)\varphi(x)\,dx = f(a)\int_a^\xi \varphi(x)\,dx + f(b)\int_\xi^b \varphi(x)\,dx \qquad (a \leqq \xi \leqq b).$$

which is the Weierstrass form of Bonnet's second law of the mean.

COROLLARY 10b. *If $\alpha(x)$ is real and continuous and $f(x)$ is non-negative non-decreasing in $a \leqq x \leqq b$, then*

$$\int_a^b f(x)\,d\alpha(x) = f(b)\int_\xi^b d\alpha(x) \qquad (a \leqq \xi \leqq b);$$

*if $f(x)$ is non-negative non-increasing in $a \leqq x \leqq b$, then*

$$\int_a^b f(x)\,d\alpha(x) = f(a)\int_a^\xi \varphi(x)\,dx \qquad (a \leqq \xi \leqq b).$$

* Note that in this case we cannot appeal to Theorem 6a to show that

$$\int_a^b f(x)\,d\alpha(x) = \int_a^b f(x)\varphi(x)\,dx$$

since $f(x)$ is not continuous. That the result is true one sees easily by using Theorem 15d to change the order of integration in

$$\int_a^b df(t)\int_a^t \varphi(u)\,du.$$

Another closely related result* is contained in:

**THEOREM 10c.** *If $f(x)$ is non-negative, non-decreasing and continuous, and if $\alpha(x)$ is real and of bounded variation in $a \leqq x \leqq b$, then*

$$\int_a^b f(x)\,d\alpha(x) = Af(b),$$

*where*

(5) $\qquad \underset{a \leqq x \leqq b}{\text{l.b.}} \int_x^b d\alpha(t) \leqq A \leqq \underset{a \leqq x \leqq b}{\text{u.b.}} \int_x^b d\alpha(t).$

To prove this set

$$\beta(x) = \alpha(b) - \alpha(x) \qquad (a \leqq x \leqq b)$$

$$\underset{a \leqq x \leqq b}{\text{u.b.}} \beta(x) = U, \qquad \underset{a \leqq x \leqq b}{\text{l.b.}} \beta(x) = L.$$

Then

$$I = \int_a^b f(x)\,d\alpha(x) = -\int_a^b f(x)\,d\beta(x)$$

$$= f(a)\beta(a) + \int_a^b \beta(x)\,df(x).$$

Since $f(a)$ is not negative and $f(x)$ is non-decreasing,

$$I \leqq f(a)U + [f(b) - f(a)]U = Uf(b)$$
$$I \geqq f(a)L + [f(b) - f(a)]L = Lf(b).$$

Hence

$$I = Af(b),$$

where $A$ is between $U$ and $L$ and hence satisfies (5). This completes the proof of the theorem.

**COROLLARY 10c.** *If $f(x)$ is non-negative, non-increasing, and continuous, then*

$$\int_a^b f(x)\,d\alpha(x) = f(a)B,$$

*where*

$$\underset{a \leqq x \leqq b}{\text{l.b.}} \int_a^x d\alpha(t) \leqq B \leqq \underset{a \leqq x \leqq b}{\text{u.b.}} \int_a^x d\alpha(t).$$

It is easy to extend the theorem to include improper integrals. For example, if $f(x)$ is non-negative, non-decreasing, continuous, and

---

* See R. P. Boas [1937a] p. 8, or R. P. Boas and D. V. Widder [1939] p. 6.

bounded in $a \leqq x < \infty$; if $\alpha(x)$ is a real function of bounded variation in $a + \epsilon \leqq x \leqq \epsilon^{-1}$ for every sufficiently small positive $\epsilon$ and of such a nature that

$$\int_{a+}^{\infty} f(x)\, d\alpha(x)$$

and $\alpha(\infty)$ exist, then

$$\int_{a+}^{\infty} f(x)\, d\alpha(x) = f(\infty)A,$$

where

$$\underset{a \leqq x < \infty}{\text{l.b.}} \int_{x}^{\infty} d\alpha(t) \leqq A \leqq \underset{a \leqq x < \infty}{\text{u.b.}} \int_{x}^{\infty} d\alpha(t).$$

## 11. Change of Variable

We discuss here the formula for the change of variable of a Stieltjes integral.

**THEOREM 11a.** *If $f(x)$ is continuous, $\alpha(x)$ is of bounded variation, and $\beta(x)$ is continuous increasing with no points of invariability in $a \leqq x \leqq b$, then*

(1) $$\int_{a}^{b} f(x)\, d\alpha(x) = \int_{c}^{d} f(\beta(x))\, d\alpha(\beta(x)),$$

*where*

$$a = \beta(c) \qquad b = \beta(d).$$

The proof follows at once from the definition of the integral. Clearly both integrals (1) exist by Theorem 4a. But

$$\sum_{i=0}^{n-1} f(x_i)[\alpha(x_{i+1}) - \alpha(x_i)] = \sum_{i=0}^{n-1} f(\beta(t_i))[\alpha(\beta(t_{i+1})) - \alpha(\beta(t_i))],$$

where

$$x_i = \beta(t_i) \qquad (i = 0, 1, \cdots, n-1).$$

Since $\beta(t)$ is increasing the sequences $\{x_i\}_0^n$ and $\{t_i\}_0^n$ are increasing together; and, by the continuity of $\beta(t)$, norms of the corresponding subdivisions approach zero together. Hence equation (1) must hold.

**THEOREM 11b.** *If $f(x)$ is continuous, $\alpha(x)$ is continuous increasing with no points of invariability, and if $\gamma(x)$ is the inverse of $\alpha(x)$, then*

(2) $$\int_{a}^{b} f(x)\, d\alpha(x) = \int_{\alpha(a)}^{\alpha(b)} f(\gamma(x))\, dx.$$

This follows from Theorem 11a by setting $\beta(x)$ of that theorem equal to $\gamma(x)$ of Theorem 11b.

It may happen that the integral on the right-hand side of (2) exists as a Lebesgue integral even though $f(x)$ is not continuous. In this case its value is taken as the definition of the left-hand side. The integral on the left is then called a Lebesgue-Stieltjes integral. In fact it is possible to relax further the conditions imposed on $\alpha(x)$, in which case some care in the definition of the inverse $\gamma(x)$ must be exercised. Since we shall have no need for this degree of generality we omit the details.

As an example in which (2) exists as a Lebesgue-Stieltjes integral but not in the Riemann-Stieltjes sense, define $\alpha(x)$ as $e^x$ and $f(x)$ as zero or unity according as $x$ is rational or irrational.

## 12. Variation of the Indefinite Integral

We next consider the variation of an indefinite Stieltjes integral

$$\beta(x) = \int_a^x f(t)\, d\alpha(t) \tag{1}$$

If $\alpha(x)$ is of bounded variation and $f(x)$ is continuous in $a \leq x \leq b$, it is evident that $\beta(x)$ is also of bounded variation there and that its total variation does not exceed

$$\int_a^b |f(t)|\, |d\alpha(t)| \tag{2}$$

This follows from Theorem 5c. We wish to show in fact that it is precisely equal to this integral.

**THEOREM 12.** *Let $\alpha(x)$ and $f(x)$ be respectively of bounded variation and continuous in $a \leq x \leq b$. Then the variation of $\beta(x)$, defined by (1), is (2).*

Let $u(x)$ be the variation of $\alpha(t)$ in the interval $a \leq t \leq x$. Denote the integral (2) by $J$; then

$$J = \int_a^b |f(t)|\, du(t)$$

For a given subdivision $\Delta$ of norm $\delta$,

$$a = x_0 < x_1 < x_2 < \cdots < x_n = b,$$

set

$$S_\delta = \sum_{k=0}^{n-1} \left| \int_{x_k}^{x_{k+1}} f(t)\, d\alpha(t) \right|$$

Then

$$0 \leq S_\delta \leq J$$

We wish to show that by choice of $\Delta$ the sum $S_\delta$ may be brought arbitrarily near to $J$.

Let $\epsilon$ be an arbitrary positive number. Choose $\delta_0$ and a subdivision $\Delta$ of norm $\delta$ less than $\delta_0$ such that

(3) $$0 \leq u(b) - \sum_{k=0}^{n-1} |\alpha(x_{k+1}) - \alpha(x_k)| < \epsilon$$

and

(4) $$|f(t') - f(t'')| < \epsilon \qquad (|t' - t''| < \delta_0 \,; a \leq t' \leq t'' \leq b).$$

This is possible by the definition of $u(b)$ and by the uniform continuity of $f(x)$.

By the first law of the mean

$$J = \sum_{k=0}^{n-1} |f(\xi_k)| \int_{x_k}^{x_{k+1}} du(t) \qquad (x_k \leq \xi_k \leq x_{k+1}).$$

Hence

$$J - S_\delta = \sum_{k=0}^{n-1} |f(\xi_k)| \left\{ \int_{x_k}^{x_{k+1}} du(t) - \left| \int_{x_k}^{x_{k+1}} d\alpha(t) \right| \right\}$$
$$+ \sum_{k=0}^{n-1} \left| \int_{x_k}^{x_{k+1}} f(\xi_k)\, d\alpha(t) \right| - \left| \int_{x_k}^{x_{k+1}} f(t)\, d\alpha(t) \right|.$$

The first of these sums does not exceed

$$\max_{a \leq t \leq b} |f(t)| \left\{ u(b) - \sum_{k=0}^{n-1} |\alpha(x_{k+1}) - \alpha(x_k)| \right\} \leq \max_{a \leq t \leq b} |f(t)| \epsilon$$

by (3). The second is not greater than

$$\sum_{k=0}^{n-1} \left| \int_{x_k}^{x_{k+1}} \{f(\xi_k) - f(t)\}\, d\alpha(t) \right| \leq \epsilon u(b)$$

by (4). This establishes our theorem.*

COROLLARY 12. *Let $\alpha(x)$ and $f(x)$ be respectively of bounded variation and continuous in $(-\infty, \infty)$ and let*

(5) $$\int_{-\infty}^{\infty} |f(t)|\,|d\alpha(t)| < \infty.$$

---

*The essentials of this proof were concocted orally by J. D. Tamarkin and S. E. Warschawski during the course of one of their nocturnal promenades. It could of course be obtained from the more familiar corresponding theorem for Lebesgue integrals.

*Then the variation of*

$$\beta(x) = \int_{-\infty}^{x} f(t)\, d\alpha(t)$$

*in* $(-\infty, \infty)$ *is the integral* (5).

## 13. Stieltjes Integrals as Infinite Series; Second Method

We saw in section 9 that an improper Stieltjes integral becomes an infinite series if the function $\alpha(t)$ is chosen as a step-function with infinitely many jumps. Here we shall show that the same result may be achieved for a finite interval and a continuous $\alpha(t)$ by choice of the function integrated.

THEOREM 13a. *Let $f(x)$ be defined by the equations*

$$f(x) = \begin{cases} 0 & x < c_1 \\ u_1 & c_1 \leq x \end{cases} \qquad (a \leq c_1 \leq b),$$

*and let $\alpha(x)$ be non-decreasing in $(a, b)$ and continuous at $x = c_1$. Then*

$$\int_a^b f(x)\, d\alpha(x) = u_1[\alpha(b) - \alpha(c_1)].$$

This result is obtained by integration by parts or directly from the definition of the integral.

THEOREM 13b. *If $c_1, c_2, \cdots, c_m$ are distinct points of $(a \leq x \leq b)$, if $\alpha(x)$ is non-decreasing in $(a, b)$ and continuous at these points, and if\**

$$f(x) = \sum_{c_k \leq x} u_k \qquad (-\infty < x < \infty),$$

*then*

$$\int_a^b f(x)\, d\alpha(x) = \sum_{k=1}^{m} [\alpha(b) - \alpha(c_k)] u_k.$$

This follows from Theorem 13a, for the present function $f(x)$ is the sum of $m$ functions of the type integrated in that theorem.

THEOREM 13c. *If $c_1, c_2, \cdots$ are points of $(a, b)$, if the series*

$$\sum_{k=1}^{\infty} u_k$$

*converges absolutely, if*

$$f(x) = \sum_{c_k \leq x} u_k \qquad (-\infty < x < \infty),$$

---

\* In this notation it is to be understood that $f(x)$ is zero if no $c_k$ lies between $a$ and $x$.

and if $\alpha(x)$ is non-decreasing in $(a, b)$ and continuous at the points $c_1, c_2, \cdots,$ then

$$\int_a^b f(x)\, d\alpha(x) = \sum_{k=1}^{\infty} [\alpha(b) - \alpha(c_k)] u_k.$$

Let $\epsilon$ be an arbitrary positive number. Determine $m$ so that

$$\sum_{k=m+1}^{\infty} |u_k| < \epsilon.$$

Write

$$f(x) = \sideset{}{'}\sum_{a \leq c_k \leq x} u_k + \sideset{}{''}\sum_{a \leq c_k \leq x} u_k = f_1(x) + f_2(x),$$

where in the first sum $k$ runs through the integers $1, 2, \cdots, m$; through all larger integers in the second.

Introduce a subdivision $\Delta$ of $(a, b)$ of norm $\delta$. Set

$$\sigma_n = \sum_{k=0}^{n-1} f_1(\xi_k)[\alpha(x_{k+1}) - \alpha(x_k)] + \sum_{k=0}^{n-1} f_2(\xi_k)[\alpha(x_{k+1}) - \alpha(x_k)]$$

$$= \sigma_n' + \sigma_n'',$$

where

$$x_k \leq \xi_k \leq x_{k+1}.$$

Clearly

$$|f_2(x)| < \epsilon$$

$$|\sigma_n''| \leq \epsilon[\alpha(b) - \alpha(a)].$$

We may choose $\delta$ so small that

$$\left| \sigma_n' - \sum_{k=1}^{m} u_k[\alpha(b) - \alpha(c_k)] \right| < \epsilon$$

by Theorem 13b. Also

$$\left| \sum_{k=m+1}^{\infty} u_k[\alpha(b) - \alpha(c_k)] \right| \leq \epsilon[\alpha(b) - \alpha(a)].$$

Hence

$$\left| \sum_{k=1}^{\infty} u_k[\alpha(b) - \alpha(c_k)] - \sigma_n \right| < \epsilon[1 + 2\alpha(b) - 2\alpha(a)],$$

if $\delta$ is sufficiently small. This shows that

$$\lim_{\delta \to 0} \sigma_n = \sum_{k=1}^{\infty} u_k[\alpha(b) - \alpha(c_k)],$$

and the proof is complete.

## 14. Further Conditions of Integrability

The results of the previous section enable us to discuss the integrability of one function of bounded variation with respect to another. Preliminary to this discussion we introduce two lemmas.

**LEMMA 14a.** *Let $f(x)$ be of bounded variation in an interval $a \leqq x \leqq b$; let $f(x)$ be defined outside the interval in any way so that $f(a-)$ and $f(b+)$ exist; and let the points of discontinuity of $f(x)$ in $a \leqq x \leqq b$ be $c_1, c_2, \cdots$. Then the functions*

$$\varphi(x) = \sum_{c_k \leqq x} [f(c_k) - f(c_k-)] \qquad (-\infty < x < \infty)$$

$$\psi(x) = \sum_{c_k < x} [f(c_k+) - f(c_k)] \qquad (-\infty < x < \infty)$$

*are continuous in $(-\infty, \infty)$ except at the points $c_1, c_2, \cdots$, where*

(1) $\qquad \varphi(c_k) - \varphi(c_k-) = f(c_k) - f(c_k-)$

(2) $\qquad \varphi(c_k+) - \varphi(c_k) = 0 = \psi(c_k) - \psi(c_k-)$

(3) $\qquad \psi(c_k+) - \psi(c_k) = f(c_k+) - f(c_k).$

The proof will be sufficiently clear if we prove (1). Since $f(x)$ is of bounded variation the series

$$\sum_{k=1}^{\infty} |f(c_k) - f(c_k-)|$$

converges. Corresponding to a given positive $\epsilon$ we determine $m$ so that

$$\sum_{k=m+1}^{\infty} |f(c_k) - f(c_k-)| < \epsilon.$$

If $x_0$ is not one of the $c_k$ we can find a positive $\delta$ so small that the points $c_1, c_2, \cdots, c_m$ are not in the interval $(x_0 - \delta, x_0 + \delta)$. For $h$ in absolute value less than this $\delta$ we have

$$|\varphi(x_0 + h) - \varphi(x_0)| \leqq \sum_{x_0 - \delta < c_k \leqq x_0 + \delta} |f(c_k) - f(c_k-)| < \epsilon.$$

This proves that $\varphi(x)$ is continuous except perhaps at points $c_k$.

For a point $c_j$ we have

$$\varphi(x) = f(c_j) - f(c_j-) + \sum_{c_k \leqq x}{}' [f(c_k) - f(c_k-)] \qquad (c_j \leqq x)$$

$$= \sum_{c_k \leqq x}{}' [f(c_k) - f(c_k-)], \qquad (x > c_j)$$

where the primed summation is taken over all the $c_k$ indicated except $c_j$. The function defined by the primed summation is continuous at $x = c_j$ by the first part of our proof. Hence (1) is established.

LEMMA 14b. *Under the conditions of Lemma 14a the function*

$$g(x) = f(x) - \varphi(x) - \psi(x)$$

*is continuous in* $a \leq x \leq b$.

The proof follows in an obvious way from equations (1), (2) and (3).

THEOREM 14. *If $f(x)$ and $\alpha(x)$ are of bounded variation in $a \leq x \leq b$ and have no common points of discontinuity, then each is integrable with respect to the other from a to b.*

We may assume that $\alpha(x)$ is non-decreasing. Then since $g(x)$ is continuous it is integrable with respect to $\alpha(x)$. The same is true of $\varphi(x)$ by Theorem 13c and of $\psi(x)$ by an obvious modification of that theorem. It is consequently true of $f(x)$ and our result is established.

## 15. Iterated Integrals

In this section we state without proof several theorems regarding the change in the order of iterated integrals.

THEOREM 15a. *If $f(x, y)$ is continuous in the rectangle $a \leq x \leq b$, $c \leq y \leq d$ and if $\alpha(x)$ and $\beta(y)$ are of bounded variation in the intervals $a \leq x \leq b$, $c \leq y \leq d$, respectively, then*

$$\int_a^b d\alpha(x) \int_c^d f(x, y) \, d\beta(y) = \int_c^d d\beta(y) \int_a^b f(x, y) \, d\alpha(x).$$

THEOREM 15b. *If $f(x)$ is continuous in $a \leq x \leq b$, if $\beta(y)$ is of bounded variation in $c \leq y \leq d$, and if $\alpha(x, y)$ is a continuous function of $y$ in $c \leq y \leq d$ for each $x$ in $a \leq x \leq b$ and of uniformly bounded variation in $x$ for all values of $y$ in $c \leq y \leq d$, then*

$$\int_a^b f(x) \, d_x \int_c^d \alpha(x, y) \, d\beta(y) = \int_c^d d\beta(y) \int_a^b f(x) \, d_x \alpha(x, y).$$

A proof of this will be found in H. E. Bray [1919]. Compare also N. Wiener [1933] p. 26.

THEOREM 15c. *Let $f(x, y)$ be continuous for all $x$ and $y$; let $\alpha(x)$ and $\beta(y)$ be functions of bounded variation in every finite interval; let the integrals*

$$\int_{-\infty}^{\infty} f(x, y) \, d\beta(y), \qquad \int_{-\infty}^{\infty} f(x, y) \, d\alpha(x)$$

*be continuous functions of $x$ and of $y$ respectively for all values of these variables; and let at least one of the iterated integrals*

$$\int_{-\infty}^{\infty} |\, d\alpha(x) \,| \int_{-\infty}^{\infty} |f(x, y)| \, |\, d\beta(y) \,|$$

$$\int_{-\infty}^{\infty} |\, d\beta(y) \,| \int_{-\infty}^{\infty} |f(x, y)| \, |\, d\alpha(x) \,|$$

be finite. Then

(1) $$\int_{-\infty}^{\infty} d\alpha(x) \int_{-\infty}^{\infty} f(x, y) \, d\beta(y) = \int_{-\infty}^{\infty} d\beta(y) \int_{-\infty}^{\infty} f(x, y) \, d\alpha(x).$$

By an integral of the form

$$\int_{-\infty}^{\infty} |f(x, y)| \, |d\alpha(x)|$$

we mean

$$\int_{0}^{\infty} |f(x, y)| \, dV_1(x) - \int_{-\infty}^{0} |f(x, y)| \, dV_2(x),$$

where $V_1(x)$ is the variation of $\alpha(t)$ in $0 \leq t \leq x$ and $V_2(x)$ is the variation of $\alpha(t)$ in $x \leq t \leq 0$.

This theorem is a special case of the following more general one in which the integrals involved are Lebesgue-Stieltjes integrals. We shall refer to it as the Fubini theorem.

THEOREM 15d. *Let $\alpha(x)$ and $\beta(y)$ be functions of bounded variation in every finite interval and let $f(x, y)$ be such that at least one of the iterated integrals*

$$\int_{-\infty}^{\infty} |d\alpha(x)| \int_{-\infty}^{\infty} |f(x, y)| \, |d\beta(y)|,$$

$$\int_{-\infty}^{\infty} |d\beta(y)| \int_{-\infty}^{\infty} |f(x, y)| \, |d\alpha(x)|$$

*is finite. Then* (1) *holds.*

For a proof of this theorem see S. Saks [1933], pp. 76–81. In the application we shall have to make of this theorem, each of the simple integrals involved in (1) will exist as Riemann-Stieltjes integrals at least for all but a countable number of values of the parameter.

## 16. The Selection Principle

In this section we discuss the selection principle or diagonal process as applied to an important result of E. Helly [1921] concerning bounded sequences of functions of bounded variation.*

16.1. We begin by considering bounded sequences of constants.

THEOREM 16.1. *Let the complex constants $a_{m,n}$ and the positive number $A$ be such that*

$$|a_{m,n}| < A \qquad (m, n = 0, 1, 2, \cdots).$$

* Compare also A. Wintner [1939] p. 81.

*Then there exists a set of positive integers*

$$n_0 < n_1 < n_2 < \cdots$$

*and a set of numbers* $a_0, a_1, \cdots$ *such that*

$$\lim_{i \to \infty} a_{m,n_i} = a_m \qquad (m = 0, 1, 2, \cdots).$$

Since the sequence $\{a_{0,n}\}_{n=0}^{\infty}$ is bounded, there is at least one limit point, which we may denote by $a_0$. Hence we may find integers

(1) $\qquad\qquad\qquad n_0^0 < n_1^0 < n_2^0 < \cdots$

such that

$$\lim_{i \to \infty} a_{0,n_i^0} = a_0.$$

Likewise the set $\{a_{1,n_i^0}\}_{i=0}^{\infty}$ has a limit point $a_1$, so that we may find integers

$$n_0^1 < n_1^1 < n_2^1 < \cdots$$

all included in (1) such that

$$\lim_{i \to \infty} a_{1,n_i^1} = a_1.$$

Proceeding in this way we obtain for any integer $m$ a set of integers

$$n_0^m < n_1^m < n_2^m < \cdots$$

all included in the set

$$n_0^{m-1} < n_1^{m-1} < n_2^{m-1} < \cdots$$

and a number $a_m$ such that

$$\lim_{i \to \infty} a_{m,n_i^m} = a_m.$$

If now we set

$$n_i = n_i^i \qquad\qquad (i = 0, 1, 2, \cdots)$$

it is clear that the sequence $\{n_i\}_0^{\infty}$ is increasing and that

$$\lim_{i \to \infty} a_{m,n_i} = a_m \qquad (m = 0, 1, 2, \cdots).$$

16.2. We turn next to sequences of monotonic functions.

THEOREM 16.2. *Let the real non-decreasing functions* $\alpha_n(x)$ *and the positive constant* $A$ *be such that*

$$|\alpha_n(x)| < A \qquad (n = 0, 1, 2, \cdots; a \leq x \leq b).$$

*Then there exists a set of integers*

$$n_0 < n_1 < n_2 < \cdots$$

*and a non-decreasing bounded function $\alpha(x)$ such that*

$$\lim_{i \to \infty} \alpha_{n_i}(x) = \alpha(x) \qquad (a \leq x \leq b).$$

Arrange all the rational numbers of $(a, b)$ in a sequence $\{x_m\}_0^\infty$. Then apply Theorem 16.1 to the set of numbers

$$a_{m,n} = \alpha_n(x_m) \qquad (n, m = 0, 1, 2, \cdots).$$

We thus obtain a function $\alpha(x)$ defined at the rational points and a sequence of integers

(1) $$n_0^0 < n_1^0 < n_2^0 < \cdots$$

such that

$$\lim_{i \to \infty} \alpha_{n_i^0}(x_m) = \alpha(x_m) \qquad (m = 0, 1, 2, \cdots)$$

Now set

$$\varlimsup_{i \to \infty} \alpha_{n_i^0}(x) = \bar{\alpha}(x)$$

$$\varliminf_{i \to \infty} \alpha_{n_i^0}(x) = \underline{\alpha}(x).$$

Then clearly

$$\alpha(x_m) = \bar{\alpha}(x_m) = \underline{\alpha}(x_m) \qquad (m = 0, 1, 2, \cdots).$$

Moreover, from their very definitions it is evident that $\bar{\alpha}(x)$ and $\underline{\alpha}(x)$ are non-decreasing functions. Their common points of continuity are consequently dense in $(a, b)$. Let $y$ be such a point. Then

$$\bar{\alpha}(y) = \underline{\alpha}(y)$$

since one can find a sequence of rational numbers $\{x_{m_i}\}_{i=0}^\infty$ for which

$$y = \lim_{i \to \infty} x_{m_i},$$

and since $\bar{\alpha}(x) = \underline{\alpha}(x)$ for each of these rational numbers. We define $\alpha(y)$ as

$$\alpha(y) = \bar{\alpha}(y) = \underline{\alpha}(y).$$

Thus $\alpha(x)$ has been defined at all but a countable set of points $\{z_m\}_{m=0}^\infty$, the points of discontinuity of $\bar{\alpha}(x)$ or $\underline{\alpha}(x)$. Apply Theorem 16.1 to the set of numbers

$$\alpha_{n_i^0}(z_m) \qquad (i, m = 0, 1, 2, \cdots).$$

We obtain a set of integers
$$n_0 < n_1 < n_2 < \cdots$$
all included in the sequence (1) and a sequence $\{a_m\}_0^\infty$ such that
$$\lim_{i \to \infty} \alpha_{n_i}(z_m) = a_m \qquad (m = 0, 1, 2, \cdots).$$
Complete the definition of $\alpha(x)$ by the equations
$$\alpha(z_m) = a_m \qquad (m = 0, 1, 2, \cdots).$$
We then see that
$$\lim_{i \to \infty} \alpha_{n_i}(x) = \alpha(x) \qquad (a \leqq x \leqq b),$$
that $\alpha(x)$ is non-decreasing in $(a, b)$, and that
$$|\alpha(x)| \leqq A.$$
This completes the proof of the theorem.

16.3. The results of the previous section are easily extended to sequences of functions of bounded variation as follows.

THEOREM 16.3. *Let the sequence of functions $\{\alpha_n(x)\}_0^\infty$ be of uniformly bounded variation in $a \leqq x \leqq b$ and such that*
$$|\alpha_n(a)| < A \qquad (n = 0, 1, 2, \cdots)$$
*for some constant $A$. Then there exists a set of integers*
$$n_0 < n_1 < n_2 < \cdots$$
*and a function $\alpha(x)$ of bounded variation in $a \leqq x \leqq b$ such that*
$$\lim_{i \to \infty} \alpha_{n_i}(x) = \alpha(x) \qquad (a \leqq x \leqq b).$$

It will be sufficient to prove the result for real functions. For, if $\beta_n(x)$ and $\gamma_n(x)$ are the real and imaginary parts of $\alpha_n(x)$ respectively, the sequences $\{\beta_n(x)\}_0^\infty$ and $\{\gamma_n(x)\}_0^\infty$ both satisfy the hypothesis of the theorem. For a suitable sequence of integers $\{n_i^0\}_{i=0}^\infty$ and a function $\beta(x)$ of bounded variation we should have
$$\lim_{i \to \infty} \beta_{n_i^0}(x) = \beta(x) \qquad (a \leqq x \leqq b).$$
From the sequence $\{n_i^0\}_{i=0}^\infty$ we could then pick another $\{n_i^1\}_{i=0}^\infty$, and we could determine a function $\gamma(x)$ of bounded variation in $(a, b)$ such that
$$\lim_{i \to \infty} \gamma_{n_i^1}(x) = \gamma(x). \qquad (a \leqq x \leqq b).$$

The complex function $\alpha(x)$ whose real and imaginary parts are $\beta(x)$ and $\gamma(x)$ respectively would then be the function of bounded variation sought, $\{n_i^1\}_0^\infty$ the sequence sought.

To prove the theorem in the real case we write $\alpha_n(x)$ as the difference of two non-decreasing functions,

$$\alpha_n(x) = \eta_n(x) - \zeta_n(x) + \alpha_n(a)$$
$$V[\alpha_n]_a^x = \eta_n(x) + \zeta_n(x).$$

By hypothesis there exists a constant $M$ such that

$$0 \leq V[\alpha_n]_a^b < M \qquad (n = 0, 1, 2, \cdots),$$

so that the sequences $\{\eta_n(x)\}_0^\infty$ and $\{\zeta_n(x)\}_0^\infty$ satisfy the conditions of Theorem 16.2. Moreover, the sequence $\{\alpha_n(a)\}_0^\infty$ is bounded by hypothesis. Hence by three applications of the selection principle we obtain the desired result.

COROLLARY 16.3. *The theorem holds if the interval $a \leq x \leq b$ is replaced by $a \leq x < \infty$ throughout.*

For, let $M$ be a uniform upper bound for the total variation of the sequence $\{\alpha_n(x)\}_0^\infty$ on $a \leq x < \infty$. By the theorem there exists a sequence of integers

(1) $$n_0^1 < n_1^1 < n_2^1 < \cdots$$

and a function $\alpha(x)$ of total variation not greater than $M$ such that

(2) $$\lim_{i \to \infty} \alpha_{n_i^1}(x) = \alpha(x) \qquad (a \leq x \leq a + 1).$$

Moreover, it is possible by a second application of the theorem to pick from the sequence (1) a subsequence

$$n_0^2 < n_1^2 < n_2^2 < \cdots$$

and to determine a function $\beta(x)$ of variation not exceeding $M$ in $a \leq x \leq a + 2$ such that

$$\lim_{i \to \infty} \alpha_{n_i^2}(x) = \beta(x) \qquad (a \leq x \leq a + 2).$$

But by (2), $\beta(x) = \alpha(x)$ in $a \leq x \leq a + 1$, so that $\beta(x)$ may be regarded as extending the domain of definition of $\alpha(x)$. Thus by use of successive subsequences we successively increase the domain of definition of $\alpha(x)$. Finally, by use of the diagonal process we obtain a subsequence of the integers

$$n_0 < n_1 < n_2 < \cdots$$

such that
$$\lim_{i \to \infty} \alpha_{n_i}(x) = \alpha(x) \qquad (a \leq x < \infty).$$

It is clear from the way in which $\alpha(x)$ was defined that its variation in $a \leq x < \infty$ does not exceed $M$. This completes the proof of the corollary.

16.4. Another result, which we shall need, was proved independently by E. Helly [1921] and H. E. Bray [1919]; compare also G. C. Evans [1927] p. 15.

THEOREM 16.4. *Let the sequence of functions $\{\alpha_n(x)\}_0^\infty$ be of uniformly bounded variation in $a \leq x \leq b$. let*
$$\lim_{n \to \infty} \alpha_n(x) = \alpha(x) \qquad (a \leq x \leq b),$$
*and let $f(x)$ be continuous in $a \leq x \leq b$. Then*
$$\lim_{n \to \infty} \int_a^b f(x) \, d\alpha_n(x) = \int_a^b f(x) \, d\alpha(x).$$

The hypothesis implies that $\alpha(x)$ is of bounded variation. Hence there exists a number $T$ which is not less than the variation of any of the functions $\alpha_n(x)$ or $\alpha(x)$. Let $\epsilon$ be an arbitrary positive number. Choose a subdivision of $(a, b)$ of norm so small that the oscillation of $f(x)$ is less than $\epsilon$ in any of the sub-intervals. If the points of subdivision are

(1) $$a = x_0 < x_1 < \cdots < x_m = b.$$

we have
$$H_n = \int_a^b f(x) \, d\alpha_n(x) - \int_a^b f(x) \, d\alpha(x)$$
$$= \sum_{i=0}^{m-1} \int_{x_i}^{x_{i+1}} [f(x) - f(x_i)] \, d\alpha_n(x) - \sum_{i=0}^{m-1} \int_{x_i}^{x_{i+1}} [f(x) - f(x_i)] \, d\alpha(x)$$
$$+ \sum_{i=0}^{m-1} f(x_i) \int_{x_i}^{x_{i+1}} d[\alpha_n(x) - \alpha(x)].$$

Hence
$$|H_n| \leq \epsilon T + \epsilon T + \max_{a \leq x \leq b} |f(x)| \sum_{i=0}^{m-1} \left| \int_{x_i}^{x_{i+1}} d[\alpha_n(x) - \alpha(x)] \right|.$$

By letting $n$ become infinite we obtain
$$\varlimsup_{n \to \infty} |H_n| \leq 2\epsilon T,$$
from which the result is immediately obvious.

32    THE STIELTJES INTEGRAL    [Ch. I

COROLLARY 16.4. *Under the conditions of the theorem*

$$\int_a^b |d\alpha(x)| \leq \lim_{n \to \infty} \int_a^b |d\alpha_n(x)|.$$

For if $\Delta$ is an arbitrary subdivision (1) of $(a, b)$, then

$$\sum_{i=0}^{m-1} \left| \int_{x_i}^{x_{i+1}} d\alpha_n(x) \right| \leq \sum_{i=0}^{m-1} \int_{x_i}^{x_{i+1}} |d\alpha_n(x)|$$

$$= \int_a^b |d\alpha_n(x)|.$$

Allowing $n$ to become infinite we obtain

$$\sum_{i=0}^{m-1} |\alpha(x_{i+1}) - \alpha(x_i)| \leq \lim_{n \to \infty} \int_a^b |d\alpha_n(x)|.$$

But by a suitable choice of $\Delta$ the left-hand side of this inequality may be brought arbitrarily near to the variation of $\alpha(x)$ in $(a, b)$. This establishes the corollary.

16.5. It should be observed that the results of the previous section do not hold for an infinite interval. In fact Theorem 16.4 fails if the hypothesis holds in $a \leq x < b$ and the integral is replaced by a Cauchy-value

$$\int_a^{b-} f(x) \, d\alpha_n(x).$$

For, take $a = 0$, $b = 1$, $f(x) = 1$, and define $\alpha_n(x)$ as follows

$$\alpha_n(x) = 0 \qquad \left(0 \leq x \leq 1 - \frac{1}{n}\right)$$

$$= nx - n + 1 \qquad \left(1 - \frac{1}{n} \leq x < 1\right).$$

Then the sequence $\{\alpha_n(x)\}_1^\infty$ is of uniformly bounded variation in $(0, 1)$ and

$$\lim_{n \to \infty} \alpha_n(x) = \alpha(x) = 0 \qquad (0 \leq x < 1).$$

But

$$\lim_{n \to \infty} \int_0^{1-} d\alpha_n(x) = \lim_{n \to \infty} \alpha_n(1-) = 1$$

$$\int_0^{1-} d\alpha(x) = \alpha(1-) = 0.$$

By use of an exponential change of variable this example also shows that the Helly-Bray theorem, Theorem 16.4, does not hold on an infinite interval.

## 17. Weak Compactness

In later work we shall have occasion to use several theorems concerning *weak compactness* of a set of functions. We state them here for convenience of reference and refer the student to a proof.*

THEOREM 17a. *If for some real number p greater than unity each function of the sequence $\{\alpha_n(x)\}_0^\infty$ belongs to the class $L^p$ in $(a, b)$, and if there exists a constant $M_p$ such that*

$$\int_a^b |\alpha_n(t)|^p \, dt < M_p \qquad (n = 0, 1, 2, \cdots).$$

*then it is possible to find a function $\alpha(x)$ of $L^p$ in $(a, b)$ and a sequence of integers*

$$n_0 < n_1 < n_2 < \cdots$$

*such that*

$$\lim_{i \to \infty} \int_a^b \alpha_{n_i}(t)\beta(t) \, dt = \int_a^b \alpha(t)\beta(t) \, dt$$

*for every function $\beta(x)$ of $L^{p/(p-1)}$ in $(a, b)$. Moreover,*

$$\left\{ \int_a^b |\alpha(t)|^p \, dt \right\}^{1/p} \leq \lim_{i \to \infty} \left\{ \int_a^b |\alpha_{n_i}(t)|^p \, dt \right\}^{1/p}$$

THEOREM 17b. *If the functions of the sequence $\{\alpha_n(x)\}_0^\infty$ are uniformly bounded in $a \leq x \leq b$, then there exists a function $\alpha(x)$ bounded there and a sequence of integers*

$$n_0 < n_1 < n_2 < \cdots$$

*such that*

$$\lim_{i \to \infty} \int_a^b \alpha_{n_i}(t)\beta(t) \, dt = \int_a^b \alpha(t)\beta(t) \, dt$$

*for every function $\beta(x)$ of $L$ in $(a, b)$. Moreover,*

$$\text{true max}_{a \leq x \leq b} |\alpha(x)| \leq \lim_{i \to \infty} \text{true max}_{a \leq x \leq b} |\alpha_{n_i}(x)|$$

* See, for example S. Banach [1932] p. 130.

In the latter theorem the symbol

$$\text{true} \max_{a \leq x \leq b} |\alpha(x)|$$

means the greatest lower bound of numbers $w$ for which

$$|\alpha(x)| \leq w$$

almost everywhere in $(a, b)$.

It is easily seen by an exponential change of variable that these theorems hold equally well on an infinite interval.

## CHAPTER II

## THE LAPLACE TRANSFORM

### 1. Region of Convergence

Let $\alpha(t)$ be a complex function of the real variable $t$ defined on the interval $0 \leq t < \infty$. Denote its real and imaginary parts by $\alpha'(t)$ and $\alpha''(t)$ respectively,

$$\alpha(t) = \alpha'(t) + i\alpha''(t).$$

Let $\alpha(t)$ be of bounded variation in the interval $0 \leq t \leq R$ for every positive $R$. Clearly this will be the case if and only if $\alpha'(t)$ and $\alpha''(t)$ have the same property.

Let $s$ be a complex variable with real and imaginary parts $\sigma$ and $\tau$ respectively,

$$s = \sigma + i\tau.$$

It follows from Theorem 4a of Chapter I that the integral

$$\int_0^R e^{-st}\, d\alpha(t)$$

exists for each positive $R$ and for every complex $s$. It has the value

$$\int_0^R e^{-\sigma t} \cos \tau t\, d\alpha'(t) - \int_0^R e^{-\sigma t} \sin \tau t\, d\alpha''(t)$$

$$+ i\int_0^R e^{-\sigma t} \sin \tau t\, d\alpha'(t) + i\int_0^R e^{-\sigma t} \cos \tau t\, d\alpha''(t).$$

We now define the improper integral

(1) $$\int_0^\infty e^{-st}\, d\alpha(t) = \lim_{R \to \infty} \int_0^R e^{-st}\, d\alpha(t).$$

If this limit exists for a given value of $s$ we say the integral (1) converges for that value of $s$. Or, we say that the right-hand side of (1) is the *Cauchy value* of the integral at its upper limit. If the limit (1) does not exist, the integral diverges.

It is sometimes useful to consider the Cauchy value of the integral at its lower limit also. If $\alpha(t)$ is of bounded variation in the interval

$\epsilon \leq t \leq R$ for every positive $\epsilon$ and every positive $R$, then we use the notation

$$(2) \qquad \int_{0+}^{\infty} e^{-st}\, d\alpha(t) = \lim_{\substack{\epsilon \to 0 \\ R \to \infty}} \int_{\epsilon}^{R} e^{-st}\, d\alpha(t)$$

when this limit exists. For example, if $\alpha(0) = 0$, $\alpha(t) = t \sin(1/t)$ ($t > 0$), then $\alpha(t)$ is not of bounded variation in any interval including the origin. Yet the integral (2) exists for $\sigma > 0$, and

$$\int_{0+}^{\infty} e^{-st}\, d\alpha(t) = s \int_{0}^{\infty} e^{-st} t \sin(1/t)\, dt \qquad (\sigma > 0).$$

We shall understand throughout when the integral (1) is written that $\alpha(t)$ is of bounded variation in $(0, R)$ for every positive $R$; when (2) is written that $\alpha(t)$ is of bounded variation in $(\epsilon, R)$ for all positive numbers $\epsilon$ and $R$. We now prove*

THEOREM 1. *If*

$$(3) \qquad \underset{0 \leq u < \infty}{\text{u.b.}} \left| \int_{0}^{u} e^{-s_0 t}\, d\alpha(t) \right| = M < \infty$$

*then* (1) *converges for every $s$ for which $\sigma > \sigma_0$, and*

$$(4) \qquad \int_{0}^{\infty} e^{-st}\, d\alpha(t) = (s - s_0) \int_{0}^{\infty} e^{-(s-s_0)t} \beta(t)\, dt,$$

*where*

$$(5) \qquad \beta(u) = \int_{0}^{u} e^{-s_0 t}\, d\alpha(t) \qquad (u \geq 0)$$

*the integral on the right-hand side of* (4) *converging absolutely.*

For, if $\beta(u)$ is defined by (5) we have

$$\int_{0}^{R} e^{-st}\, d\alpha(t) = \int_{0}^{R} e^{-(s-s_0)t}\, d\beta(t)$$

by Theorem 6b of Chapter I. Integration by parts gives

$$\int_{0}^{R} e^{-st}\, d\alpha(t) = e^{-(s-s_0)R} \beta(R) + (s - s_0) \int_{0}^{R} e^{-(s-s_0)t} \beta(t)\, dt.$$

By (3)

$$\lim_{R \to \infty} e^{-(s-s_0)R} \beta(R) = 0$$

* See D. V. Widder [1929]

for $\sigma > \sigma_0$. The same hypothesis shows that

$$\int_0^\infty e^{-(s-s_0)t} \beta(t)\, dt$$

converges absolutely. In fact

$$\left| \int_0^\infty e^{-(s-s_0)t} \beta(t)\, dt \right| \leq M \int_0^\infty e^{-(\sigma-\sigma_0)t}\, dt = \frac{M}{\sigma - \sigma_0} \qquad (\sigma > \sigma_0).$$

Hence the theorem is established.

COROLLARY 1a. *If the integral* (1) *converges for* $s = \sigma_0 + i\tau_0$, *it converges for all* $s = \sigma + i\tau$ *for which* $\sigma > \sigma_0$.

For, the convergence of (1) at $s = s_0$ implies (3).

COROLLARY 1b. *The region of convergence of* (1) *is a half-plane.*

For, Theorem 1 shows that the divergence of (1) at a point $s = s_0$ implies its divergence at all points $s$ for which $\sigma < \sigma_0$. Hence three possibilities arise:

(a) the integral converges for no point;
(b) it converges for every point;
(c) it converges for $\sigma > \sigma_c$ and diverges for $\sigma < \sigma_c$.

In case (c) we define the number $\sigma_c$ as the *abscissa of convergence*, the line $\sigma = \sigma_c$ as the *axis of convergence*. In case (a) we write $\sigma_c = +\infty$; in case (b), $\sigma_c = -\infty$. The following examples show that all three cases may actually arise:

(a) $\qquad\qquad \int_0^\infty e^{-st} e^{e^t}\, dt \qquad\qquad (\sigma_c = \infty)$

(b) $\qquad\qquad \int_0^\infty e^{-st} \cdot e^{-e^t}\, dt \qquad\qquad (\sigma_c = -\infty)$

(c) $\qquad\qquad \int_0^\infty e^{-st}\, dt \qquad\qquad (\sigma_c = 0).$

When the integral (1) converges it defines a function of $s$ which we denote by $f(s)$. This function is called the Laplace-Stieltjes transform of $\alpha(t)$. If

$$f(s) = \int_0^\infty e^{-st} \varphi(t)\, dt$$

we refer to $f(s)$ as the Laplace transform of $\varphi(t)$. In either case $f(s)$ is called the *generating* function. The function $\varphi(t)$ is usually* referred to

---

* This is the terminology adopted by most authors. But S. Pincherle [1905] reverses the names of these functions. The complex inversion formula for the Laplace transform reveals a certain reciprocity between the two functions which

as the *determining* function, but we shall sometimes use this name for $\alpha(t)$ also.

## 2. Abscissa of Convergence

In this section we shall derive a formula for the abscissa of convergence of a Laplace integral in terms of its determining function. This formula is the analogue of the familiar one in the theory of power series

$$\frac{1}{\rho} = \varlimsup_{n \to \infty} \sqrt[n]{|a_n|}.$$

Here $\rho$ is the radius of convergence of the power series

$$\sum_{n=0}^{\infty} a_n z^n.$$

2.1. We first establish certain relations between the order properties of the determining function and the convergence properties of the corresponding Laplace integral.

THEOREM 2.1. *If*

$$\alpha(t) = O(e^{\gamma t}) \qquad (t \to \infty)$$

*for some real number $\gamma$, then the integral*

(1) $$\int_0^\infty e^{-st} \, d\alpha(t)$$

*converges for $\sigma > \gamma$.*

The hypothesis implies the existence of a constant $M$ such that

$$|\alpha(t)| \leq M e^{\gamma t} \qquad (0 \leq t < \infty),$$

since we are assuming that $\alpha(t)$ is of bounded variation in every finite interval. Hence*

$$\int_0^\infty e^{-st} \alpha(t) \, dt \ll M \int_0^\infty e^{-(\sigma-\gamma)t} \, dt = \frac{M}{\sigma - \gamma} \qquad (\sigma > \gamma),$$

---

makes the confusion of terms understandable. The terminology adopted here seems the more natural, since (1) reduces to a power series if $\alpha(t)$ is a suitable step-function. In that case the sum of the series is always referred to as the generating function of the coefficients of the series.

\* The symbol $\ll$ means that the integral on the left is dominated by the one on the right, or that the integrand of the first is in absolute value not greater than that of the second over the whole range of integration.

so that the integral on the left-hand side of the inequality converges absolutely for $\sigma > \gamma$. Moreover

$$\int_0^R e^{-st} d\alpha(t) = \alpha(R)e^{-sR} - \alpha(0) + s\int_0^R e^{-st} \alpha(t)\, dt$$

$$\alpha(R)e^{-sR} = o(1) \qquad (R \to \infty\,;\, \sigma > \gamma),$$

so that the integral (1) is seen to converge when $\sigma > \gamma$. In fact

$$\int_0^\infty e^{-st} d\alpha(t) = s\int_0^\infty e^{-st}\alpha(t)\, dt - \alpha(0) \qquad (\sigma > \gamma).$$

This establishes our result.

COROLLARY 2.1. *If $\alpha(\infty)$ exists and if*

$$\alpha(t) - \alpha(\infty) = O(e^{\gamma t}) \qquad (t \to \infty)$$

*for some real number $\gamma$, then the integral (1) converges for $\sigma > \gamma$.*

This follows at once since

$$\int_0^\infty e^{-st} d[\alpha(t) - \alpha(\infty)] = \int_0^\infty e^{-st} d\alpha(t).$$

If $\gamma < 0$ and $\alpha(\infty) \neq 0$ this corollary yields a larger region of convergence than the theorem, for $\alpha(t)$ itself is no better than $O(1)$ and by Theorem 2.1 $\sigma_c \leq 0$. By Corollary 2.1 $\sigma_c \leq \gamma$.

Note that the exact converse of Theorem 2.1 is not true, as example (c) of §1 shows. That integral converges for $\sigma > 0$, but $\alpha(t) = t$ is not bounded.

2.2. We now proceed from the convergence properties of the integral to the order properties of $\alpha(t)$.

THEOREM 2.2a. *If the integral*

(1) $$\int_0^\infty e^{-st} d\alpha(t)$$

*converges for $s = s_0 = \gamma + i\delta$ with $\gamma > 0$, then*

$$\alpha(t) = o(e^{\gamma t}) \qquad (t \to \infty).$$

For, by Theorem 6b of Chapter I

$$\alpha(t) - \alpha(0) = \int_0^t e^{s_0 u} d\beta(u),$$

where

(2) $$\beta(t) = \int_0^t e^{-s_0 u} d\alpha(u) \qquad (0 < t < \infty).$$

Integration by parts gives

$$\alpha(t) - \alpha(0) = \beta(t)e^{s_0 t} - s_0 \int_0^t e^{s_0 u} \beta(u)\, du.$$

By hypothesis $\beta(\infty)$ exists, so that

$$\lim_{t\to\infty} [\alpha(t) - \alpha(0)]e^{-s_0 t} = \beta(\infty) - \lim_{t\to\infty} s_0 e^{-s_0 t} \int_0^t e^{s_0 u} \beta(u)\, du$$

$$= \lim_{t\to\infty} s_0 e^{-s_0 t} \int_0^t e^{s_0 u} [\beta(\infty) - \beta(u)]\, du.$$

It is easily seen that this last limit is zero if the real part of $s_0$ is positive, whence

$$\alpha(t) - \alpha(0) = o(e^{\gamma t}) \qquad (t \to \infty)$$

$$\alpha(t) = o(e^{\gamma t}) \qquad (t \to \infty).$$

THEOREM 2.2b. *If the integral* (1) *converges for* $s = s_0 = \gamma + i\delta$ *with* $\gamma < 0$, *then* $\alpha(\infty)$ *exists and*

$$\alpha(t) - \alpha(\infty) = o(e^{\gamma t}) \qquad (t \to \infty).$$

Since $\gamma$ is negative the integral (1) converges for $s = 0$, so that the existence of $\alpha(\infty)$ is assured. But

$$\alpha(\infty) - \alpha(t) = \int_t^\infty e^{s_0 u}\, d\beta(u),$$

where $\beta(t)$ is defined by (2). Integration by parts gives

$$\alpha(\infty) - \alpha(t) = -e^{s_0 t}\beta(t) - s_0 \int_t^\infty e^{s_0 u} \beta(u)\, du$$

$$\lim_{t\to\infty} [\alpha(\infty) - \alpha(t)]e^{-s_0 t} = -\beta(\infty) - \lim_{t\to\infty} s_0 e^{-s_0 t} \int_t^\infty e^{s_0 u} \beta(u)\, du$$

$$= \lim_{t\to\infty} s_0 e^{-s_0 t} \int_t^\infty e^{s_0 u}[\beta(\infty) - \beta(u)]\, du$$

$$= 0.$$

This proves the theorem.

It is to be noticed that both theorems are false when $\gamma = 0$. For if $\alpha(0) = 0$, $\alpha(t) = 1$ $(t > 0)$, the integral (1) converges for all $s$. Yet $\alpha(t) \neq o(1)$ as $t$ becomes infinite. Also if $\alpha(t) = 2$ $(t \leq 1)$, $\alpha(t) = 2\sqrt{t}$ $(t > 1)$, (1) becomes at $s = i$

$$\int_1^\infty \frac{e^{-iu}}{\sqrt{u}}\, du,$$

a convergent integral. Yet $\alpha(\infty)$ does not exist.

2.3. By use of these results we may frequently express a Laplace-Stieltjes integral in terms of an ordinary **Laplace integral**.

THEOREM 2.3a. *If the integral*

$$(1) \qquad f(s_0) = \int_0^\infty e^{-s_0 t}\, d\alpha(t)$$

*converges with* $\sigma_0 > 0$, *then*

$$(2) \qquad f(s_0) = s_0 \int_0^\infty e^{-s_0 t} \alpha(t)\, dt - \alpha(0).$$

*The integral* (2) *converges absolutely if* $s_0$ *is replaced by any number with larger real part.*

For,

$$\int_0^R e^{-s_0 t}\, d\alpha(t) = e^{-s_0 R} \alpha(R) + s_0 \int_0^R e^{-s_0 t} \alpha(t)\, dt - \alpha(0)$$

By letting $R$ become infinite and making use of Theorem 2.2a we obtain (2) That the integral

$$(3) \qquad \int_0^\infty e^{-st} \alpha(t)\, dt$$

converges absolutely for $\sigma > \sigma_0$ follows from the fact that

$$\alpha(t) = o(e^{\sigma_0 t}) \qquad (t \to \infty).$$

But it must not be supposed that (2) converges absolutely or even that (3) converges on the whole line $\sigma = \sigma_0$. For example, if

$$\alpha(t) = \int_0^t \frac{e^u}{\sqrt{u}}\, du \qquad (t \geq 0)$$

and if $\sigma_0 = 1$, then (3) converges for $s_0 = 1 + i$ since (1) becomes

$$\int_0^\infty \frac{e^{-it}}{\sqrt{t}}\, dt.$$

But (3) does not converge absolutely at $s = 1 + i$, nor does it converge at $s = 1$, since by Fubini's theorem

$$\int_0^\infty e^{-t}\, dt \int_0^t \frac{e^u}{\sqrt{u}}\, du = \int_0^\infty \frac{e^u}{\sqrt{u}}\, du \int_u^\infty e^{-t}\, dt = \int_0^\infty \frac{du}{\sqrt{u}} = \infty.$$

THEOREM 2.3b. *If* (1) *converges with* $\sigma_0 < 0$, *then* $\alpha(\infty)$ *exists and*

$$(4) \qquad f(s_0) = \alpha(\infty) - \alpha(0) + s_0 \int_0^\infty e^{-s_0 t} [\alpha(t) - \alpha(\infty)]\, dt.$$

*The integral* (4) *converges absolutely if $s_0$ is replaced by any number with larger real part.*

For, we have

$$f(s_0) = \int_0^\infty e^{-s_0 t} d[\alpha(t) - \alpha(\infty)].$$

Integrating by parts and using Theorem 2.2b we obtain (4).

**2.4.** We are now in a position to establish the formula for the abscissa of convergence.

THEOREM 2.4a. *If*

$$\varlimsup_{t \to \infty} \frac{\log |\alpha(t)|}{t} = k \neq 0,$$

*then $\sigma_c = k$ for the integral*

(1) $$\int_0^\infty e^{-st} d\alpha(t).$$

First consider the case $k > 0$. We show that (1) converges for $\sigma > k$. Let $\epsilon$ be an arbitrary positive constant. The hypothesis implies that

$$\alpha(t) = O(e^{(k+\epsilon)t}) \qquad (t \to \infty).$$

Hence by Theorem 2.1 the integral (1) converges for $\sigma > k + \epsilon$. That is, (1) converges for $\sigma > k$.

We show next that (1) diverges for $\sigma < k$. Suppose it converged for $s = \gamma$ where

(2) $$0 < \gamma < k.$$

By Theorem 2.2a we should have

$$\alpha(t) = o(e^{\gamma t}) \qquad (t \to \infty).$$

But this implies the existence of constants $M$ and $t_0$ such that

$$|\alpha(t)| < Me^{\gamma t}$$

for all $t$ greater than $t_0$. Hence

$$\log |\alpha(t)| < \log M + \gamma t \qquad (t_0 < t < \infty),$$

whence

$$\varlimsup_{t \to \infty} \frac{\log |\alpha(t)|}{t} = k \leqq \gamma.$$

This contradicts (2) so that our theorem is established for positive $k$.

If $k < 0$ the same argument used for positive $k$ shows that (1) converges for $\sigma > k$. The hypothesis now clearly implies that $\alpha(\infty) = 0$. If (1) converged for $\sigma = \gamma < k$, we should have by Theorem 2.2b that

$$\alpha(t) = o(e^{\gamma t}) \qquad (t \to \infty),$$

whence

$$\varlimsup_{t \to \infty} \frac{\log |\alpha(t)|}{t} = k \leqq \gamma < k.$$

The contradiction shows that (1) must diverge for $\sigma < k$, and our result is completely established.

COROLLARY 2.4a. *If*

$$\varlimsup_{t \to \infty} \frac{\log |\alpha(t)|}{t} = +\infty \, (-\infty),$$

*then* $\sigma_c = \infty \, (-\infty)$.

THEOREM 2.4b. *If the integral (1) has a non-negative abscissa of convergence $\sigma_c$, then*

$$(3) \qquad \sigma_c = \varlimsup_{t \to \infty} \frac{\log |\alpha(t)|}{t}.$$

First suppose $\sigma_c = 0$. Then if the limit superior in (3) were different from zero we should have a contradiction by Theorem 2.4a. If $\sigma_c > 0$ and if the limit superior (3) were different from $\sigma_c$ and different from zero we obtain a contradiction again by use of Theorem 2.4a. If this limit superior were zero we could conclude only that $\sigma_c \leqq 0$; but this also contradicts the hypothesis.

THEOREM 2.4c. *If*

$$\varlimsup_{t \to \infty} \frac{\log |\alpha(t)|}{t} = 0,$$

*and if $\alpha(t)$ approaches no limit as $t$ becomes infinite, then $\sigma_c = 0$.*

For, the hypothesis clearly implies that (1) converges for $\sigma > 0$ and diverges for $s = 0$. Note that the theorem is false without the hypothesis that $\alpha(t)$ approaches no limit. For, if $\alpha(t) = 1 - e^{-t}$ then $k = 0$ but $\sigma_c = -1$.

THEOREM 2.4d. *If $\alpha(\infty)$ exists and if*

$$(4) \qquad \varlimsup_{t \to \infty} \frac{\log |\alpha(t) - \alpha(\infty)|}{t} = l \leqq 0,$$

*then* $\sigma_c = l$.

First note that $k = 0$ implies (4) whenever $\alpha(\infty)$ exists. For, $k = 0$ implies for every positive $\epsilon$ that

$$\alpha(t) = O(e^{\epsilon t}) \qquad (t \to \infty)$$

$$\alpha(t) - \alpha(\infty) = O(e^{\epsilon t}) \qquad (t \to \infty),$$

and this implies (4).

Now Corollary 2.1 shows that (1) converges for $\sigma > l$. On the other hand if (1) converged for $s = \gamma < l$, then $\gamma$ would be negative and we should have by Theorem 2.2b that

$$\alpha(\infty) - \alpha(t) = o(e^{\gamma t}) \qquad (t \to \infty)$$

$$l = \varlimsup_{t \to \infty} \frac{\log |\alpha(\infty) - \alpha(t)|}{t} \leq \gamma < l.$$

The contradiction shows that (1) diverges for $\sigma < l$. Hence $\sigma_c = l$.

**THEOREM 2.4e.** *If* (1) *has a negative abscissa of convergence* $\sigma_c$, *then* $\alpha(\infty)$ *exists and*

(5) $$\sigma_c = \varlimsup_{t \to \infty} \frac{\log |\alpha(\infty) - \alpha(t)|}{t}$$

Clearly (1) must converge at $s = 0$ so that $\alpha(\infty)$ exists. Also $|\alpha(\infty) - \alpha(t)|$ is bounded so that the limit superior (5) is less than or equal to zero. Hence Theorem 2.4d is applicable, giving a contradiction if the limit superior (5) were different from $\sigma_c$.

2.5. All the theorems proved thus far in this chapter apply equally well to Dirichlet series

(1) $$f(s) = \sum_{n=1}^{\infty} a_n e^{-\lambda_n s}$$

$$0 \leq \lambda_1 < \lambda_2 < \lambda_3 < \cdots, \qquad \lim_{n \to \infty} \lambda_n = \infty$$

For, if $\alpha(t)$ is defined by the equations

$$\alpha(t) = a_1 + a_2 + \cdots + a_n \qquad (\lambda_n < t < \lambda_{n+1})$$

(2) $$\alpha(0) = 0$$

$$\alpha(t) = \frac{\alpha(t+) + \alpha(t-)}{2} \qquad (t > 0),$$

we have

$$\int_0^\infty e^{-st} d\alpha(t) = \sum_{n=1}^\infty a_n e^{-\lambda_n s}$$

whenever the integral or series converges.

For example, we have

**THEOREM 2.5.** *If*

(3) $$\varlimsup_{n\to\infty} \frac{\log \left|\sum_{k=1}^{n} a_k\right|}{\lambda_n} = k \neq 0,$$

*then the abscissa of convergence of* (2.4) *is* $k$.

To prove this we must show that if $\alpha(t)$ is defined by (2) then (3) implies

(4) $$\varlimsup_{t\to\infty} \frac{\log |\alpha(t)|}{t} = k.$$

Setting $s_n$ equal to the sum of the first $n$ coefficients $a_k$, we have

$$\frac{\log |\alpha(t)|}{t} = \frac{\log |s_n|}{t} < \frac{\log |s_n|}{\lambda_n} \qquad (\lambda_n < t < \lambda_{n+1}),$$

so that

$$\varlimsup_{t\to\infty} \frac{\log |\alpha(t)|}{t} \leq k.$$

On the other hand if

$$\varlimsup_{t\to\infty} \frac{\log |\alpha(t)|}{t} < k,$$

it must be possible to find $k' < k$ such that

$$\frac{\log |\alpha(t)|}{t} < k'$$

for all $t$ sufficiently large. That is, for $n$ sufficiently large

$$\frac{\log |s_n|}{t} < k' \qquad (\lambda_n < t < \lambda_{n+1}).$$

Letting $t$ approach $\lambda_n$, we obtain

$$\frac{\log |s_n|}{\lambda_n} \leq k',$$

whence

$$\varlimsup_{n\to\infty} \frac{\log |s_n|}{\lambda_n} \leq k'$$

contrary to the hypothesis (3). Consequently (3) implies (4).

## 3. Absolute Convergence

Denote the total variation of the function $\alpha(t)$ in the interval $0 \leq t \leq x$ by $u(x)$. The integral (1.1) is said to converge absolutely at a point $s = \sigma + i\tau$ if the integral

$$(1) \qquad \int_0^\infty e^{-\sigma t} |d\alpha(t)| = \int_0^{\infty^*} e^{-\sigma t} du(t)$$

converges. In particular, if $\alpha(t)$ is an integral of a function $\varphi(t)$,

$$\alpha(t) = \int_0^t \varphi(y)\, dy,$$

then

$$u(t) = \int_0^t |\varphi(y)|\, dy,$$

and (1) becomes

$$\int_0^\infty e^{-\sigma t} |\varphi(t)|\, dt.$$

It is also easy to see that when $\alpha(t)$ is defined by §2.5 (2) the integral (1) becomes

$$\sum_{n=1}^\infty |a_n| e^{-\lambda_n \sigma},$$

so that our definition in this case conforms with the usual notion of the **absolute convergence of a series**.

3.1. We now show the existence in some cases of a half-plane of absolute convergence.

THEOREM 3.1. *If the integral*

$$(1) \qquad \int_0^\infty e^{-st} d\alpha(t)$$

*converges absolutely for* $s = \sigma_0 + i\tau_0$, *then it converges uniformly and absolutely in the half-plane* $\sigma \geq \sigma_0$.

For,

$$\int_0^\infty e^{-\sigma t} du(t) \ll \int_0^\infty e^{-\sigma_0 t} du(t) \qquad\qquad \sigma \geq \sigma_0.$$

This result enables us to define an axis of absolute convergence $\sigma = \sigma_a$, proceeding as in the definition of $\sigma_c$. The following example shows that $\sigma_c$ and $\sigma_a$ need not be coincident:

$$(2) \qquad \int_0^\infty e^{-st} e^{kt} \sin e^{kt}\, dt \qquad\qquad (k > 0).$$

Here $\sigma_c = 0$ and $\sigma_a = k$. The integral converges absolutely for $\sigma > k$ since

$$| e^{-st}e^{kt} \sin e^{kt} | \leqq e^{-(\sigma-k)t} \qquad (0 \leqq t < \infty).$$

It does not converge absolutely for $s = k$ since

$$\int_0^\infty | \sin e^{kt} | \, dt = \frac{1}{k} \int_1^\infty \frac{| \sin u |}{u} du = \infty.$$

The relation

$$\int_0^\infty e^{-st} e^{kt} \sin e^{kt} \, dt = \frac{1}{k} \int_1^\infty \frac{\sin u}{u^{s/k}} du$$

shows clearly that $\sigma_c = 0$.

That the axis of absolute convergence may disappear completely for a convergent integral is seen by the example

(3) $$\int_0^\infty e^{-(s-1)t} e^{e^t} \sin e^{e^t} \, dt = \int_e^\infty \frac{\sin u}{(\log u)^s} du.$$

Here $\sigma_c = 0$ and $\sigma_a = \infty$.

3.2. It is clear that the results of section 2 are applicable to the axis of absolute convergence mutatis mutandis. For example, we state

THEOREM 3.2. *If*

$$\varlimsup_{t \to \infty} \frac{\log u(t)}{t} = k \neq 0,$$

where $u(t)$ is the total variation of $\alpha(x)$ in the interval $0 \leqq x \leqq t$, then the integral §3.1 (1) has the abscissa of absolute convergence $\sigma_a = k$.

3.3. We next prove a result concerning the relation between $\sigma_c$ and $\sigma_a$.

THEOREM 3.3. *If $\alpha(t)$ is monotonic in the intervals*

$$\lambda_n < t < \lambda_{n+1} \qquad (n = 0, 1, 2, \cdots),$$

*where*

$$0 = \lambda_0 < \lambda_1 < \lambda_2 < \cdots$$

$$\lim_{n \to \infty} \lambda_n = \infty,$$

*then*

$$\sigma_a - \sigma_c \leqq \varlimsup_{n \to \infty} \frac{\log n}{\lambda_n}$$

*for the integral*

(1) $$\int_0^\infty e^{-st} \, d\alpha(t).$$

We may assume that

$$\varlimsup_{n\to\infty} \frac{\log n}{\lambda_n} = l < \infty,$$

for otherwise there is nothing to prove. Clearly $l$ is not negative. It follows from Theorem 2.5 that the series

$$\sum_{n=0}^{\infty} e^{-\lambda_n s}$$

has abscissa of convergence equal to $l$ if $l$ is positive. The same is true if $l = 0$, for the series diverges when $s = 0$, and Theorem 2.1 can be used to show that it converges for all positive values of $s$.

Suppose that (1) converges for $s = s_0$. Then there exists a positive number $R_0$ such that

$$\left| \int_{R'}^{R''} e^{-s_0 t}\, d\alpha(t) \right| \leqq 1 \qquad (R', R'' > R_0).$$

In particular

(2) $$\left| \int_{\lambda_n+}^{t} e^{-s_0 y}\, d\alpha(y) \right| \leqq 1 \qquad (t > \lambda_n > R_0).$$

By Theorem 5c of Chapter I we have

(3) $$| a'_{n+1} e^{-s_0 \lambda_{n+1}} | \leqq 1 \qquad (\lambda_n > R_0)$$

(4) $$| a''_n e^{-s_0 \lambda_n} | \leqq 1 \qquad (\lambda_n > R_0),$$

where

$$\alpha(\lambda_n +) - \alpha(\lambda_n) = a''_n$$

$$\alpha(\lambda_{n+1}) - \alpha(\lambda_{n+1} -) = a'_{n+1} \qquad (n = 1, 2, \cdots).$$

To prove the theorem we must show that the integral 3 (1) converges for $\sigma > \sigma_0 + l$, or that the series

$$\sum_{n=0}^{\infty} \int_{\lambda_n}^{\lambda_{n+1}} e^{-\sigma t}\, du(t)$$

converges in the same region. On account of the monotonic character of $\alpha(t)$ in the interval $\lambda_n < t < \lambda_{n+1}$ it follows that

$$\int_{\lambda_n}^{\lambda_{n+1}} e^{-\sigma t}\, du(t) = \left| \int_{\lambda_n+}^{\lambda_{n+1}-} e^{-\sigma t}\, d\alpha(t) \right| + | a''_n | e^{-\lambda_n \sigma} + | a'_{n+1} | e^{-\lambda_{n+1} \sigma}.$$

But by (3) and (4) the series

$$\sum_{n=1}^{\infty} \{ | a''_n | e^{-\lambda_n \sigma} + | a'_{n+1} | e^{-\lambda_{n+1} \sigma} \}$$

will converge whenever the series

(5) $$\sum_{n=1}^{\infty} e^{-\tau \lambda_n (\sigma - \sigma_0)}$$

does. But we have seen that (5) converges when $\sigma - \sigma_0 > l$. It remains to consider the series

$$\sum_{n=1}^{\infty} \left| \int_{\lambda_n+}^{\lambda_{n+1}-} e^{-\sigma t} \, d\alpha(t) \right|$$

Set

$$\beta_n(t) = \int_{\lambda_n+}^{t} e^{-s_0 y} \, d\alpha(y) \qquad (t > \lambda_n > R_0).$$

Then

$$\int_{\lambda_n+}^{\lambda_{n+1}-} e^{-\sigma t} \, d\alpha(t) = \int_{\lambda_n+}^{\lambda_{n+1}-} e^{-(\sigma-s_0)t} \, d\beta_n(t)$$

$$= \beta_n(\lambda_{n+1}-)e^{-(\sigma-s_0)\lambda_{n+1}} + (\sigma - s_0) \int_{\lambda_n}^{\lambda_{n+1}} e^{-(\sigma-s_0)t} \beta_n(t) \, dt.$$

By (2) we see that $|\beta_n(t)| \leq 1$ when $t > \lambda_n$ for all $n$ sufficiently large. Hence we need only show that

$$\sum_{1}^{\infty} e^{-(\sigma-\sigma_0)\lambda_{n+1}} + \int_{0}^{\infty} e^{-(\sigma-\sigma_0)t} \, dt < \infty$$

when $\sigma - \sigma_0 > l$. This is clearly the case, so that the theorem is established.

COROLLARY 3.3.* *For any Dirichlet series*

$$\sum_{n=1}^{\infty} a_n e^{-\lambda_n s}$$

*we have*

$$\sigma_a - \sigma_c \leq \varlimsup_{n \to \infty} \frac{\log n}{\lambda_n}$$

For, in this case $\alpha(t)$ is constant in each of the intervals

$$\lambda_n < t < \lambda_{n+1}$$

As further examples of the theorem consider the integrals §3.1(2) and §3.1(3). In the first, $\lambda_n = \dfrac{\log n\pi}{k}$ and $l = k$. In the second, $\lambda_n = \log \log n\pi$ and $l = \infty$.

* See E. Landau [1909], p. 732.

## 4. Uniform Convergence

We have seen that the general Laplace-Stieltjes integral with abscissa of absolute convergence $\sigma_a$ converges uniformly when $\sigma \geq \sigma_a + \delta$ for every positive $\delta$. The conclusion does not remain true, however, if $\sigma_a$ is replaced by $\sigma_c$. We are thus led to define an *axis of uniform convergence*.

**4.1.** We first establish:

**THEOREM 4.1.** *If the integral*

$$(1) \qquad \int_0^\infty e^{-st}\,d\alpha(t)$$

*converges uniformly for* $s = \sigma_0 + i\tau$, $-\infty < \tau < \infty$, *then it converges uniformly in the half-plane* $\sigma \geq \sigma_0$.

Given an arbitrary positive $\epsilon$, we must show the existence of a number $R_0$ independent of $\sigma$ and $\tau$ for $\sigma \geq \sigma_0$, $-\infty < \tau < \infty$ and such that

$$(2) \qquad \left| \int_R^\infty e^{-(\sigma+i\tau)t}\,d\alpha(t) \right| < \epsilon \qquad (R \geq R_0).$$

We choose $R_0$ so that

$$(3) \qquad \left| \int_R^\infty e^{-(\sigma_0+i\tau)t}\,d\alpha(t) \right| < \frac{\epsilon}{2} \qquad (R \geq R_0,\ -\infty < \tau < \infty).$$

This is possible by the hypothesis of uniform convergence on the line $\sigma = \sigma_0$. Define a function $\beta(t, \tau)$ by the equation

$$\beta(t, \tau) = \int_t^\infty e^{-(\sigma_0+i\tau)y}\,d\alpha(y) \qquad (t \geq 0,\ -\infty < \tau < \infty).$$

Then (3) becomes

$$(4) \qquad |\beta(t, \tau)| < \frac{\epsilon}{2} \qquad (t \geq R_0,\ -\infty < \tau < \infty).$$

If $R'$ is an arbitrary number greater than $R$, then*

$$\int_R^{R'} e^{-(\sigma+i\tau)t}\,d\alpha(t) = -\int_R^{R'} e^{-(\sigma-\sigma_0)t}\,d_t\beta(t, \tau)$$

for every real $\tau$. Hence

$$-\int_R^{R'} e^{-(\sigma+i\tau)t}\,d\alpha(t) = e^{-(\sigma-\sigma_0)R'}\beta(R', \tau) - e^{-(\sigma-\sigma_0)R}\beta(R, \tau)$$

$$+ (\sigma - \sigma_0)\int_R^{R'} e^{-(\sigma-\sigma_0)t}\beta(t, \tau)\,dt$$

$$\int_R^\infty e^{-(\sigma+i\tau)t}\,d\alpha(t) = e^{-(\sigma-\sigma_0)R}\beta(R, \tau) - (\sigma - \sigma_0)\int_R^\infty e^{-(\sigma-\sigma_0)t}\beta(t, \tau)\,dt$$

---
* The notation $d_t\beta(t, \tau)$ indicates that $\tau$ is to be held constant during the integration.

for $\sigma > \sigma_0$. By use of (4) we obtain

$$\left| \int_R^\infty e^{-(\sigma+i\tau)t} \, d\alpha(t) \right| \leq \frac{\epsilon}{2} + \frac{\epsilon}{2} (\sigma - \sigma_0) \int_R^\infty e^{-(\sigma-\sigma_0)t} \, dt < \epsilon$$

for $\sigma > \sigma_0$. This relation is also true if $\sigma = \sigma_0$, as one sees directly from (3). Hence (2) is established.

4.2. We now introduce the following

DEFINITION 4.2. *If for every positive $\epsilon$ the integral §4.1 (1) converges uniformly in the half-plane $\sigma \geqq \sigma_u + \epsilon$ and fails to do so in the half-plane $\sigma \geqq \sigma_u - \epsilon$, then $\sigma_u$ is called the abscissa of uniform convergence and the line $\sigma = \sigma_u$ is called the axis of uniform convergence.*

It is clear that $\sigma_c \leqq \sigma_u$, and Theorem 3.1 shows that $\sigma_u \leqq \sigma_a$. In order to obtain a formula for $\sigma_u$ in terms of $\alpha(t)$ we first introduce the functions

$$\beta(x, \tau) = \int_0^x e^{-i\tau t} \, d\alpha(t) \qquad (x \geqq 0; -\infty < \tau < \infty)$$

$$T(x) = \underset{-\infty < \tau < \infty}{\text{u.b.}} |\beta(x, \tau)| \qquad (x \geqq 0).$$

We shall need two preliminary results.

LEMMA 4.2a. *If*

(1) $$T(x) = O(e^{\gamma x}) \qquad (x \to \infty)$$

*for some real number $\gamma$, then the integral*

(2) $$\int_0^\infty e^{-st} \, d\alpha(t)$$

*converges uniformly in the half-plane $\sigma \geqq \gamma + \delta$ for every positive $\delta$.*

Since

$$\int_0^R e^{-(\sigma+i\tau)t} \, d\alpha(t) = \int_0^R e^{-\sigma t} \, d_t \beta(t, \tau) = e^{-\sigma R} \beta(R, \tau) + \sigma \int_0^R e^{-\sigma t} \beta(t, \tau) \, dt,$$

it will be sufficient to show that

$$\lim_{R \to \infty} e^{-\sigma R} \beta(R, \tau) = 0$$

uniformly for $\sigma \geqq \gamma + \delta$, $-\infty < \tau < \infty$, and that

$$\lim_{R \to \infty} \sigma \int_R^\infty e^{-\sigma t} \beta(t, \tau) \, dt = 0$$

uniformly in the same region. But in that region

$$|e^{-\sigma R} \beta(R, \tau)| \leqq e^{-(\gamma+\delta)R} T(R) = O(e^{-\delta R}) \qquad (R \to \infty)$$

$$\left| \sigma \int_R^\infty e^{-\sigma t} \beta(t, \tau) \, dt \right| \leqq \sigma \int_R^\infty e^{-\sigma t} T(t) \, dt \leqq M \frac{(\gamma + \delta)}{\delta} e^{-\delta R}.$$

Here $M$ is a constant independent of $\sigma$ and $\tau$, so that the lemma is established.

LEMMA 4.2b. *If for a fixed positive number $\gamma$ the integral*

$$(3) \qquad \int_0^\infty e^{-(\gamma+i\tau)t}\, d\alpha(t)$$

*converges uniformly in the interval $-\infty < \tau < \infty$, then*

$$T(x) = O(e^{\gamma x}) \qquad (x \to \infty).$$

Since (3) converges uniformly we can find a number $R_0$ independent of $\tau$ in $(-\infty, \infty)$ such that

$$\left| \int_{R_0}^R e^{-(\gamma+i\tau)t}\, d\alpha(t) \right| \leq 1 \qquad (R \geq R_0).$$

Set

$$\zeta(x, \tau) = \int_0^x e^{-(\gamma+i\tau)t}\, d\alpha(t) \qquad (x \geq 0, -\infty < \tau < \infty).$$

Then for $x > 0$

$$|\zeta(x, \tau)| \leq \int_0^{R_0} e^{-\gamma t}|d\alpha(t)| + 1 = K.$$

Furthermore, we have

$$\beta(x, t) = \int_0^x e^{\gamma t}\, d_t \zeta(t, \tau) = \zeta(x, \tau) e^{\gamma x} - \gamma \int_0^x e^{\gamma t} \zeta(t, \tau)\, dt$$

$$|\beta(x, \tau)| \leq Ke^{\gamma x} + \gamma \int_0^x e^{\gamma t} K\, dt \leq Ke^{\gamma x} + K(e^{\gamma x} - 1)$$

$$\leq 2Ke^{\gamma x}$$

That is,

$$T(x) = \underset{-\infty < \tau < \infty}{\text{u.b.}} |\beta(x, \tau)| = O(e^{\gamma x}),$$

and the proof of the lemma is complete.

By use of these results we are now in a position to prove:

THEOREM 4.2. *If*

$$\varlimsup_{x \to \infty} \frac{\log T(x)}{x} = k > 0,$$

*then*

$$\sigma_u = k.$$

We show first that the integral (2) converges uniformly in the half-plane $\sigma \geq k + \delta$ for every positive $\delta$. It follows from the hypothesis that
$$T(x) = O(e^{(k+\delta/2)x}) \qquad (x \to \infty),$$
and then by Lemma 4.2a that the integral converges uniformly for $\sigma \geq k + \delta$.

On the other hand (2) must fail to converge uniformly in the half-plane $\sigma \geq k - \delta$ for every positive $\delta$. Otherwise it would converge uniformly on some line $\sigma = \gamma$ where $0 < \gamma < k$. Then by Lemma 4.2b we should have
$$T(x) = O(e^{\gamma x}) \qquad (x \to \infty),$$
whence
$$\varlimsup_{x \to \infty} \frac{\log T(x)}{x} \leq \gamma < k.$$

This contradicts the hypothesis. Hence the proof of our theorem is complete.

COROLLARY 4.2.* *If $\sigma_u > 0$, then*
$$\varlimsup_{x \to \infty} \frac{\log T(x)}{x} = \sigma_u.$$

If
$$\varlimsup_{x \to \infty} \frac{\log T(x)}{x} > 0,$$
the proof follows from Theorem 4.2. It is impossible to have
$$\varlimsup_{x \to \infty} \frac{\log T(x)}{x} \leq 0,$$
for this would imply that
$$T(x) = O(e^{x\sigma_u/2}) \qquad (x \to \infty).$$
This, by virtue of Lemma 4.2a, would imply that (2) converges uniformly for $\sigma \geq \tfrac{3}{4}\sigma_u$, contrary to hypothesis.

4.3. It may be shown by example that it is possible for $\sigma_u$ to be greater than $\sigma_c$. In this case we have thus far proved nothing about the uniform convergence of the Laplace integral in the neighborhood of

---

* This theorem was first proved by H. Bohr [1913] in the special case in which (2) is an ordinary Dirichlet series.

the axis of convergence. In this connection the following result will be useful.

**THEOREM 4.3** *If the integral*

(1) $$\int_0^\infty e^{-st}\,d\alpha(t)$$

*converges at* $s = \sigma_0 + i\tau_0$, *and if $H$ and $K$ are any constants for which $H > 0$, $K > 1$, then the integral (1) converges uniformly in the region $\Delta$ defined by the inequalities*

(2) $$|s - s_0| \leq K(\sigma - \sigma_0)e^{H(\sigma-\sigma_0)}, \qquad (\sigma \geq \sigma_0)$$

Note that if $s$ is in $\Delta$ we must have $\sigma > \sigma_0$ or else $s = s_0$. Hence by Corollary 1a the integral (1) converges in $\Delta$. If $\epsilon$ is an arbitrary positive number we wish to show that we can determine a number $R_0$ independent of $s$ in $\Delta$ such that for $R > R_0$

(3) $$\left|\int_R^\infty e^{-st}\,d\alpha(t)\right| < \epsilon.$$

Set

$$\beta(t) = \int_0^t e^{-s_0 u}\,d\alpha(u) \qquad (0 \leq t < \infty)$$

and determine $R_0$ greater than $H$ and such that

$$|\beta(t) - \beta(t')| < \epsilon/K$$

for all values of $t$ and $t'$ greater than $R_0$. This is possible by the convergence of (1) at $s_0$.

Then

$$\int_R^\infty e^{-st}\,d\alpha(t) = \int_R^\infty e^{-(s-s_0)t}\,d[\beta(t) - \beta(R)]$$

$$= (s - s_0)\int_R^\infty e^{-(s-s_0)t}[\beta(t) - \beta(R)]\,dt \qquad (\sigma > \sigma_0)$$

for any positive $R$. If $R > R_0$

$$\left|\int_R^\infty e^{-st}\,d\alpha(t)\right| \leq \frac{\epsilon}{K}\frac{|s - s_0|}{\sigma - \sigma_0}e^{-(\sigma-\sigma_0)R} \qquad (\sigma > \sigma_0).$$

Hence if $s$ is any point of $\Delta$ not $s_0$

$$\left|\int_R^\infty e^{-st}\,d\alpha(t)\right| \leq \epsilon e^{-(\sigma-\sigma_0)(R-H)} < \epsilon.$$

If $s = s_0$

$$\left|\int_R^\infty e^{-st}\,d\alpha(t)\right| = |\beta(\infty) - \beta(R)| \leq \frac{\epsilon}{K} < \epsilon,$$

so that (3) holds for all $s$ in $\Delta$, and the theorem is proved.

We now investigate the nature of the region $\Delta$. Setting $\sigma - \sigma_0 = x$ and $\tau - \tau_0 = y$ we see that the boundary curve $C$ of $\Delta$ is symmetric in the $x$-axis and that the equation of the part above that axis is

$$y = x(K^2 e^{2Hx} - 1)^{1/2} \qquad (x \geq 0).$$

The first and second derivatives of $y$ with respect to $x$ are

$$y' = \frac{K^2 e^{2Hx}(1 + Hx) - 1}{(K^2 e^{2Hx} - 1)^{1/2}}$$

$$y'' = \frac{HK^2 e^{2Hx}[K^2 e^{2Hx}(2 + Hx) - 2 - 2Hx]}{(K^2 e^{2Hx} - 1)^{3/2}}.$$

Since

$$e^{2Hx} \geq 1 + 2Hx$$

$$K^2 e^{2Hx}(2 + Hx) \geq 2 + 5Hx + 2H^2 x^2 \geq 2 + 2Hx, \quad (x \geq 0)$$

we see that the curve is concave upward above the $y$ axis. It passes through the origin and increases monotonically as $x$ increases. Since

$$\lim_{x \to 0+} y' = (K^2 - 1)^{1/2} > 0,$$

the curve $C$ has a cusp at the origin, the angle between the two tangents being

$$2 \tan^{-1}(K^2 - 1)^{1/2}.$$

As $x$ becomes infinite $y$ becomes infinite. The region $\Delta$ consists of the curve $C$ and the part of the plane lying to its right. The half-line

$$y = (K^2 - 1)^{1/2} x \qquad (x \geq 0)$$

lies entirely in $\Delta$, but no other half-line through the origin with greater slope lies entirely in $\Delta$.

Set

$$\tan \tau = (K^2 - 1)^{1/2}, \qquad \sec \tau = K.$$

Then the angular region

(4) $\qquad |s - s_0| \leq K(\sigma - \sigma_0), \qquad (\sigma \geq \sigma_0)$

lies entirely in $\Delta$, a result which is also clear from the fact that (4) evidently implies (2). We have thus proved

COROLLARY 4.3. *If* (1) *converges for* $s = \sigma_0 + i\tau_0$ *and if* $K > 1$ *it converges uniformly in the region*

$$|s - s_0| \leq K(\sigma - \sigma_0), \qquad (\sigma \geq \sigma_0)$$

4.4 We prove next the following result:*

THEOREM 4.4. *If the integral* §4.3 (1) *converges at* $s = s_0$ *and if* $H$ *is any positive number, it converges uniformly in the region* $\Delta'$ *defined by the inequality*

$$|\tau - \tau_0| \leq e^{H(\sigma - \sigma_0)} - 1$$

For, if $s$ is a point of $\Delta'$, then

$$|s - s_0| \leq \sigma - \sigma_0 + e^{H(\sigma - \sigma_0)} - 1$$
$$< \left(\frac{1}{H} + 1\right) e^{H(\sigma - \sigma_0)}$$

Let $\delta$ be an arbitrary positive number. If $\sigma - \sigma_0 \geq \delta$ we can determine $H' > H$ so that

$$\frac{e^{H(\sigma - \sigma_0)}}{\sigma - \sigma_0} < e^{H'(\sigma - \sigma_0)}$$

Hence if $s$ belongs to $\Delta'$ and is such that $\sigma - \sigma_0 \geq \delta$ it also belongs to a region $\Delta$ of Theorem 4.3:

$$|s - s_0| \leq \left(\frac{1}{H} + 1\right)(\sigma - \sigma_0) e^{H'(\sigma - \sigma_0)} \qquad (\sigma > \sigma_0)$$

The remaining part of $\Delta'$ where $\sigma - \sigma_0 < \delta$ can be included in a triangular region of the type described in Corollary 4.3, so that the theorem is established.

COROLLARY 4.4 *If* $H$ *and* $\delta$ *are any positive numbers and if* §4.3 (1) *converges at* $s = s_0$, *then it converges uniformly in the region*

(1)  $\qquad \sigma \geq \sigma_0 + \delta, \qquad |\tau| \leq e^{H\sigma}.$

By Theorem 4.4 the integral converges uniformly in the region

$$|\tau - \tau_0| < e^{2H(\sigma - \sigma_0)} - 1.$$

If $s$ satisfies (1) then

$$|\tau - \tau_0| \leq |\tau_0| + e^{H\sigma} < e^{2H(\sigma - \sigma_0)} - 1$$

for all $\sigma$ sufficiently large. The remainder of the region (1) can again be included in a triangular region.

* Compare E. Landau [1909] p. 739.

## 5. Analytic Character of the Generating Function

It is now easy to see that a Laplace integral represents an analytic function in its region of convergence.

**LEMMA 5.** *If $0 \leq a \leq b$, then the function*

$$f(s) = \int_a^b e^{-st} \, d\alpha(t)$$

*is entire, and*

$$f^{(k)}(s) = (-1)^k \int_a^b e^{-st} t^k \, d\alpha(t) \quad (k = 1, 2, 3, \cdots).$$

By the uniform convergence of the exponential series we have

$$f(s) = \int_a^b e^{-st} \, d\alpha(t) = \sum_{n=0}^{\infty} \frac{(-s)^n}{n!} \int_a^b t^n \, d\alpha(t)$$

Clearly this series converges uniformly for $s$ in any bounded region, so that its sum, $f(s)$, is entire and

$$f^{(k)}(s) = \sum_{n=k}^{\infty} \frac{s^{n-k}}{(n-k)!} \int_a^b (-t)^n \, d\alpha(t) \quad (k = 1, 2, \cdots)$$

$$= \int_a^b e^{-st} (-t)^k \, d\alpha(t).$$

This completes the proof of the lemma.

By use of this result we establish:

**THEOREM 5a.** *If the integral*

(1)  $$f(s) = \int_0^\infty e^{-st} \, d\alpha(t)$$

*converges for $\sigma > \sigma_c < \infty$, then $f(s)$ is analytic for $\sigma > \sigma_c$, and*

$$f^{(k)}(s) = \int_0^\infty e^{-st} (-t)^k \, d\alpha(t).$$

For if $s_0$ is an arbitrary point in the half-plane $\sigma > \sigma_c$, we can surround it by a circle $K$ which also lies in that half-plane. By Theorem 4.3 the integral (1), and hence the series

$$f(s) = \sum_{n=0}^{\infty} \int_n^{n+1} e^{-st} \, d\alpha(t),$$

converges uniformly in $K$. Since each term of the series is entire, we may apply the theorem of Weierstrass concerning the term by term differentiation of series to obtain the desired result.

From the analogy with power series one might be tempted to suppose that a function defined by a Laplace integral would have at least one singularity on the axis of convergence of the integral. This, however, is not the case. The integral

$$f(s) = \int_0^\infty e^{-st} e^t \sin(e^t)\, dt = \int_1^\infty \frac{\sin x}{x^s}\, dx$$

clearly has $\sigma_c = 0$ for its abscissa of convergence. But for $\sigma > 0$ we have, after integrating by parts

$$f(s) = \cos 1 - s \int_1^\infty \frac{\cos x}{x^{s+1}}\, dx.$$

This integral converges for $\sigma > -1$, so that this equation serves to extend the function $f(s)$ analytically into the half-plane $\sigma > -1$. That is, $f(s)$ certainly has no singularities on the imaginary axis, the axis of convergence of the original Laplace integral. In fact one could show easily by successive integration by parts that $f(s)$ is entire.

A case in which there is certainly a singularity on the axis of convergence is described in the following theorem.*

THEOREM 5b. *If $\alpha(t)$ is monotonic, then the real point of the axis of convergence of*

$$f(s) = \int_0^\infty e^{-st}\, d\alpha(t)$$

*is a singularity of $f(s)$.*

The statement of the theorem assumes that the abscissa of convergence, $\sigma_c$, is finite. It is no restriction to assume it zero. For

(2) $$f(s + \sigma_c) = \int_0^\infty e^{-st}\, d\beta(t),$$

where

$$\beta(t) = \int_0^t e^{-\sigma_c u}\, d\alpha(u) \qquad (0 \leq t < \infty).$$

The integral (2) has abscissa of convergence zero and its determining function $\beta(t)$ is also monotonic. For definiteness we take $\alpha(t)$ nondecreasing.

Assuming that $\sigma_c = 0$ we must prove the origin a singular point of $f(s)$. Suppose the contrary: then the series

$$f(s) = \sum_{k=0}^\infty \frac{f^{(k)}(1)}{k!} (s - 1)^k$$

* Compare H. Hamburger [1921] p. 306.

converges for some real negative value of $s$, say $s = -\delta$. That is,

$$f(-\delta) = \sum_{k=0}^{\infty} (-1)^k \frac{(\delta + 1)^k}{k!} \int_0^{\infty} e^{-t}(-t)^k\, d\alpha(t)$$

(3)
$$= \int_0^{\infty} e^{-t} \sum_{k=0}^{\infty} \frac{1}{k!}(\delta + 1)^k (t)^k\, d\alpha(t)$$

$$= \int_0^{\infty} e^{\delta t}\, d\alpha(t),$$

if the interchange of summation and integral signs is permissible. To justify this we have

$$e^{t(\delta+1)} = \sum_{k=0}^{\infty} (-1)^k \frac{(\delta + 1)^k}{k!}(-t)^k \ll \sum_{k=0}^{\infty} \frac{(\delta + 1)^k}{k!} R^k \quad (0 \leq t \leq R),$$

so that

(4)
$$\int_0^{R} e^{\delta t}\, d\alpha(t) = \sum_{k=0}^{\infty} (-1)^k \frac{(\delta + 1)^k}{k!} \int_0^{R} e^{-t}(-t)^k\, d\alpha(t)$$

by Theorem 5d of Chapter I.

Now series (4) converges uniformly for $1 \leq R < \infty$ since it is dominated by the series

$$\sum_{k=0}^{\infty} \frac{(\delta + 1)^k}{k!} \int_0^{\infty} e^{-t} t^k\, d\alpha(t) = \sum_{k=0}^{\infty} (-1)^k f^{(k)}(1) \frac{(\delta + 1)^k}{k!},$$

which we have assumed converges to $f(-\delta)$. Hence we may let $R$ become infinite in (4) to obtain (3). That is, the Laplace integral representation of $f(s)$ converges for $s = -\delta$ contrary to the hypothesis $\sigma_c = 0$. The contradiction shows that $f(s)$ must have a singularity at $s = 0$, and our proof is complete.

## 6. Uniqueness of Determining Function

We shall show that the determining function $\alpha(t)$, if it is normalized, is uniquely determined by its generating function $f(s)$,

(1)
$$f(s) = \int_0^{\infty} e^{-st}\, d\alpha(t).$$

We shall also consider the integral

(2)
$$f(s) = \int_{0+}^{\infty} e^{-st}\, d\alpha(t).$$

In the case of the integral (2) we note first that its convergence* for

* See §1 for our conventions concerning the integral (2).

60    THE LAPLACE TRANSFORM    [Ch. II

any fixed $s = s_0$ implies the existence of $\alpha(0+)$. For, if we set

$$\beta(t) = \int_t^1 e^{-s_0 u}\, d\alpha(u) \qquad (0 \leq t \leq 1),$$

we have

(3)
$$\alpha(1) - \alpha(\epsilon) = \int_\epsilon^1 e^{s_0 t} e^{-s_0 t}\, d\alpha(t) = -\int_\epsilon^1 e^{s_0 t}\, d\beta(t)$$
$$= e^{s_0 \epsilon}\beta(\epsilon) + s_0 \int_\epsilon^1 e^{s_0 t}\beta(t)\, dt \qquad (0 < \epsilon < 1).$$

Since $\beta(0+)$ exists by virtue of the assumption that (2) converges at $s = s_0$, it follows from equation (3) that $\alpha(0+)$ also exists. We say that the function $\alpha(t)$ of (2) is normalized if $\alpha(0+)$ is zero and if

$$\alpha(t) = \frac{\alpha(t+) + \alpha(t-)}{2} \qquad (t > 0)$$

6.1. To prove uniqueness theorems for 6(1) or 6(2) we need:

THEOREM 6.1. *If $\alpha(t)$ is a normalized function of bounded variation in* (0, 1) *such that*

(1)
$$\int_0^1 t^n\, d\alpha(t) = 0 \qquad (n = 0, 1, 2, \cdots),$$

*then $\alpha(t)$ is identically zero in $0 \leq t \leq 1$.*

Since $\alpha(0) = 0$ by hypothesis, equation (1) with $n$ equal to zero shows that $\alpha(1)$ is zero. Hence (1) becomes after integration by parts

(2)
$$\int_0^1 t^n \alpha(t)\, dt = 0 \qquad (n = 0, 1, 2, \cdots).$$

If

(3)
$$\beta(t) = \int_0^t \alpha(u)\, du \qquad (0 \leq t \leq 1),$$

then $\beta(1) = 0$ by (2) with $n = 0$, and (2) becomes

(4)
$$\int_0^1 t^n \beta(t)\, dt = 0 \qquad (n = 0, 1, 2, \cdots).$$

If $\epsilon$ is an arbitrary positive number, it is possible by virtue of the approximation theorem of Weierstrass to determine a polynomial $P(t)$ which approximates to the continuous conjugate $\overline{\beta(t)}$ of $\beta(t)$ with an error less than $\epsilon$,

(5)
$$|\beta(t) - P(t)| < \epsilon \qquad (0 \leq t \leq 1)$$

Hence by (4) and (5)

$$\int_0^1 |\beta(t)|^2 dt = \int_0^1 \beta(t)[\overline{\beta(t)} - P(t)] dt$$

$$\int_0^1 |\beta(t)|^2 dt \leqq \epsilon \int_0^1 |\beta(t)| dt.$$

Since $\epsilon$ was arbitrary this shows that $\beta(t)$ is identically zero in $0 \leqq t \leqq 1$.

By (3) we then see that $\alpha(t)$ is zero at all its points of continuity. Since these are dense in $(0, 1)$ and since for every $t$ between zero and unity $\alpha(t+)$ and $\alpha(t-)$ both exist, it is evident that

$$\alpha(t+) = \alpha(t-) = \alpha(t) = 0 \qquad (0 < t < 1).$$

Hence $\alpha(t)$ is identically zero in the closed interval $(0, 1)$, as was stated in the theorem.

COROLLARY 6.1a. *If*

(6) $$\int_{0+}^{1-} t^n d\alpha(t) = 0 \qquad (n = 0, 1, 2, \cdots),$$

*and if $\alpha(t)$ is normalized, then*

$$\alpha(t) = 0 \qquad (0 < t < 1).$$

By our conventions the existence of (6) implies that $\alpha(t)$ is of bounded variation in $(\epsilon, 1 - \epsilon)$ for every positive $\epsilon$. Moreover, equations (6) with $n = 0$ shows that $\alpha(0+)$ and $\alpha(1-)$ exist. The first of these limits is zero since $\alpha(t)$ is normalized, the second is also zero by (6). Hence (6) implies

(7) $$\int_{0+}^{1-} t^n \alpha(t) dt = 0 \qquad (n = 0, 1, 2, \cdots)$$

Since $\alpha(t)$ is integrable in $(0, 1)$, equations (7) imply (2), and the conclusion is obvious.

COROLLARY 6.1b. *If $\varphi(t)$ belongs to $L$ in $(0, 1)$, and if*

$$\int_0^1 t^n \varphi(t) dt = 0 \qquad (n = 0, 1, 2, \cdots),$$

*then $\varphi(t)$ is zero almost everywhere.*

6.2. We now prove a result which was due to M. Lerch [1903] in its original form.

THEOREM 6.2. *If $s_0$ is a point in the region of convergence of §6(1) and $l$ is a positive number, if $\alpha(t)$ is normalized,\* and if*

---

\* Compare Definition 8 of Chapter I. Here we mean, of course, that $\alpha(t)$ is normalized in the interval $(0, R)$ for every positive $R$

(1) $\quad f(s_0 + nl) = \int_0^\infty e^{-(s_0+nl)t} d\alpha(t) = 0 \qquad (n = 0, 1, 2, \cdots),$

**then**
$$\alpha(t) = 0 \qquad (0 \leq t < \infty).$$

For, set
$$\beta(t) = \int_t^\infty e^{-s_0 u} d\alpha(u) \qquad (0 \leq t < \infty).$$

Then $\beta(t)$ is of bounded variation in $(0, R)$ for every positive $R$ and $\beta(\infty) = 0$. In terms of $\beta(t)$ equations (1) become

$$\int_0^\infty e^{-nlt} d\beta(t) = 0 \qquad (n = 0, 1, 2, \cdots).$$

Set $e^{-lt}$ equal to a new variable $u$ and obtain

(2) $\quad \int_{0+}^1 u^n d\beta(l^{-1} \log u^{-1}) = 0 \qquad (n = 0, 1, 2, \cdots).$

Clearly $\beta(l^{-1} \log u^{-1})$ is normalized, so that we may apply Corollary 6.1a to show it indentically zero. But by Theorem 5c of Chapter I

$$\beta(t+) - \beta(t-) = e^{-s_0 t}[\alpha(t-) - \alpha(t+)] \quad (0 < t < \infty).$$

Hence $\alpha(t)$ is identically zero in $(0 \leq t < \infty)$.

COROLLARY 6.2a. *If in Theorem 6.2 equations* (1) *are replaced by*

$$\int_{0+}^\infty e^{-(s_0+nl)t} d\alpha(t),$$

**then**
$$\alpha(t) = 0 \qquad (0 < t < \infty).$$

For, we may define $\beta(t)$ as before. It is no longer of bounded variation in an interval including the origin. Equations (2) become

$$\int_{0+}^{1-} u^n d\beta(l^{-1} \log u^{-1}) = 0 \qquad (n = 0, 1, 2, \cdots),$$

but Corollary 6.1a is still applicable.

COROLLARY 6.2b. *If in Theorem 6.2 equations* (1) *are replaced by*

$$\int_0^\infty e^{-(s_0+nl)t} \varphi(t) dt = 0 \qquad (n = 0, 1, \cdots),$$

*where* $\varphi(t)$ *belongs to* $L$ *in* $(0, R)$ *for every positive* $R$, *then* $\varphi(t)$ *is zero almost everywhere*.

§7]  COMPLEX INVERSION FORMULA  63

For, by Theorem 6.2 the function

$$\alpha(t) = \int_0^t \varphi(u)\, du$$

vanishes identically. Its derivative is $\varphi(t)$ almost everywhere.

6.3. We may now prove the desired uniqueness theorem.

THEOREM 6.3. *There cannot exist two different normalized determining functions corresponding to the same generating function.*

The proof is obvious. More precisely the theorem means that if $\alpha_1(t)$ and $\alpha_2(t)$ are normalized and such that

$$\int_0^\infty e^{-st}\, d\alpha_1(t) = \int_0^\infty e^{-st}\, d\alpha_2(t)$$

for all $s$ in some common region of convergence, then

$$\alpha_1(t) = \alpha_2(t) \qquad (0 \leq t < \infty).$$

Or, if

$$\int_0^\infty e^{-st} \varphi_1(t)\, dt = \int_0^\infty e^{-st} \varphi_2(t)\, dt$$

for all $s$ in some common region of convergence, then

$$\varphi_1(t) = \varphi_2(t)$$

for almost all positive values of $t$.

## 7. Complex Inversion Formula*

We shall next obtain a formula which gives the determining function $\alpha(t)$ in terms of the generating function $f(s)$, an *inversion* formula for the Laplace transform. The Cauchy formula for the coefficients of a power series in terms of its sum furnishes an analogue for the result which we shall obtain.

The Laplace integral with determining function $\alpha(t)$ reduces to a series in powers of $e^{-s}$ if

$$\alpha(0) = 0$$

$$\alpha(t) = s_n = a_1 + a_2 + \cdots + a_n \cdot (n-1 \leq t < n;\ n = 1, 2, 3, \cdots).$$

Then

(1) $$f(s) = F(z) = \sum_{n=1}^\infty a_n z^n \qquad (z = e^{-s}).$$

---
* See J. D. Tamarkin [1926] where a historical introduction to this subject is given.

Since
$$\frac{F(z)}{1-z} = \sum_{n=1}^{\infty} s_n z^n,$$
we have
$$s_n = \frac{1}{2\pi i} \int_C \frac{F(z)}{(1-z)z^{n+1}} dz,$$
where the integration is taken in the positive sense over a circle $C$, $|z| = \gamma$, the radius $\gamma$ being less than unity and less than $\rho$, the radius of convergence of (1). A change of variable gives

(2) $$s_n = \frac{1}{2\pi i} \int_\pi^{c+i\pi} \frac{f(s)e^{ns}}{1 - e^{-s}} ds$$

$$c = \log\frac{1}{\gamma}, \quad c > \log\frac{1}{\rho}, \quad c > 0$$

By the theory of residues, one could show* that the integral (2) can be replaced by

$$\alpha(t) = s_n = \frac{1}{2\pi i} \int_{c-i\infty}^{c+i\infty} \frac{f(s)e^{st}}{s} ds \quad \left(n - 1 < t < n, c > 0, c > \log\frac{1}{\rho}\right)$$

We do not give the details, since we shall give a rigorous proof for the general Laplace transform. The present approach merely shows the **type of formula to expect, and shows why one must expect the restriction** $c > 0$ in the general formula below.

**7.1.** For the inversion of the integral we shall need a result concerning the familiar Dirichlet integral, the proof of which we insert for completeness.

THEOREM 7.1. *If $\alpha(t)$ is of bounded variation in $0 \leq t \leq \delta$, $\delta > 0$, then*

(1) $$\lim_{T \to \infty} \frac{1}{\pi} \int_0^\delta \alpha(t) \frac{\sin Tt}{t} dt = \frac{\alpha(0+)}{2}$$

Clearly if equation (1) is true for each of two functions $\alpha(t)$ it is true for their sum. Hence it is no restriction to suppose $\alpha(t)$ real. In fact it is no restriction to suppose that $\alpha(t)$ is non-negative, non-decreasing. As a final reduction we may suppose that $\alpha(0+) = 0$ since

$$\lim_{T \to \infty} \frac{1}{\pi} \int_0^\delta \frac{\sin Tt}{t} dt = \frac{1}{\pi} \int_0^\infty \frac{\sin x}{x} dx = \tfrac{1}{2}.$$

---

* See, for example E. Landau [1909] chapter 77

Then if $\epsilon$ is an arbitrary positive constant we can choose $\eta$ so that $\alpha(t) < \epsilon$ when $0 \leq t \leq \eta < \delta$. By the second law of the mean

$$\frac{1}{\pi} \int_0^\delta \alpha(t) \frac{\sin Tt}{t} dt = \frac{1}{\pi} \int_\eta^\delta \alpha(t) \frac{\sin Tt}{t} dt$$
$$+ \frac{1}{\pi} \alpha(\eta) \int_\xi^\eta \frac{\sin Tt}{t} dt \qquad (0 \leq \xi \leq \eta).$$

By the Riemann-Lebesgue theorem

$$\varlimsup_{T \to \infty} \left| \frac{1}{\pi} \int_0^\delta \alpha(t) \frac{\sin Tt}{t} dt \right| \leq \epsilon A,$$

where $A$ is a constant such that

$$\left| \int_\xi^\eta \frac{\sin tT}{t} dt \right| = \left| \int_{\xi T}^{\eta T} \frac{\sin x}{x} dx \right| \leq A\pi.$$

Since $\epsilon$ was arbitrary, our result is established.

7.2. An immediate consequence of this result is:

THEOREM 7.2. *If $\varphi(u)$ belongs to $L$ in $(-\infty, \infty)$ and is of bounded variation in some two-sided neighborhood of a point $t$, then*

(1) $$\lim_{T \to \infty} \frac{1}{\pi} \int_{-\infty}^\infty \frac{\varphi(u) \sin T(t-u)}{t-u} du = \frac{\varphi(t+) + \varphi(t-)}{2}$$

For, if $\epsilon$ is an arbitrary positive number, we can determine $R$ greater than $t$, greater than $-t$, and so large that

$$\left| \frac{1}{\pi} \int_R^\infty \varphi(u) \frac{\sin T(t-u)}{t-u} du \right| \leq \frac{1}{\pi} \frac{1}{R-t} \int_R^\infty |\varphi(u)| du \leq \epsilon$$

$$\left| \frac{1}{\pi} \int_{-\infty}^{-R} \varphi(u) \frac{\sin T(t-u)}{t-u} du \right| \leq \frac{1}{\pi} \frac{1}{R+t} \int_{-\infty}^{-R} |\varphi(u)| du \leq \epsilon.$$

Also

$$\frac{1}{\pi} \int_{-R}^R \frac{\varphi(u) \sin T(t-u)}{t-u} du = \frac{1}{\pi} \int_0^{R-t} \varphi(t+y) \frac{\sin Ty}{y} dy$$
$$+ \frac{1}{\pi} \int_0^{t+R} \varphi(t-y) \frac{\sin Ty}{y} dy = \frac{\varphi(t+) + \varphi(t-)}{2} + o(1) \quad (T \to \infty)$$

by an application of the Riemann-Lebesgue theorem and of Theorem 7.1. Hence

$$\varlimsup_{T \to \infty} \left| \frac{1}{\pi} \int_{-\infty}^\infty \frac{\varphi(u) \sin T(t-u)}{t-u} du - \frac{\varphi(t+) + \varphi(t-)}{2} \right| \leq 2\epsilon,$$

from which (1) follows at once.

COROLLARY 7.2. *If $\varphi(u)$ belongs to $L$ in $(R, \infty)$, and if $t < R$, then*

$$\lim_{T \to \infty} \int_R^\infty \frac{\varphi(u) \sin T(t-u)}{t-u} du = 0.$$

7.3. We first develop an inversion formula* for the case in which the determining function $\alpha(t)$ is the integral of a function $\varphi(t)$. That is,

(1) $$f(s) = \int_0^\infty e^{-st} \varphi(t) \, dt,$$

and we seek to determine $\varphi(t)$ in terms of $f(s)$.

THEOREM 7.3. *If $\varphi(u)$ belongs to $L$ in $(0, R)$ for every positive $R$ and if the integral (1) converges absolutely on the line $\sigma = c$, then*

$$\lim_{T \to \infty} \frac{1}{2\pi i} \int_{c-iT}^{c+iT} f(s) e^{st} \, ds = 0 \qquad (t < 0).$$

*If in addition $\varphi(u)$ is of bounded variation in a neighborhood** of $u = t$ $(t \geqq 0)$, then*

(2) $$\lim_{T \to \infty} \frac{1}{2\pi i} \int_{c-iT}^{c+iT} f(s) e^{st} \, ds = \frac{\varphi(t+) + \varphi(t-)}{2} \qquad (t > 0)$$

$$= \frac{\varphi(0+)}{2} \qquad (t = 0).$$

For any value of $t$ we have

$$\frac{1}{2\pi i} \int_{c-iT}^{c+iT} f(s) e^{st} \, ds = \frac{1}{2\pi i} \int_{c-iT}^{c+iT} e^{st} \, ds \int_0^\infty e^{-su} \varphi(u) \, du$$

$$= \frac{1}{\pi} \int_0^\infty \frac{\varphi(u) e^{c(t-u)}}{t-u} \sin T(t-u) \, du.$$

The interchange of the order of integration here effected is justified by use of Theorem 3.1, the integral (1) being uniformly convergent along the vertical line segment from $c - iT$ to $c + iT$ in the complex $s$-plane. If $\varphi(u)$ is defined to be zero in $(-\infty, 0)$, the function $\varphi(u) e^{c(t-u)}$ belongs to $L$ in $(-\infty, \infty)$ so that Theorem 7.2 is applicable to it. The conclusion of that theorem yields Theorem 7.3.

7.4. The condition imposed on $\varphi(u)$ in the neighborhood of $u = t$ is precisely Jordan's condition for the convergence of the Fourier series corresponding to $\varphi(u)$. All the familiar convergence tests for the

---

* B. Riemann [1876].
** This is a two-sided neighborhood if $t > 0$, a right-hand neighborhood if $t = 0$.

§7]  COMPLEX INVERSION FORMULA  67

convergence of these series have their analogues in the present theory. The following corresponds to Dini's condition. To simplify the statement of the theorem let us define $\varphi(u)$ to be zero for negative values of $u$.

THEOREM 7.4. *If $\varphi(u)$ belongs to $L$ in $(0, R)$ for every positive $R$ and if the integral §7.3 (1) converges absolutely on the line $\sigma = c$, then for any negative $t$*

$$\lim_{T \to \infty} \frac{1}{2\pi i} \int_{c-iT}^{c+iT} f(s)e^{st} \, ds = 0.$$

*If for some fixed non-negative $t$ the limits $\varphi(t+)$ and $\varphi(t-)$ exist and for some positive $\delta$*

(1) $$\int_0^\delta \frac{|\varphi(t+u) - \varphi(t+)|}{u} \, du < \infty$$

(2) $$\int_0^\delta \frac{|\varphi(t-u) - \varphi(t-)|}{u} \, du < \infty,$$

*then*

$$\lim_{T \to \infty} \frac{1}{2\pi i} \int_{c-iT}^{c+iT} f(s)e^{st} \, ds = \frac{\varphi(t+) + \varphi(t-)}{2}.$$

By use of Corollary 7.2 it is easy to see that we need only prove the equation

$$\lim_{T \to \infty} \frac{1}{\pi} \int_{t-\delta}^{t+\delta} \frac{\varphi(u) e^{c(t-u)} \sin T(t-u)}{t-u} \, du$$

$$= \frac{\varphi(t+) + \varphi(t-)}{2} \quad (-\infty < t < \infty),$$

where $\delta$ is positive but as small as we like. By obvious changes of variable we see that it will be sufficient to prove

(3) $$\lim_{T \to \infty} \frac{1}{\pi} \int_0^\delta \frac{\varphi(t+u)e^{-cu} - \varphi(t+)}{u} \sin Tu \, du = 0$$

(4) $$\lim_{T \to \infty} \frac{1}{\pi} \int_0^\delta \frac{\varphi(t-u)e^{cu} - \varphi(t-)}{u} \sin Tu \, du = 0.$$

Since

(5) $$\frac{\varphi(t+u)e^{-cu} - \varphi(t+)}{u} = \frac{[\varphi(t+u) - \varphi(t+)]e^{-cu}}{u}$$

$$+ \frac{\varphi(t+)[e^{-cu} - 1]}{u}$$

it is clear that the function on the left-hand side of the equation belongs to $L$ in $(0, \delta)$ for $\delta$ sufficiently small. Consequently we may apply the Riemann-Lebesgue theorem to establish (3) A similar argument applies to (4)

7.5. It must not be supposed that the integral

$$(1) \qquad \frac{1}{2\pi i} \int_{c-i\infty}^{c+i\infty} f(s)e^{st}\, ds \qquad (c > \sigma_a)$$

necessarily converges. Thus if $\varphi(t) = 1$ in §7.3 (1), then $f(s) = s^{-1}$ for $\sigma > 0$. The conditions of Theorems 7.3 and 7.4 are both satisfied so that the principal value of (1) exists. But if $t = 0$, $c = 1$ it is clear that (1) diverges. In this connection we prove:

**THEOREM 7.5.** *Under the conditions of Theorem 7.4 the integral (1) converges when $t < 0$, or when $t \geq 0$ if $\varphi(u)$ is continuous\* at $u = t$.*

Let $T$ and $U$ be arbitrary positive constants. We wish to show that

$$\lim_{T\to\infty,\, U\to\infty} \int_{c+iT}^{c+iU} e^{st} f(s)\, ds = 0$$

when $T$ and $U$ become infinite independently This will prove that

$$\frac{1}{2\pi i} \int_{c}^{c+i\infty} f(s)e^{st}\, ds$$

converges. When this is established it will be evident to the reader how to modify the argument to treat the integral

$$\frac{1}{2\pi i} \int_{c-i\infty}^{c} f(s)e^{st}\, ds$$

It will be sufficient to show that

$$\lim_{T\to\infty,\, U\to\infty} \frac{1}{2\pi i} \int_0^\infty \frac{\varphi(u)e^{c(t-u)}}{t-u} [e^{iU(t-u)} - e^{iT(t-u)}]\, du = 0 \qquad (t > 0)$$

or that

$$\lim_{T\to\infty,\, U\to\infty} \int_{t-\delta}^{t+\delta} \frac{\varphi(u)e^{c(t-u)}}{t-u} [e^{iU(t-u)} - e^{iT(t-u)}]\, du = 0$$

for some small positive $\delta$. Making use of hypotheses §7.4 (1) and §7.4 (2), of equation §7.4 (5) and of the Riemann-Lebesgue theorem we see that we need only prove that

$$\lim_{T\to\infty,\, U\to\infty} \left\{ \int_0^\delta \frac{\varphi(t-)}{y}[e^{iUy} - e^{iTy}]\, dy - \int_0^\delta \frac{\varphi(t+)}{y}[e^{-iUy} - e^{-iTy}]\, dy \right\} = 0$$

---
\* Recall that $\varphi(u)$ has been defined as zero for $u < 0$

Since $\varphi(t+) = \varphi(t-)$ this is equivalent to

$$\lim_{T\to\infty, U\to\infty} \int_0^\delta \frac{\sin Uy - \sin Ty}{y}\, dy = 0.$$

But this follows from Theorem 7.1. The modifications necessary if $t$ is negative or zero are apparent, so that the theorem is established.

7.6. An inversion formula for the general Laplace-Stieltjes integral is easily derived from the foregoing results.

THEOREM 7.6a. *If $\alpha(t)$ is a normalized function of bounded variation in $(0, R)$ for every positive $R$, and if the integral*

(1) $$f(s) = \int_0^\infty e^{-st}\, d\alpha(t)$$

*has an abscissa of convergence $\sigma_c$, then for $c > 0$, $c > \sigma_c$*

(2) $$\lim_{T\to\infty} \frac{1}{2\pi i} \int_{c-iT}^{c+iT} \frac{f(s)}{s} e^{st}\, ds = \begin{cases} \alpha(t) & (t > 0) \\ \dfrac{\alpha(0+)}{2} & (t = 0) \\ 0 & (t < 0). \end{cases}$$

Note that if $\alpha(t)$ is an integral of $\varphi(t)$, then (1) reduces to §7.3 (1). If (2) is formally differentiated with respect to $t$ it reduces to §7.3 (2). It is important to observe the condition $c > 0$ here imposed.

If $\sigma > 0$ and $\sigma > \sigma_c$, then by Theorem 2.3a

(3) $$f(s) = s \int_0^\infty e^{-st}\alpha(t)\, dt$$

the integral converging absolutely. We have now only to apply Theorem 7.3 to the integral (3) to obtain our result.

COROLLARY 7.6a. *For any negative $t$*

$$\frac{1}{2\pi i}\int_{c-i\infty}^{c+i\infty} \frac{f(s)e^{st}}{s}\, ds = 0 \qquad (c > \sigma_c, c > 0)$$

*If $t$ is not negative but is a point of continuity of $\alpha(t)$ for which\**

$$\int_0^\delta \{|\alpha(t+u) - \alpha(t)| + |\alpha(t-u) - \alpha(t)|\} u^{-1}\, du < \infty$$

*for some positive $\delta$, then*

$$\frac{1}{2\pi i}\int_{c-i\infty}^{c+i\infty} \frac{f(s)}{s} e^{st}\, ds = \begin{cases} \alpha(t) & (t > 0) \\ \dfrac{\alpha(0+)}{2} & (t = 0) \end{cases}$$

\* We are assuming that $\alpha(t)$ is zero for $t \leq 0$.

This follows at once from Theorem 7.5.

**THEOREM 7.6b.** *Under the conditions of Theorem 7.6a with* $\sigma_c < c < 0$

$$\lim_{T \to \infty} \frac{1}{2\pi i} \int_{c-iT}^{c+iT} \frac{f(s)}{s} e^{st} \, ds = \begin{cases} \alpha(t) - \alpha(\infty) & (t > 0) \\ \dfrac{\alpha(0+)}{2} - \alpha(\infty) & (t = 0) \\ -\alpha(\infty) & (t < 0). \end{cases}$$

This theorem is only applicable when the integral has a negative abscissa of convergence, in which case $\alpha(\infty)$ certainly exists. By Theorem 2.3b we have for all $s$ with real part greater than $\sigma_c$

(4) $$f(s) = \alpha(\infty) + s \int_0^\infty e^{-st}[\alpha(t) - \alpha(\infty)] \, dt.$$

But it is a familiar fact that for any negative $c$

(5) $$\lim_{T \to \infty} \frac{1}{2\pi i} \int_{c-iT}^{c+iT} \frac{e^{st}}{s} \, ds = \begin{cases} 0 & (t > 0) \\ -\dfrac{1}{2} & (t = 0) \\ -1 & (t < 0). \end{cases}$$

Applying Theorem 7.3 to the integral (4) we have

$$\lim_{T \to \infty} \frac{1}{2\pi i} \int_{c-iT}^{c+iT} \frac{f(s) - \alpha(\infty)}{s} e^{st} \, ds = \begin{cases} \alpha(t) - \alpha(\infty) & (t > 0) \\ \dfrac{\alpha(0+) - \alpha(\infty)}{2} & (t = 0) \\ 0 & (t < 0). \end{cases}$$

Combining this result with (5) gives the result of the theorem.

We observe that Theorem 7.6a, or Theorem 7.6b when applicable, may be used to give a new proof of the uniqueness theorem for the Laplace-Stieltjes integral, that is of Theorem 6.3.

## 8. Integrals of the Determining Function

Equation §7.6 (2) provides a formula for determining an integral of $\varphi(u)$ when $f(s)$ is defined by §7.3 (1). We may generalize this to obtain all the successive integrals of $\varphi(u)$. Set

$$\varphi_1(t) = \int_0^t \varphi(u) \, du \qquad (t > 0)$$

$$\varphi_n(t) = \int_0^t \varphi_{n-1}(u) \, du \qquad (t > 0, n = 2, 3, \ldots).$$

§8] INTEGRALS OF DETERMINING FUNCTION 71

Then

(1) $$\varphi_n(t) = \int_0^t \frac{(t-u)^{n-1}}{\Gamma(n)} \varphi(u)\, du \qquad (t > 0).$$

In accordance with the ideas of Liouville and Riemann equation (1), with $n$ replaced by an arbitrary positive number $\rho$, is taken as the definition of the integral of $\varphi(u)$ of order $\rho$. Note that for negative $t$ the integrand need not be real for non-integral values of $\rho$. Consequently we consider only non-negative values of $t$. We shall need several preliminary results concerning these integrals.*

**LEMMMA 8a.** *If $\varphi(u)$ belongs to $L$ in $(0, R)$ for every positive $R$, the same is true of $\varphi_\rho(u)$ when $\rho > 0$.*

For,

$$\int_0^R |\varphi_\rho(t)|\, dt \leqq \int_0^R dt \int_0^t \frac{(t-u)^{\rho-1}}{\Gamma(\rho)} |\varphi(u)|\, du$$

$$= \int_0^R |\varphi(u)| \frac{(R-u)^\rho}{\Gamma(\rho+1)}\, du \leqq \frac{R^\rho}{\Gamma(\rho+1)} \int_0^R |\varphi(u)|\, du.$$

We have here made use of the Fubini theorem to change the order of integration.

**LEMMA 8b.** *If $\varphi(u)$ belongs to $L$ in $(0, R)$ for every positive $R$, then for $\rho \geqq 1$ $\varphi_\rho(u)$ is of bounded variation there.*

It is no restriction to assume $\varphi(u)$ real. Then write $\varphi(u)$ as the difference of two non-negative functions,

$$\varphi(u) = |\varphi(u)| - \{|\varphi(u)| - \varphi(u)\}.$$

It is clear that $\varphi_\rho(u)$ is the difference of two non-decreasing bounded functions.

**LEMMA 8c.** *If $\varphi(u)$ belongs to $L$ in $(0, R)$ for every positive $R$ and is of bounded variation in a neighborhood\* of a point $u = t$ ($t \geqq 0$), then for $0 < \rho < 1$, $\varphi_\rho(u)$ is of bounded variation in a neighborhood† of $u = t$ and satisfies a Lipschitz condition of order $\rho$ at $u = t$,*

$$\varphi_\rho(t+h) - \varphi_\rho(t) = O(|h|^\rho) \qquad (|h| \to 0,\ t+h > 0).$$

We may assume without loss of generality that $\varphi(u)$ is real. Let $\varphi(u)$ be of bounded variation in an interval $(a, b)$ where $a \geqq 0$. Let $t$ and $x$ be interior points of the open interval $(a, b)$. Then

$$\Gamma(\rho)\varphi_\rho(x) = \int_0^a (x-u)^{\rho-1} \varphi(u)\, du + \int_a^x (x-u)^{\rho-1} \varphi(u) du$$

$$= I_1(x) + I_2(x).$$

\* Compare G. H. Hardy [1918], G. H. Hardy and J. E. Littlewood [1928].
† This is a two-sided neighborhood if $t > 0$, a right-hand neighborhood if $t = 0$.

Since $I_1(x)$ has a bounded derivative, obtained by differentiation under the integral sign, in a neighborhood of $x = t$ it is of bounded variation there and satisfies a Lipschitz condition of order unity (and hence of order $\rho$) at $x = t$. Write $I_2(x)$ as a Stieltjes integral

$$(2) \qquad I_2(x) = \frac{(x-a)^\rho}{\rho} \varphi(a) + \frac{1}{\rho} \int_a^x (x-u)^\rho \, d\varphi(u) \qquad (a < x < b)$$

$$= \frac{(x-a)^\rho}{\rho} \varphi(a) + I_3(x)$$

It is only necessary to consider $I_3(x)$. If $t = a = 0$ this equation makes the result stated obvious. For, since $\varphi(u)$ is the difference of two non-negative, non-decreasing functions in $(a, b)$, it is no restriction to assume at once that $\varphi(u)$ is itself such a function. Then $I_3(x)$ is non-decreasing in a neighborhood of $t = 0$ and

$$I_3(h) = O(h^\rho) \qquad (h \to 0+)$$

If $t > 0$, and if $a \leqq t_1 \leqq t_2 \leqq b$, we have

$$I_3(t_2) - I_3(t_1) = \int_a^{t_2} (t_2 - u)^\rho \, d\varphi(u) - \int_a^{t_1} (t_1 - u)^\rho \, d\varphi(u)$$

$$= \int_a^{t_1} [(t_2 - u)^\rho - (t_1 - u)^\rho] \, d\varphi(u)$$

$$+ \int_{t_1}^{t_2} (t_2 - u)^\rho \, d\varphi(u) = I_4 + I_5$$

Clearly $I_4$ and $I_5$ are non-negative, so that $I_3(x)$ is non-decreasing in a neighborhood of $x = t$. Moreover, for $t_2 = t + h$ and $t_1 = t$, $1 > h > 0$, we have

$$I_4 \leqq \int_a^t (t + h - u)^\rho \left[ 1 - \frac{t - u}{t + h - u} \right] d\varphi(u) \leqq h^\rho \int_a^b d\varphi(u)$$

$$I_5 \leqq h^\rho \int_a^b d\varphi(u)$$

If $h < 0$ take $t_1 = t + h$ and $t_2 = t$. The above argument shows that $I_4$ and $I_5$ are again $O(|h|^\rho)$. Consequently we have completed the proof that $\varphi_\rho(u)$ is of bounded variation in a neighborhood of $u = t$ and that $\varphi_\rho(u)$ satisfies a Lipschitz condition of order $\rho$ at $u = t$.

8.1. By use of the foregoing results concerning generalized (fractional) integrals we now prove easily:

## §8] INTEGRALS OF DETERMINING FUNCTION

**Theorem 8.1.** *If $\varphi(u)$ belongs to $L$ in $(0, R)$ for every positive $R$, and if the integral*

$$f(s) = \int_0^\infty e^{-st} \varphi(t)\, dt$$

*has an abscissa of convergence $\sigma_c$, then for $\rho \geq 1$*

$$\lim_{T \to \infty} \frac{1}{2\pi i} \int_{c-iT}^{c+iT} \frac{f(s)}{s^\rho} e^{st}\, ds = \begin{cases} \varphi_\rho(t) & t \geq 0 \\ 0 & t \leq 0, \end{cases}$$

*where $c > \sigma_c$, $c > 0$*

If $\rho$ is a positive integer it is easy to see by integration by parts that

$$(1) \qquad f(s) = s^\rho \int_0^\infty e^{-st} \varphi_\rho(t)\, dt \qquad (\sigma > 0,\ \sigma > \sigma_c).$$

This equation also holds for all $\rho \geq 1$. For, consider the integral on the right, with $\varphi_\rho(t)$ replaced by its integral expression §8 (1). Integration by parts, if $\rho > 1$, gives

$$(2) \qquad \int_0^\infty e^{-st} \varphi_\rho(t)\, dt = \int_0^\infty e^{-st}\, dt \int_0^t \frac{(t-u)^{\rho-2}}{\Gamma(\rho-1)} \varphi_1(u)\, du.$$

After changing the order of integration this becomes

$$(3) \qquad \begin{aligned}\int_0^\infty e^{-st} \varphi_\rho(t)\, dt &= \int_0^\infty \varphi_1(u)\, du \int_u^\infty \frac{e^{-st}(t-u)^{\rho-2}}{\Gamma(\rho-1)}\, dt \\ &= \frac{1}{s^{\rho-1}} \int_0^\infty \varphi_1(u) e^{-su}\, du.\end{aligned}$$

This interchange is justified by Fubini's theorem provided the iterated integral

$$\int_0^\infty |\varphi_1(u)|\, du \int_u^\infty \frac{e^{-\sigma t}}{\Gamma(\rho-1)} (t-u)^{\rho-2}\, dt = \frac{1}{\sigma^{\rho-1}} \int_0^\infty |\varphi_1(u)|\, e^{-\sigma u}\, du$$

exists. This follows from Theorem 2.2a when $\sigma > 0$, $\sigma > \sigma_c$.

Integrating (3) by parts now gives

$$(4) \qquad \int_0^\infty e^{-st} \varphi_\rho(t)\, dt = s^{-\rho} \int_0^\infty e^{-su} \varphi(u)\, du = f(s) s^{-\rho}$$

which is equation (1). By Lemma 8b $\varphi_\rho(t)$ is of bounded variation in $(0, R)$ for every positive $R$. Moreover, since

$$\varphi_\rho(t) = \int_0^t \frac{(t-u)^{\rho-2}}{\Gamma(\rho-1)} \varphi_1(u)\, du,$$

and since, by Theorem 2.2a,

$$\varphi_1(t) = O(e^{\sigma_0 t}) \qquad (t \to \infty, \sigma_0 > \sigma_c, \sigma_0 > 0),$$

we have

$$\varphi_\rho(t) = O(e^{\sigma_0 t} t^{\rho-1}) \qquad (t \to \infty, \rho \geq 1),$$

so that the integral (1) converges absolutely for $\sigma > 0$, $\sigma > \sigma_c$.

By Lemma 8b $\varphi_\rho(u)$ is of bounded variation in any finite interval of the positive axis. Hence Theorem 7.3 is applicable to the function $f(s)s^{-\rho}$ defined by the integral (1). The conclusion is precisely Theorem 8.1.

8.2. For the case in which $0 < \rho < 1$ we must impose the further restriction on $\varphi(u)$ that its Laplace transform should have an axis of absolute convergence.

THEOREM 8.2. *If $\varphi(u)$ satisfies the conditions of Theorem 7.3, then for $0 < \rho < 1$*

(1) $$\lim_{T\to\infty} \frac{1}{2\pi i} \int_{c-iT}^{c+iT} \frac{f(s)}{s^\rho} e^{st} ds = \begin{cases} \varphi_\rho(t) & (t \geq 0) \\ 0 & (t \leq 0), \end{cases}$$

where $c > \sigma_a$.

For,

$$\int_0^\infty e^{-st} \varphi_\rho(t) dt = \int_0^\infty e^{-st} dt \int_0^t \frac{(t-u)^{\rho-1}}{\Gamma(\rho)} \varphi(u) du$$

$$= \int_0^\infty \varphi(u) du \int_u^\infty e^{-st} \frac{(t-u)^{\rho-1}}{\Gamma(\rho)} dt$$

$$= \frac{1}{s^\rho} \int_0^\infty e^{-su} \varphi(u) du.$$

The interchange of the order of integration is justified by Fubini's theorem for $\sigma > \sigma_a$. For then

$$\int_0^\infty e^{-\sigma u} |\varphi(u)| du < \infty.$$

Hence §8.1 (4) is still valid. Moreover, it is clear by Fubini's theorem that the integral on the left of §8.1 (4) is absolutely convergent for $\sigma > \sigma_a$. Under our present hypothesis it is also clear by Lemma 8c that $\varphi_\rho(u)$ is of bounded variation in a neighborhood of $u = t$. Hence Theorem 7.3 is again applicable. The conclusion is (1) since by Lemma 8c the function $\varphi_\rho(u)$ is continuous at $u = t$.

8.3. In Theorems 8.1 and 8.2 we made use of the principal values of the inversion integrals employed. In this paragraph we give a sufficient condition for the convergence of these integrals.

THEOREM 8.3a. *Under the conditions of Theorem 8.1 the integral*

$$\int_{c-i\infty}^{c+i\infty} \frac{f(s)}{s^\rho} e^{st} \, ds \qquad (c > 0, c > \sigma_c)$$

*converges for all $t$ if $\rho > 1$.*

For, we showed that

(1) $$f(s) = s^\rho \int_0^\infty e^{-st} \varphi_\rho(t) \, dt \qquad (\sigma > 0, \sigma > \sigma_c)$$

and that

(2) $$\varphi_\rho(t) = \int_0^t \frac{(t-u)^{\rho-2}}{\Gamma(\rho-1)} \varphi_1(u) \, du.$$

The integral (1) converges absolutely. Since $\varphi_1(u)$ is of bounded variation we see by Lemma 8c applied to (2) that $\varphi_\rho(u)$ satisfies a Lipschitz condition of order $\rho - 1$ at every point $u \neq t$. Hence $\varphi_\rho(u)$ is continuous and the functions

$$[\varphi_\rho(t+u) - \varphi_\rho(t)]u^{-1}, \qquad [\varphi_\rho(t-u) - \varphi_\rho(t)]u^{-1}$$

belong to $L$ for $0 \leq u \leq \delta$ for some positive $\delta$. Our result is consequently established by use of Theorem 7.5.

THEOREM 8.3b. *If $\varphi(u)$ satisfies the conditions of Theorem 7.3, then the integral*

$$\int_{c-i\infty}^{c+i\infty} \frac{f(s)}{s^\rho} e^{st} \, ds \qquad (c > \sigma_a)$$

*converges for $0 < \rho \leq 1$.*

Again we obtain (1) as in the previous section. The integral (1) converges absolutely. For $0 < \rho < 1$ we see by Lemma 8c that $\varphi_\rho(u)$ satisfies a Lipschitz condition of order $\rho$ at $u = t$. The remainder of the proof follows as in Theorem 8.3a.

## 9. Summability of Divergent Integrals

In section 7.5 it was pointed out that the integral §7.5 (1) may be divergent. For this reason we were obliged to consider the principal value of the integral. If we relax the conditions imposed on $\varphi(u)$, for example Jordan's or Dini's condition, this principal value may not exist. To treat this case we introduce the notion of the summability of an integral.

DEFINITION 9.* *The integral*

$$\int_{-\infty}^{\infty} f(u)\, du$$

*is summable by arithmetic means, or by Cesàro's means of order* 1, $(C, 1)$, *to the value* $A$ *if*

(1) $$\lim_{T \to \infty} \int_{-T}^{T} \left(1 - \frac{|u|}{T}\right) f(u)\, du = A.$$

It is easy to see that

$$\frac{1}{T} \int_0^T dt \int_{-t}^t f(u)\, du = \int_{-T}^T \left(1 - \frac{|u|}{T}\right) f(u)\, du,$$

so that the term arithmetic mean is justified. This equation makes evident the consistency of the method. That is, if

$$\lim_{T \to \infty} \int_{-T}^T f(u)\, du = A,$$

then (1) follows. For, it is a familiar fact that if a function approaches a limit its arithmetic mean approaches the same limit

9.1. LEMMA 9.1a. *If $\varphi(t)$ belongs to $L$ in $(0, \delta)$ and if $\varphi(0+)$ exists, then*

(1) $$\lim_{T \to \infty} \frac{2}{\pi T} \int_0^\delta \varphi(t) \frac{\sin^2 (Tt/2)}{t^2} dt = \frac{\varphi(0+)}{2}.$$

Since

(2) $$\frac{2}{\pi T} \int_0^\delta \frac{\sin^2 (Tt/2)}{t^2} dt = \frac{1}{\pi} \int_0^{T\delta/2} \frac{\sin^2 u}{u^2} du,$$

the limit of the left-hand side is $1/2$, so that (1) is valid when $\varphi(t)$ is constant. If $\epsilon$ is given there exists $\zeta = \zeta(\epsilon)$ such that

$$|\varphi(t) - \varphi(0+)| < \epsilon \qquad (0 \leq t \leq \zeta)$$

Then

$$\left| \frac{2}{\pi T} \int_0^\delta [\varphi(t) - \varphi(0+)] \frac{\sin^2 (Tt/2)}{t^2} dt \right| < \epsilon + \frac{2}{\pi T} \int_\zeta^\delta \frac{|\varphi(t) - \varphi(0+)|}{t^2} dt$$

$$\overline{\lim_{T \to \infty}} \frac{2}{\pi T} \int_0^\delta [\varphi(t) - \varphi(0+)] \frac{\sin^2 \frac{Tt}{2}}{t^2} dt \leq \epsilon,$$

from which (1) is immediate.

* See E. W. Hobson [1926] vol. 2, p. 384.

## §9] SUMMABILITY OF DIVERGENT INTEGRALS

**LEMMA 9.1b.** *If $\beta(u)$ is bounded in $(0, \infty)$, then*

$$I(T) = \frac{1}{T}\int_\delta^\infty \beta(u)\, d\left(\frac{\sin^2(Tu/2)}{u^2}\right) = o(1) \qquad (T \to \infty)$$

*for any positive $\delta$*

For,

$$I(T) = -\frac{1}{T}\int_\delta^\infty \beta(u)\frac{1-\cos Tu}{u^3}\,du + \int_\delta^\infty \beta(u)\frac{\sin Tu}{2u^2}\,du = I_1 + I_2.$$

Clearly

$$I_1(T) = O\left(\frac{1}{T}\right) \qquad (T \to \infty),$$

and

$$I_2(T) = o(1) \qquad (T \to \infty),$$

by the Riemann-Lebesgue theorem.

**9.2.** The lemmas of the previous section permit us to establish easily the following result.*

**THEOREM 9.2.** *If $\varphi(u)$ belongs to $L$ in $(0, R)$ for every $R > 0$, and if*

$$f(s) = \int_0^\infty e^{-st}\varphi(t)\,dt \qquad (\sigma > \sigma_c < \infty),$$

*then for $c > \sigma_c$ the integral*

$$\frac{1}{2\pi i}\int_{c-i\infty}^{c+i\infty} f(s)e^{st}\,ds$$

*is summable $(C, 1)$ to zero for $t < 0$, to $\varphi(0+)/2$ if $t = 0$, and to $[\varphi(t+) + \varphi(t-)]/2$ whenever these expressions have a meaning*

Since

(1) $$\int_0^\infty e^{-(c+iu)y}\varphi(y)\,dy$$

converges uniformly for $-T \leq u \leq T$, where $T$ is arbitrary, we have

$$I = \frac{1}{2\pi}\int_{-T}^T \left(1 - \frac{|u|}{T}\right)e^{(c+iu)t}\,du \int_0^\infty e^{-(c+iu)y}\varphi(y)\,dy$$

$$= \frac{2}{\pi T}\int_0^\infty \varphi(y)e^{c(t-y)}\frac{\sin^2[(t-y)T/2]}{(t-y)^2}\,dy.$$

---
* Compare G. H. Hardy [1921].

If $t > 0$,

$$I = \frac{2}{\pi T} \int_0^t \varphi(t - u) e^{cu} \frac{\sin^2 (Tu/2)}{u^2} du$$
$$+ \frac{2}{\pi T} \int_0^\infty \varphi(t + u) e^{-cu} \frac{\sin^2 (Tu/2)}{u^2} du = I_1 + I_2.$$

By Lemma 9.1a

$$\lim_{T \to \infty} I_1 = \frac{\varphi(t-)}{2}.$$

Let $\delta$ be an arbitrary positive number. Then

$$I_2 = \frac{2}{\pi T} \left( \int_0^\delta + \int_\delta^\infty \right) \varphi(t + u) e^{-cu} \frac{\sin^2 (Tu/2)}{u^2} du = I_3 + I_4.$$

By Lemma 9.1a

$$\lim_{T \to \infty} I_3 = \frac{\varphi(t+)}{2}.$$

To treat $I_4$ we introduce the function

$$\beta(y) = \int_\delta^y \varphi(t + u) e^{-cu} du \qquad (y \geq \delta)$$
$$= e^{ct} \int_{t+\delta}^{t+y} \varphi(u) e^{-cu} du.$$

Since (1) converges it follows that $\beta(y)$ is a bounded function ($t$ being fixed). But

$$I_4 = \frac{-2}{\pi T} \int_\delta^\infty \beta(u) \, d\left( \frac{\sin^2 (Tu/2)}{u^2} \right).$$

By Lemma 9.1b we see that $I_4$ tends to zero with $1/T$ so that our theorem is established.

9.3. Following a familiar terminology, we say that the set of values $x$ for which

$$\int_0^h |f(x + t) - f(x)| \, dt = o(h) \qquad (h \to 0)$$

is the Lebesgue set for the function $f(x)$. The following result* will now be established.

* Compare G. H. Hardy [1921].

§9] SUMMABILITY OF DIVERGENT INTEGRALS 79

THEOREM 9.3. *If $\varphi(u)$ belongs to $L$ in $(0, R)$ for every positive $R$ and if*

$$f(s) = \int_0^\infty e^{-st} \varphi(t)\, dt \qquad (\sigma > \sigma_c < \infty),$$

*then for every $t \geqq 0$ in the Lebesgue set for $\varphi(u)$ the integral*

(1) $$\frac{1}{2\pi i} \int_{c-i\infty}^{c+i\infty} f(s) e^{st}\, ds \qquad (c > \sigma_c)$$

*is summable $(C, 1)$ to $\varphi(t)$.*

It will be sufficient to prove that

$$\lim_{T \to \infty} \frac{2}{\pi T} \left[ \int_0^\delta \varphi(t-u) e^{cu} \frac{\sin^2(Tu/2)}{u^2}\, du \right.$$
$$\left. + \int_0^\delta \varphi(t+u) e^{-cu} \frac{\sin^2(Tu/2)}{u^2}\, du \right] = \varphi(t)$$

or, by §9.1 (2), that

$$\lim_{T \to \infty} \frac{1}{T} \int_0^\delta \omega(u) \frac{\sin^2(Tu/2)}{u^2}\, du = 0,$$

$$\omega(u) = \varphi(t-u) e^{cu} + \varphi(t+u) e^{-cu} - 2\varphi(t).$$

But, if $t$ is in the Lebesgue set, then

(2) $$\Omega(h) = \int_0^h |\omega(u)|\, du = o(h) \qquad (h \to 0).$$

Hence

$$\left| \frac{1}{T} \int_0^{1/T} \omega(u) \frac{\sin^2 Tu}{u^2}\, du \right| \leqq T \int_0^{1/T} |\omega(u)|\, du = o(1) \qquad (T \to \infty).$$

Also

$$\left| \frac{1}{T} \int_{1/T}^\delta \omega(u) \frac{\sin^2(Tu/2)}{u^2}\, du \right| \leqq \frac{1}{T} \int_{1/T}^\delta \frac{|\omega(u)|}{u^2}\, du$$
$$= \frac{\Omega(\delta)}{T\delta^2} - \Omega\left(\frac{1}{T}\right) T + \frac{2}{T} \int_{1/T}^\delta \frac{\Omega(u)}{u^3}\, du.$$

By virtue of equation (2) each term on the right-hand side of the last equation is clearly* $o(1)$ as $T$ becomes infinite, so that our theorem is proved.

COROLLARY 9.3a. *The integral (1) is summable $(C, 1)$ to $\varphi(t)$ for almost all positive $t$.*

* One may apply l'Hospital's rule for the last term.

COROLLARY 9.3b. *Two functions $\varphi(t)$ which differ in a set of positive measure on the positive t-axis cannot have the same Laplace transform.*

This is again the uniqueness theorem which we have proved before. The present section gives a new proof.

## 10. Inversion When the Determining Function Belongs to $L^2$

Another case in which it is desirable to discuss the inversion formula is that in which $\varphi(t)$ belongs to $L^2$ on the interval $(0, \infty)$. That is,

$$\int_0^\infty |\varphi(t)|^2 dt < \infty$$

We use the notation

$$\underset{a\to\infty}{\text{l.i.m.}}^{(p)} \varphi_a(t) = \varphi(t)$$

to mean that $\varphi_a(t)$ and $\varphi(t)$ belong to $L^p$, $(p > 0)$, in $(-\infty, \infty)$ and that

$$\lim_{a\to\infty} \int_{-\infty}^\infty |\varphi_a(t) - \varphi(t)|^p dt = 0.$$

We must first see if the generating function is defined when $\varphi(t)$ belongs to $L^2$.

THEOREM 10. *If $\varphi(t)$ belongs to $L^2$ in $(0, \infty)$ then*

(1) $$\underset{R\to\infty}{\text{l.i.m.}}^{(2)} \int_0^R e^{-st} \varphi(t)\, dt$$

*exists for $\sigma \geqq 0$ and defines a function $f(s)$ there which is analytic for $\sigma > 0$. Moreover,*

(2) $$f(s) = \int_0^\infty e^{-st} \varphi(t)\, dt \qquad (\sigma > 0),$$

*the integral converging absolutely for $\sigma > 0$, and*

(3) $$\underset{\sigma\to 0+}{\text{l.i.m.}}^{(2)} f(\sigma + i\tau) = f(i\tau)$$

For, since $\varphi(t)$ belongs to $L^2$ in $(0, \infty)$, the same is true of $e^{-\sigma t}\varphi(t)$ for any positive $\sigma$. The existence of (1) follows from the Plancherel theorem, and $f(\sigma + i\tau)$ considered as a function of $\tau$ is the Fourier transform of a function which is zero for negative $t$ and is $\sqrt{2\pi}\, e^{-\sigma t}\varphi(t)$ for positive $t$. That (2) converges absolutely follows from Schwarz's inequality

$$\left[\int_0^\infty e^{-\sigma t} |\varphi(t)|\, dt\right]^2 \leqq \int_0^\infty e^{-2\sigma t} dt \int_0^\infty |\varphi(t)|^2 dt.$$

The analyticity of $f(s)$ follows from Theorem 5a. That it is the same function defined by (1) results from familiar facts concerning mean convergence.*

Finally, since $f(\sigma + i\tau) - f(i\tau)$ is the Fourier transform of $[e^{-\sigma t} - 1]\varphi(t)$, we see by Parseval's equation that

$$\int_{-\infty}^{\infty} |f(\sigma + i\tau) - f(i\tau)|^2 d\tau = 2\pi \int_{0}^{\infty} (e^{-\sigma t} - 1)^2 |\varphi(t)|^2 dt.$$

As $\sigma$ tends to zero through positive values, the right-hand side of this equation tends to zero by Theorem 4.3, so that (3) is established.

10.1. For the present case the inversion formula now takes the following form.

THEOREM 10.1. *If $f(s)$ is defined as in Theorem* 10, *then*

$$\underset{T\to\infty}{\text{l.i.m.}}^{(2)} \frac{1}{2\pi i} \int_{-iT}^{iT} f(s)e^{st} ds = \begin{cases} \varphi(t) & (t \geq 0) \\ 0 & (t < 0), \end{cases}$$

*or for $c > 0$*

(1) $$\frac{e^{ct}}{2\pi} \underset{T\to\infty}{\text{l.i.m.}}^{(2)} \int_{-T}^{T} f(c + i\tau)e^{i\tau t} d\tau = \begin{cases} \varphi(t) & (t \geq 0) \\ 0 & (t < 0), \end{cases}$$

*and*

$$\int_{0}^{\infty} e^{-2ct} |\varphi(t)|^2 dt = \frac{1}{2\pi} \int_{-\infty}^{\infty} |f(c + i\tau)|^2 d\tau.$$

This is an immediate consequence of the Plancherel and Parseval theorems. We observe that (1) is equivalent to the two equations

$$\lim_{T\to\infty} \int_{0}^{\infty} \left| \varphi(t)e^{-ct} - \frac{1}{2\pi} \int_{-T}^{T} f(c + i\tau)e^{i\tau t} d\tau \right|^2 dt = 0$$

$$\lim_{T\to\infty} \int_{-\infty}^{0} \left| \int_{-T}^{T} f(c + i\tau)e^{i\tau t} d\tau \right|^2 dt = 0$$

10.2. From §10.1 (1) we could conclude that

(1) $$\frac{1}{2\pi i} \int_{c-iT}^{c+iT} f(s)e^{st} ds \qquad (c > 0)$$

converges in the mean to $\varphi(t)$ over every *finite* interval. But we can show that (1) cannot converge in the mean to $\varphi(t)$ in the *infinite* interval $(-\infty, \infty)$ unless $f(s)$ is identically zero. To establish this we shall need

* See, for example, E. C. Titchmarsh [1932], p. 390 (iii).

the following result, which is a special case of the general uniqueness theorem in the theory of Fourier transforms.

**LEMMA 10.2.** *If $\varphi(t)$ belongs to $L^2$, and if*

$$\int_0^\infty e^{-ct}\varphi(t) \sin xt\, dt = 0 \qquad (c > 0)$$

*for all $x$, then $\varphi(t)$ is zero for almost all positive values of $t$.*

Clearly $e^{-ct}\varphi(t)$ and $te^{-ct}\varphi(t)$ belongs to $L$ in $(0, \infty)$. The integral

$$\int_0^\infty e^{-ct} t\varphi(t) \cos xt\, dt$$

converges uniformly for all $x$, so that its value is also zero for all $x$. Hence

$$\int_0^\infty \frac{dx}{a^2 + x^2} \int_0^\infty e^{-ct} t\varphi(t) \cos xt\, dt = 0 \qquad (a > 0, -\infty < x < \infty)$$

$$\int_0^\infty e^{-ct} t\varphi(t)\, dt \int_0^\infty \frac{\cos xt}{a^2 + x^2}\, dx = \frac{\pi}{2a} \int_0^\infty \varphi(t) e^{-(a+c)t} t\, dt = 0 \qquad (a > 0).$$

The interchange in the order of integration is justified by Fubini's theorem. An appeal to Corollary 6.2b concludes the proof.

We are now able to prove the following result.*

**THEOREM 10.2.** *If $f(s)$ is defined as in Theorem 10 and is not identically zero, then for $c > 0$ the function*

(2) $$\psi_T(t) = \frac{1}{2\pi i} \int_{c-iT}^{c+iT} f(s) e^{st}\, ds$$

*cannot belong to $L^2$ in the interval $-\infty < t < \infty$ for all large positive values of $T$.*

For, set

$$\varphi_T(t) = 2\pi e^{-ct} \psi_T(t).$$

Then

$$\varphi_T(t) = \int_{-T}^T f(c + i\tau) e^{i\tau t}\, d\tau.$$

We must show that

(3) $$\int_{-\infty}^\infty |\psi_T(t)|^2\, dt = \infty$$

---

* Compare D. V. Widder and N. Wiener [1938].

for values of $T$ arbitrarily large. An integration by parts gives

$$\varphi_T(t) = \frac{f(c + iT)e^{itT} - f(c - iT)e^{-itT}}{it} - \frac{1}{t} \int_{-T}^{T} e^{i\tau t} f'(c + i\tau) \, d\tau.$$

Now $f(c + iT)$ cannot be identically equal to $f(c - iT)$ for all large $T$. For this would imply, by use of §10 (2), that

(4) $$\int_0^\infty e^{-ct} \varphi(t) \sin Tt \, dt = 0$$

for large $T$ and hence* for all $T$. By Lemma 10.2 this would imply that $\varphi(t)$ is zero for almost all positive $t$ and hence that $f(s)$ is identically zero, contrary to hypothesis.

Choose $T$ one of the values for which (4) is not zero and set

$$A = |f(c + iT) - f(c - iT)| \neq 0.$$

Then by virtue of the Riemann-Lebesgue theorem

$$it\varphi_T(t) = [f(c + iT) - f(c - iT)]e^{itT}$$
$$+ 2if(c - iT) \sin Tt + o(1) \quad (t \to \infty),$$

so that

$$|t\varphi_T(t)| \geq \frac{A}{2} - 2|f(c - iT) \sin Tt| \qquad (t > t_0),$$

if $t_0$ is sufficiently large. Since the last term is zero infinitely often, we have

$$|t\varphi_T(t)| \geq \frac{A}{4}$$

in infinitely many intervals to the right of $t = t_0$ all of the same length. This insures that

$$\int_{-\infty}^{\infty} e^{2ct} |\varphi_T(t)|^2 \, dt = \infty,$$

so that (3) is established for the value of $T$ chosen. Since $T$ may be chosen arbitrarily large, the proof of the theorem is complete.

## 11. Stieltjes Resultant

Let $\alpha(t)$ and $\beta(t)$ be two functions defined for $t \geq 0$. We make the following definition:

---

* The integral represents an analytic function of $T$, since it is equal to $f(c + iT) - f(c - iT)$ except for a constant factor.

DEFINITION 11. *The Stieltjes resultant\* of $\alpha(t)$ and $\beta(t)$ is the function*

(1) $$\gamma(t) = \int_0^t \alpha(t-u)\,d\beta(u) = \int_0^t \beta(t-u)\,d\alpha(u) \qquad (t \geq 0)$$

*when these two integrals exist and are equal.*

If one of the functions is continuous and the other is of bounded variation, both vanishing at $t = 0$, both integrals will clearly exist and be equal for all positive $t$. We shall be particularly interested in the case in which both functions are normalized functions of bounded variation. Denote by $P_\alpha$ the countable set of points where $\alpha(t)$ is discontinuous, with a similar meaning for $P_\beta$. By the set $P_{\alpha+\beta}$ we understand the set of points $x$ which have coördinates of the form

$$x = x_\alpha + x_\beta,$$

where $x_\alpha$ and $x_\beta$ are the coördinates of points of $P_\alpha$ and $P_\beta$ respectively. If at least one of the sets $P_\alpha$, $P_\beta$ is empty we make the convention† that $P_{\alpha+\beta}$ shall then be empty.

11.1. With the conventions of the previous section we prove

THEOREM 11.1. *If $\alpha(t)$ and $\beta(t)$ are normalized functions of bounded variation in $0 \leq t \leq R$ with discontinuities in the sets $P_\alpha$ and $P_\beta$ respectively, then the Stieltjes resultant*

$$\gamma(t) = \int_0^t \alpha(t-u)\,d\beta(u) = \int_0^t \beta(t-u)\,d\alpha(u)$$

*exists for every $t$ of $(0, R)$ not in the set $P_{\alpha+\beta}$.*

For, if $t$ is not in $P_{\alpha+\beta}$ it is clear that $\alpha(t-u)$ considered as a function of $u$ can have no point of discontinuity in common with one of $\beta(u)$ in the interval $0 \leq u \leq R$. Hence by Theorem 14 of Chapter I the resultant $\gamma(t)$ exists for all $t$ not in $P_{\alpha+\beta}$.

11.2. We now show that the Stieltjes resultant of two functions of bounded variation, which is defined by §11 (1) except in a certain countable set of points $P_{\alpha+\beta}$, may be defined there also in such a way as to become a function of bounded variation.

THEOREM 11.2a. *If $\alpha(t)$, $\beta(t)$ and $\gamma(t)$ are defined as in Theorem 11.1, then $\gamma(t)$ may be defined in points $P_{\alpha+\beta}$ so as to become a normalized function of bounded variation in $(0, R)$.*

It is no restriction to assume that $\alpha(t)$ and $\beta(t)$ are real functions. Then since $\alpha(t)$ is of bounded variation we have

---

\* Another term sometimes used in this connection is *convolution*. The German word "Faltung" is also used in English texts; cf. N. Wiener [1933] p. 71.

† Compare A. Wintner [1934] p. 9.

$$\alpha(t) = \alpha_1(t) - \alpha_2(t)$$
$$V_\alpha(t) = \alpha_1(t) + \alpha_2(t),$$

where $\alpha_1(t)$ and $\alpha_2(t)$ vanish at $t = 0$, are non-decreasing, and have the same discontinuities as $\alpha(t)$. The function $V_\alpha(t)$ is the total variation of $\alpha(u)$ in the interval $0 \leq u \leq t$. Functions $\beta_1(t)$, $\beta_2(t)$ and $V_\beta(t)$ are defined for $\beta(t)$ in a similar way. Then if $t$ is not in $P_{\alpha+\beta}$, we have

$$\begin{aligned}(1)\quad \gamma(t) &= \int_0^t \alpha_1(t-u)\,d\beta_1(u) - \int_0^t \alpha_2(t-u)\,d\beta_1(u) \\ &\quad - \int_0^t \alpha_1(t-u)\,d\beta_2(u) + \int_0^t \alpha_2(t-u)\,d\beta_2(u)\end{aligned}$$

If $t_1$ and $t_2$ are points not in $P_{\alpha+\beta}$ such that $0 < t_1 < t_2$ and if $i$ and $j$ are either of the integers 1, 2, then

$$\int_0^{t_2} \alpha_i(t_2-u)\,d\beta_j(u) - \int_0^{t_1} \alpha_i(t_1-u)\,d\beta_j(u)$$
$$= \int_0^{t_1} [\alpha_i(t_2-u) - \alpha_i(t_1-u)]\,d\beta_j(u) + \int_{t_1}^{t_2} \alpha_i(t_2-u)\,d\beta_j(u) \geq 0.$$

That is, each of the integrals (1) defines a monotonic function on the set complementary to $P_{\alpha+\beta}$. Each is bounded there since

$$0 \leq \int_0^t \alpha_i(t-u)\,d\beta_j(u) \leq \alpha_i(t)\beta_j(t) \leq \alpha_i(R)\beta_j(R)$$

Hence $\gamma(t)$ has a right-hand limit at $t = 0$, a left-hand limit at $t = R$ and both right-hand and left-hand limits at intermediate points. Hence we may clearly complete its definition so as to make it a normalized function of bounded variation in $(0, R)$.

In subsequent work we shall assume that the Stieltjes resultant of two functions of bounded variation has been defined by the above method at all points.

THEOREM 11.2b. *If $\alpha(t)$, $\beta(t)$ and $\gamma(t)$ are defined as in Theorem 11.1, then*

$$V_\gamma(R) \leq V_\alpha(R)V_\beta(R)$$

For, define $\alpha(t)$ and $\beta(t)$ to be zero for negative $t$, to be $\alpha(R)$ and $\beta(R)$ respectively for $t > R$. Then

$$\gamma(t) = \int_{-\infty}^\infty \alpha(t-u)\,d\beta(u)$$

if $t$ is not in $P_{\alpha+\beta}$. Let
$$0 \leq a = t_0 < t_1 < \cdots < t_N = b \leq R,$$
where the $t_i$ are not in $P_{\alpha+\beta}$. Then for any positive $u$
$$\sum_{i=0}^{N-1} |\alpha(t_{i+1} - u) - \alpha(t_i - u)| \leq \sum_{i=0}^{N-1} \int_{t_i-u}^{t_{i+1}-u} |d\alpha(v)|$$
$$\leq \int_{-\infty}^{\infty} |d\alpha(v)| = V_\alpha(R).$$

Hence
$$\sum_{i=0}^{N-1} |\gamma(t_{i+1}) - \gamma(t_i)| \leq \int_{-\infty}^{\infty} V_\alpha(R) |d\beta(u)| = V_\alpha(R)V_\beta(R).$$

Since the right-hand side of this inequality is independent of the $t_i$ we may let the latter approach points of $P_{\alpha+\beta}$ and in particular we may let $a$ and $b$ approach zero and $R$ respectively. The left-hand side can then be brought arbitrarily near to the variation of $\gamma(t)$, $V_\gamma(R)$, so that the theorem is proved.

11.3. We now establish a fundamental result regarding the limiting value of a resultant function as the independent variable becomes infinite. The result contains in it the classical Cauchy theorem regarding the multiplication of absolutely convergent series.

THEOREM 11.3. *If $A(t)$ and $B(t)$ are normalized functions of bounded variation in $(0, \infty)$ which tend to limits $A$ and $B$ respectively as $t$ becomes infinite, then $C(t)$, the Stieltjes resultant of $A(t)$ and $B(t)$, tends to $AB$ as $t$ becomes infinite.*

Since
$$C(t) - AB(t) = \int_0^t [A(t-u) - A] d[B(u) - B],$$
it is no restriction to assume that $A = B = 0$ provided we relinquish the hypothesis $A(0) = B(0) = 0$ implied by the word "normalized." Change of variable and integration by parts gives
$$C(t) = \int_0^{t/2} A(t-u) dB(u) + \int_0^{t/2} B(t-u) dA(u)$$
$$+ A(0)B(t) - A(t/2)B(t/2).$$
It is clearly sufficient to show that

(1) $$\lim_{t \to \infty} \int_0^{t/2} A(t-u) dB(u) = 0.$$

If $B(u)$ is constant the result is trivial. Otherwise denote its total variation in $(0, \infty)$ by $V_B(\infty)$. To an arbitrary positive $\epsilon$ there corresponds a $t_0$ such that for $2t > t_0$

$$|A(t)| \leq \epsilon/V_B(\infty).$$

Then

$$\left|\int_0^{t/2} A(t-u)\,dB(u)\right| \leq \frac{\epsilon}{V_B(\infty)} \int_0^{t/2} |dB(u)| \leq \epsilon \quad (t > t_0),$$

so that (1), and hence the theorem, is established.

11.4. We prove next:

THEOREM 11.4. *If $\alpha(t)$, $\beta(t)$, and $\gamma(t)$ are defined as in Theorem 11.1, and if*

$$A(t) = \int_0^t e^{-s_0 u}\,d\alpha(u), \qquad B(t) = \int_0^t e^{-s_0 u}\,d\beta(u),$$

$$C(t) = \int_0^t e^{-s_0 u}\,d\gamma(u)$$

*for any complex number $s_0$, then $C(t)$ is the Stieltjes resultant of $A(t)$ and $B(t)$.*

For, let $t$ be a point at which

$$\int_0^t A(t-u)\,dB(u)$$

exists. This has the value

$$\int_0^t e^{-s_0 u}\,d\beta(u) \int_0^{t-u} e^{-s_0 x}\,d\alpha(x),$$

which becomes, after integration by parts,

$$\int_0^t e^{-s_0 u}\left[e^{-s_0(t-u)}\alpha(t-u) + s_0 \int_0^{t-u} \alpha(x)e^{-s_0 x}\,dx\right]d\beta(u)$$

$$= \int_0^t e^{-s_0 t}\alpha(t-u)\,d\beta(u) + s_0 \int_0^t e^{-s_0 u}\,d\beta(u)\int_u^t \alpha(x-u)e^{-s_0(x-u)}\,dx$$

$$= \int_0^t e^{-s_0 t}\alpha(t-u)\,d\beta(u) + s_0 \int_0^t e^{-s_0 x}\,dx \int_0^x \alpha(x-u)\,d\beta(u)$$

(1)

$$= e^{-s_0 t}\gamma(t) + s_0 \int_0^t e^{-s_0 x}\gamma(x)\,dx$$

$$= \int_0^t e^{-s_0 x}\,d\gamma(x) = C(t).$$

Since $P_{A+B} = P_{\alpha+\beta}$ by Theorem 5c of Chapter I the existence of $\gamma(t)$ for the particular $t$ chosen above is assured. For the existence of the integral (1) it is unnecessary that $\gamma(x)$ should be defined at the countable set $P_{\alpha+\beta}$. Thus $C(t)$ coincides with the resultant of $A(t)$ and $B(t)$ except perhaps in $P_{\alpha+\beta}$. But since both are normalized functions they must coincide throughout.

**11.5.** We are now able to prove the fundamental theorem concerning the product of generating functions, the analogue of Cauchy's theorem for the multiplication of power series.

**THEOREM 11.5.** *If $\alpha(t)$, $\beta(t)$, $\gamma(t)$ are defined as in Theorem 11.1* for every $R > 0$, and if the integrals*

$$f(s_0) = \int_0^\infty e^{-s_0 t} \, d\alpha(t)$$

$$g(s_0) = \int_0^\infty e^{-s_0 t} \, d\beta(t)$$

*converge absolutely, then*

(1) $$f(s_0)g(s_0) = \int_0^\infty e^{-s_0 t} \, d\gamma(t),$$

*and*

(2) $$\int_0^\infty e^{-\sigma_0 t} \, |d\gamma(t)| \leq \int_0^\infty e^{-\sigma_0 t} \, |d\alpha(t)| \int_0^\infty e^{-\sigma_0 t} \, |d\beta(t)|.$$

For, define $A(t)$, $B(t)$, $C(t)$ as in Theorem 11.4; then by Theorem 12 of Chapter I

$$V_A(t) = \int_0^t e^{-\sigma_0 u} \, |d\alpha(u)|$$

$$V_B(t) = \int_0^t e^{-\sigma_0 u} \, |d\beta(u)|,$$

so that $A(t)$ and $B(t)$ are of bounded variation in the interval $(0, \infty)$. Hence the conditions of Theorem 11.3 are satisfied by $A(t)$ and $B(t)$. By Theorem 11.4 $C(t)$ is the Stieltjes resultant of $A(t)$ and $B(t)$. Since $A = f(s_0)$, $B = g(s_0)$ we have (1) at once. By Theorem 11.2b

$$V_C(t) \leq V_A(t)V_B(t)$$

$$V_C(\infty) \leq V_A(\infty)V_B(\infty),$$

so that (2) is also established.

**COROLLARY 11.5.** *If the series*

$$f(s_0) = \sum_{n=0}^\infty a_n e^{-n s_0}, \qquad g(s_0) = \sum_{n=0}^\infty b_n e^{-n s_0}$$

converge absolutely, so does the following series whose sum is $f(s_0)g(s_0)$,

$$f(s_0)g(s_0) = \sum_{n=0}^{\infty} (a_n b_0 + a_{n-1} b_1 + \cdots + a_0 b_n)e^{-ns_0}$$

For, we have only to specialize the functions $\alpha(t)$ and $\beta(t)$ of the theorem as suitable step-functions to obtain this result.

11.6. In this section we obtain the analogue of a classical theorem of F. Mertens [1874] regarding the product of power series. This result differs from Cauchy's result only in that the condition of absolute convergence is relaxed for one of the series. The product series is then not known to converge absolutely.

THEOREM 11.6a. *If $A(t)$ and $B(t)$ are normalized functions of bounded variation in $(0, R)$ for every $R > 0$, one of which is of bounded variation in $(0, \infty)$, and if they approach limits $A$ and $B$ respectively as $t$ becomes infinite, then $C(t)$, their Stieltjes resultant, approaches $AB$ as $t$ becomes infinite.*

As in the proof of Theorem 11.3 we may assume that $A = B = 0$, that $B(t)$ is of bounded variation in $(0, \infty)$, and that

$$C(t) = \int_0^{t/2} A(t - u)\, dB(u) + \int_{t/2}^{t} A(t - u)\, dB(u) = I_1(t) + I_2(t)$$

As before

$$\lim_{t \to 0} I_1(t) = 0.$$

If we denote an upper bound of $|A(t)|$ by $M$, then

$$|I_2(t)| \leq M[V_B(t) - V_B(t/2)] = o(1) \qquad (t \to \infty).$$

This completes the proof of the theorem. By use of Theorem 11.4 one may now easily obtain:

THEOREM 11.6b. *If $\alpha(t)$, $\beta(t)$, $\gamma(t)$ are defined as in Theorem 11.1 for every $R > 0$, and if the integrals*

$$f(s_0) = \int_0^{\infty} e^{-s_0 t}\, d\alpha(t)$$

$$g(s_0) = \int_0^{\infty} e^{-s_0 t}\, d\beta(t)$$

*converge, one of them absolutely, then*

$$f(s_0)g(s_0) = \int_0^{\infty} e^{-s_0 t}\, d\gamma(t).$$

11.7. In this section we obtain an analogue of Abel's theorem* on the

* N. H. Abel [1826].

product of power series. This relaxes the condition of absolute convergence of the factor series but demands instead that these series and the product series all converge.

**THEOREM 11.7a.** *If $A(t)$ and $B(t)$ are normalized functions of bounded variations in $(0, R)$ for every $R > 0$, if $C(t)$ is their Stieltjes resultant, and if these three functions approach $A$, $B$, $C$ respectively as $t$ becomes infinite, then $C = AB$.*

Under the present hypotheses it is clear that the integrals

$$f(x) = \int_0^\infty e^{-xt} dA(t)$$

$$g(x) = \int_0^\infty e^{-xt} dB(t)$$

converge for $x \geqq 0$ and that for $x > 0$

(1) $$f(x) = x \int_0^\infty e^{-xt} dA_1(t),$$

where

$$A_1(t) = \int_0^t A(u)\, du,$$

the integral (1) converging absolutely for $x > 0$. Now

$$C_1(t) = \int_0^t C(y)\, dy = \int_0^t dy \int_0^y A(y - u)\, dB(u)$$

$$= \int_0^t dB(u) \int_u^t A(y - u)\, dy = \int_0^t A_1(t - u)\, dB(u).$$

Applying Theorem 11.6b we have

$$\frac{f(x)g(x)}{x} = \int_0^\infty e^{-xt} dC_1(t) = \int_0^\infty e^{-xt} C(t)\, dt$$

(2) $$f(x)g(x) = \int_0^\infty e^{-xt} dC(t) \qquad (x > 0).$$

This integration by parts is permissible since $C(t)$ is of bounded variation in $(0, R)$ for every $R > 0$ and is bounded in $(0, \infty)$.

By Theorem 4.3

$$\lim_{x \to 0+} f(x) = A, \qquad \lim_{x \to 0+} g(x) = B$$

$$\lim_{x \to 0+} \int_0^\infty e^{-xt} dC(t) = C.$$

Since (2) holds for all $x > 0$ we may now let $x$ approach zero and obtain $C = AB$, the result which was to be proved.

Again making use of Theorem 11.4 we can easily prove:

THEOREM 11.7b. *If $\alpha(t)$, $\beta(t)$, and $\gamma(t)$ are defined as in Theorem 11.1, then*

$$\int_0^\infty e^{-s_0 t} d\alpha(t) \int_0^\infty e^{-s_0 t} d\beta(t) = \int_0^\infty e^{-s_0 t} d\gamma(t)$$

*provided all three integrals converge.*

## 12. Classical Resultant

We define next the classical resultant of two functions $a(t)$ and $b(t)$ defined for $t \geqq 0$.

DEFINITION 12. *The resultant of $a(t)$ and $b(t)$ is the function*

$$\int_0^t a(t-u)b(u)\,du = \int_0^t b(t-u)a(u)\,du$$

*when these two integrals exist and are equal.*

Regarding the existence of the resultant we prove

THEOREM 12. *If $a(t)$ and $b(t)$ belong to $L$ in $(0, R)$, then their resultant $c(t)$ exists for almost all $t$ of $(0, R)$.*

For, define the functions $\alpha(t)$ and $\beta(t)$ as follows:

(1) $$\alpha(t) = \int_0^t a(u)\,du, \qquad \beta(t) = \int_0^t b(u)\,du.$$

These functions are continuous and of bounded variation in $(0, R)$, hence their Stieltjes resultant

$$\gamma(t) = \int_0^t \alpha(t-u)\,d\beta(u) = \int_0^t \alpha(t-u)b(u)\,du$$

exists for all positive $t$. But

$$\gamma(t) = \int_0^t b(u)\,du \int_u^t a(y-u)\,dy.$$

Interchanging the order of integration we obtain

(2) $$\gamma(t) = \int_0^t dy \int_0^y a(y-u)b(u)\,du = \int_0^t c(y)\,dy,$$

so that $c(t)$ exists for almost all $t$ and is in fact $\gamma'(t)$ for those values.

12.1. For the present case the product theorem becomes the following result.

THEOREM 12.1a. *If $c(t)$ is the resultant of $a(t)$ and $b(t)$, both of which belong to $L$ in $(0, R)$ for every $R > 0$, then*

(1) $$\int_0^\infty e^{-s_0 t} a(t)\, dt \int_0^\infty e^{-s_0 t} b(t)\, dt = \int_0^\infty e^{-s_0 t} c(t)\, dt$$

*provided both integrals on the left converge absolutely Then the integral on the right converges absolutely.*

This is a corollary of Theorem 11.5 if we define $\alpha(t)$ and $\beta(t)$ by §12 (1) Then the product (1) is equal to

(2) $$\int_0^\infty e^{-s_0 t}\, d\gamma(t)$$

where $\gamma(t)$ is the Stieltjes resultant which we showed to have the value §12 (2): hence (1) is eatablished. Since (2) converges absolutely by §11.5 (2), the same is true of the integral on the right of (1).

By use of Theorems 11.6b and 11.7b we establish in a similar way the following two results.

**THEOREM 12.1b.** *Under the conditions of Theorem 12.1a equation (1) holds if the integrals on the left both converge, one of them absolutely.*

**THEOREM 12.1c.** *Under the conditions of Theorem 12.1a equation (1) holds if all three integrals converge.*

## 13. Order on Vertical Lines

If

(1) $$f(s) = \int_0^\infty e^{-st}\, d\alpha(t)$$

has an abscissa of ordinary convergence $\sigma_c$ we can see at once that on a vertical line $\sigma = \sigma_1 > \sigma_c$ the function $f(\sigma_1 + i\tau)$ cannot increase more rapidly than $|\tau|$ as $|\tau|$ becomes infinite. For, if $\sigma_c < \sigma_0 < \sigma_1$, we have

$$f(\sigma_1 + i\tau) = (\sigma_1 + i\tau - \sigma_0) \int_0^\infty e^{-(\sigma_1 + i\tau - \sigma_0)t} \beta(t)\, dt,$$

where

(2) $$\beta(t) := \int_0^t e^{-\sigma_0 u}\, d\alpha(u) \qquad (0 \leq t)$$

Then

$$|f(\sigma_1 + i\tau)| \leq (\sigma_1 - \sigma_0 + |\tau|) \int_0^\infty e^{-(\sigma_1 - \sigma_0)t} |\beta(t)|\, dt$$

$$|f(\sigma_1 + i\tau)| = O(|\tau|) \qquad (|\tau| \to \infty)$$

However, we can prove a stronger result than this.

**THEOREM 13.** *If the integral (1) has an abscissa of convergence $\sigma_c$, then*

$$f(\sigma + i\tau) = o(|\tau|) \qquad (|\tau| \to \infty)$$

*uniformly in* $\sigma_c + \delta \leq \sigma < \infty$ *for any* $\delta > 0$.

Let $\epsilon$ be an arbitrary positive number. We wish to show that there exists a number $\tau_0$ independent of $\sigma$ in the interval $\sigma_c + \delta \leq \sigma < \infty$ such that

(3) $\qquad\qquad\qquad |f(\sigma + i\tau)| \leq \epsilon |\tau| \qquad\qquad (|\tau| \geq \tau_0).$

If $\sigma_c < \sigma_0 < \sigma + \delta$ we have

$$f(\sigma + i\tau) = \int_0^\infty e^{-(\sigma+i\tau-\sigma_0)t} d\beta(t),$$

where $\beta(t)$ is the function defined by equation (2). We show first that if the number $R$ is sufficiently large then

(4) $\qquad\qquad \left| \frac{1}{\tau} \int_R^\infty e^{-(\sigma+i\tau-\sigma_0)t} d\beta(t) \right| < \frac{\epsilon}{2}$

for all $\tau$ for which $|\tau|$ is greater than some number $\tau_1$ and for all $\sigma \geq \sigma_c + \delta$. We have

$$\int_R^\infty e^{-(\sigma-\sigma_0+i\tau)t} d\beta(t) = -\beta(R)e^{-(\sigma-\sigma_0+i\tau)R}$$
$$+ (\sigma - \sigma_0 + i\tau) \int_R^\infty e^{-(\sigma-\sigma_0+i\tau)t} \beta(t) \, dt$$

$$\left| \frac{1}{\tau} \int_R^\infty e^{-(\sigma-\sigma_0+i\tau)t} d\beta(t) \right| \leq \frac{Me^{-hR}}{\tau_1} + M\left(\frac{1}{\tau_1} + \frac{1}{h}\right) e^{-hR},$$

where $M$ is an upper bound of $\beta(t)$, and $h = \sigma_c + \delta - \sigma_0 > 0$. This inequality makes it clear that (4) will be satisfied for all numbers $R$ sufficiently large. With a value of $R$ so large that (4) holds we have

$$\left| \frac{1}{\tau} \int_0^R e^{-(\sigma-\sigma_0+i\tau)t} d\beta(t) \right| \leq \frac{1}{|\tau|} \int_0^R e^{-ht} |d\beta(t)|.$$

The right-hand side is independent of $\sigma$ and may be made less than $\epsilon/2$ by choosing $|\tau|$ sufficiently large, say greater than a number $\tau_0 > \tau_1$. Thus (3) is established.

## 14. Generating Function Analytic at Infinity

One of the simplest kinds of functions which can be represented by Laplace integrals is the class of functions analytic at infinity. We obtain first the relation between the series expansions of determining and generating functions.

THEOREM 14a. *Every function analytic at infinity is a generating function. If*

(1) $$f(s) = f(\infty) + \sum_{n=0}^{\infty} \frac{a_n n!}{s^{n+1}} \qquad |s| > c,$$

*then*

(2) $$f(s) = f(\infty) + \int_0^{\infty} e^{-st} \varphi(t) \, dt \qquad \sigma > c,$$

*where* $\varphi(t)$ *is the entire function*

(3) $$\varphi(t) = \sum_{n=0}^{\infty} a_n t^n.$$

Observe that $f(s)$ can be expressed as a Laplace-Stieltjes integral since the constant $f(\infty)$ can be so expressed. It is in this sense that $f(s)$ is a generating function.

To prove the theorem we have from (1)

$$f(s) = f(\infty) + \sum_{n=0}^{\infty} a_n \int_0^{\infty} e^{-st} t^n \, dt,$$

and the theorem is proved if we may interchange integral and summation signs. To justify this we have

$$\sum_{n=0}^{\infty} a_n \int_0^{\infty} e^{-st} t^n \, dt \ll \sum_{n=0}^{\infty} \frac{|a_n| n!}{\sigma^{n+1}},$$

the dominant series converging for $\sigma > c$ by hypothesis. This proves the theorem.

To obtain a more precise result we need several notions from the theory of entire functions.

DEFINITION 14a. *An entire function* $\varphi(t)$ *is of order* $\rho$ *if and only if*

$$M(r) = \max_{|t|=r} |\varphi(t)| = O(e^{r^{\rho+\epsilon}}) \qquad (r \to \infty)$$

*for every positive* $\epsilon$ *and for no negative* $\epsilon$.

DEFINITION 14b. *A function* $\varphi(t)$ *of order* $\rho$ *is of type* $c$ *if and only if*

$$M(r) = O(e^{(c+\epsilon)r^\rho})$$

*for every positive* $\epsilon$ *and for no negative* $\epsilon$. *The type is minimal or normal according as* $c$ *is zero or not. If no such* $c$ *exists the type is maximal.*

Clearly

$$c = \varlimsup_{r \to \infty} \frac{\log M(r)}{r^\rho}.$$

In the following lemma (see L. Bieberbach [1931] p. 238) $\varphi(t)$ is defined by (3) and

(4) $$k = \overline{\lim_{n \to \infty}} n |a_n|^{1/n}.$$

LEMMA 14. *The function $\varphi(t)$ is of order 1 and type $c \neq 0$ if and only if $k = ce$; it is of order less than 1 or of order 1 and minimal type if and only if $k = 0$.*

THEOREM 14b. *A generating function is analytic at infinity and vanishes there if and only if its determining function is of order less than 1 or of order 1 and of normal or minimal (but not maximal) type. If the order is 1 and the type $c$ then $|s| = c$ is the circle of convergence of the expansion of the generating function in powers of $1/s$.*

Consider the two related functions

(5) $$f(s) = \sum_{n=0}^{\infty} \frac{a_n n!}{s^{n+1}}$$

$$\varphi(t) = \sum_{n=0}^{\infty} a_n t^n.$$

By Stirling's formula

$$\overline{\lim_{n \to \infty}} (|a_n| n!)^{1/n} = e^{-1} \overline{\lim_{n \to \infty}} n |a_n|^{1/n} = ke^{-1}.$$

By Lemma 14 the series (5) has circle of convergence $|s| = c = ke^{-1} \geq 0$ if and only if $\varphi(t)$ is of order less than 1 or of order 1 and type $c$. But $f(s)$ is analytic at infinity and vanishes there if and only if $0 \leq c < \infty$. By Theorem 14a, $f(s)$ is the generating function of $\varphi(t)$ under these circumstances, so that the proof is complete.

COROLLARY 14b. *The entire function $\varphi(t)$ is the determining function for a generating function analytic at infinity if and only if*

(6) $$\overline{\lim_{|t| \to \infty}} \frac{\log |\varphi(t)|}{|t|} < \infty.$$

This is equivalent to the statement that there exists a positive constant $M$ such that

$$\varphi(t) = O(e^{M|t|}) \qquad (|t| \to \infty).$$

As an example take $\varphi(t) = e^{2t}$. Here $\rho = 1$ and $\sigma = 2$. The generating function $f(s) = 1/(s - 2)$ has circle of convergence $|s| = 2$ for its expansion (1).

If $\varphi(t) = \sin t$, then $\rho = 1$, $\sigma = 1$, $f(s) = 1/(s^2 + 1)$. The circle of convergence of (1) in this case is $|s| = 1$.

COROLLARY 14c. *The function*

$$f(s) = \int_0^\infty e^{-st} \varphi(t)\, dt.$$

*is an entire function of* $1/s$ *if and only if* $\varphi(t)$ *is entire and*

(7) $\qquad\qquad \varphi(t) = O(e^{\epsilon|t|}) \qquad\qquad (|t| \to \infty)$

*for every positive* $\epsilon$.

This is the case of the theorem in which $\varphi(t)$ is of minimal type. It is of particular interest since it is a generalization of a theorem of Wigert* which states that the function

$$f(z) = \sum_{n=0}^\infty a_n z^n$$

is an entire function of $1/(1 - z)$ if and only if $a_n = \varphi(n)$ where $\varphi(t)$ satisfies (7). In fact Wigert's theorem can be derived† very simply from Corollary 14c.

## 15. Periodic Determining Function

It is of interest to investigate the effect of periodicity of the determining function on the position of the singularities of the generating function. We observe that the functions

$$\frac{s}{s^2 + 1} = \int_0^\infty e^{-st} \cos t\, dt$$

$$\frac{1}{s^2 + 1} = \int_0^\infty e^{-st} \sin t\, dt$$

have their singularities at the integral points of the imaginary axis. We shall show that this is the general situation.

THEOREM 15. *If* $\alpha(t)$ *is a normalized function of bounded variation on* $0 \leq t \leq 2\pi$ *and of period* $2\pi$ *then the function*

(1) $\qquad\qquad f(s) = \int_0^\infty e^{-st}\, d\alpha(t)$

*has no singularities except perhaps poles of order unity at the integral points of the imaginary axis.*

Since $\alpha(t)$ is of bounded variation it has a Fourier development

$$\alpha(t) = \sum_{n=0}^\infty a_n \cos nt + b_n \sin nt \qquad (0 < t < \infty)$$

* See, for example, G. Faber [1903] p. 369.
† See D. V. Widder [1929] p. 732.

which converges boundedly.* By Lebesgue's limit theorem we may multiply both sides of this equation by $e^{-st}$ ($\sigma > 0$) and integrate term by term between the limits 0 and $\infty$. This gives

$$\int_0^\infty e^{-st} \alpha(t)\, dt = \sum_{n=0}^\infty \frac{a_n s + b_n n}{s^2 + n^2} \qquad (\sigma > 0).$$

Integration by parts now gives

(2) $$f(s) = \sum_{n=0}^\infty \frac{a_n s^2 + n b_n s}{s^2 + n^2}$$

But since $\alpha(t)$ is of bounded variation the sequences $\{na_n\}_0^\infty$ and $\{nb_n\}_0^\infty$ are bounded.† Hence it is easy to see that the series (2) is uniformly convergent and represents an analytic function in any finite region not including an integral point on the imaginary axis. The nature of the possible singularities at these points is evident from (2), and the proof is complete.

As an example take $\alpha(t)$ a normalized step-function which is zero in $(0, \pi)$ and unity in $(\pi, 2\pi)$ with period $2\pi$. Its expansion is

$$\alpha(t) = \frac{1}{2} - \frac{2}{\pi} \sum_{n=0}^\infty \frac{\sin (2n+1)t}{2n+1} \qquad (-\infty < t < \infty).$$

Then (2) becomes

$$f(s) = \frac{1}{1 + e^{-\pi s}} = \frac{1}{2} - \frac{2s}{\pi} \sum_{n=0}^\infty \frac{1}{s^2 + (2n+1)^2},$$

which is the familiar Mittag-Leffler development of the function.

## 16. Relation to Factorial Series

The Laplace integral enables us to obtain necessary and sufficient conditions for the representation of a function by factorial series.‡

THEOREM 16. *A necessary and sufficient condition that a function $f(s)$ can be represented by a convergent factorial series*

(1) $$f(s) = \sum_{n=0}^\infty \frac{a_n n!}{s(s+1) \cdots (s+n)}$$

*is that it be a generating function*,

(2) $$f(s) = \int_0^\infty e^{-st} \varphi(t)\, dt,$$

---

\* See E. C. Titchmarsh [1932] p. 408.
† See E. C. Titchmarsh [1932] p. 426.
‡ Compare N. E. Nörlund [1926], Chapter 6.

*with determining function*

(3) $$\varphi(t) = \sum_{n=0}^{\infty} a_n(1 - e^{-t})^n$$

*and*

(4) $$a_n = O(n^k) \qquad (n \to \infty)$$

*for some positive integer $k$.*

First supposing that (2), (3) and (4) hold, let us prove (1). By (4)

$$\varlimsup_{n \to \infty} \sqrt[n]{|a_n|} \leq 1.$$

That is, the power series

$$\sum_{n=0}^{\infty} a_n z^n$$

converges for $|z| < 1$, and (3) converges absolutely for $(0 \leq t < \infty)$. Formal integration gives

(5) $$\int_0^{\infty} e^{-st} \varphi(t)\, dt = \sum_{n=0}^{\infty} a_n \int_0^{\infty} e^{-st}(1 - e^{-t})^n\, dt$$

$$= \sum_{n=0}^{\infty} \frac{a_n n!}{s(s+1)\cdots(s+n)}.$$

To check the validity of this process we have only to note that integration over a finite range is valid by uniform convergence and that

(6) $$\int_0^{\infty} \sum_{n=0}^{\infty} |a_n|(1 - e^{-t})^n e^{-\sigma t}\, dt < \infty \qquad (\sigma > k+1).$$

To verify this we have from (4) that there exists a positive constant $M$ such that

$$\sum_{n=0}^{\infty} |a_n|(1 - e^{-t})^n < M \sum_{n=0}^{\infty} n^k(1 - e^{-t})^n.$$

Since

$$\frac{1}{(1-z)^{k+1}} = \sum_{n=0}^{\infty} \binom{n+k}{n} z^n \qquad (|z| < 1)$$

and

(7) $$\binom{n+k}{n} \sim n^k/\Gamma(k+1) \qquad (n \to \infty),$$

there must exist a positive constant $N$ such that

$$\sum_{n=0}^{\infty} |a_n| (1 - e^{-t})^n < N \sum_{n=0}^{\infty} \binom{n+k}{n} (1 - e^{-t})^n = N e^{(k+1)t}$$

$$(0 \leq t < \infty).$$

This establishes (6) and hence (5) for $\sigma > k + 1$.

Conversely let (1) converge for some $s$. It must then converge for every $s$ with larger real part.* Choose an integer $k$ for which it converges. Since its general term for $s = k$ approaches zero, we have

$$a_n = O\left(\frac{k(k+1) \cdots (n+k)}{n!}\right) \qquad (n \to \infty),$$

and by (7)

$$a_n = O(n^k) \qquad (n \to \infty).$$

Now define $\varphi(t)$ by the series (3) and substitute it in (2). Since we showed above that (2), (3) and (4) imply (1) for $\sigma > k + 1$, our proof is complete.

* Compare T. Fort [1930] p. 177, or E. Landau [1906] p. 151.

# CHAPTER III

# THE MOMENT PROBLEM

## 1. Statement of the Problem

The *moment problem of Hausdorff*, sometimes called the *little moment problem* is the following. Given a sequence of numbers

(1) $\qquad \{\mu_n\}_0^\infty : \mu_0, \mu_1, \mu_2, \cdots;$

we may ask under what conditions it is possible to determine a function $\alpha(t)$ of bounded variation in the interval $(0, 1)$ such that

(2) $\qquad \mu_n = \int_0^1 t^n \, d\alpha(t) \qquad (n = 0, 1, 2, \cdots)$

Any such sequence will be called a *moment sequence*. It is evident that not every sequence (1) has the form (2) since (2) implies that

$$|\mu_n| \leq V[\alpha(t)]_0^1$$

the quantity on the right being the variation of $\alpha(t)$ on the interval $(0, 1)$. That is, every moment sequence is bounded. It was F. Hausdorff [1921a] who first obtained necessary and sufficient conditions that a sequence should be a moment sequence.

In section 6.1 of Chapter II we showed that a sequence can have at most one representation (2) if $\alpha(t)$ is a normalized function of bounded variation. That is,

$$\alpha(0) = 0, \qquad \alpha(t) = \frac{\alpha(t+) + \alpha(t-)}{2} \qquad (0 < t < 1).$$

Since normalization of the function $\alpha(t)$ does not change the value of the integral (2) we may assume without loss of generality that $\alpha(t)$ is normalized. This we do throughout the present chapter without further repetition of the fact.

Equations (2) may be regarded as a transformation of the function $\alpha(t)$ into the sequence $\{\mu_n\}$. This transformation is closely related to the Laplace transform, is in fact the discrete analogue of the latter. For, if we replace the integer $n$ by the variable $s$ in (2) and then make the change of variable $t = e^{-u}$, we obtain

$$\mu_s = \int_0^\infty e^{-su} \, d[-\alpha(e^{-u})].$$

## 2. Moment Sequence

We introduce several definitions:

**DEFINITION 2a.**
$$\Delta^k \mu_n = \sum_{m=0}^{k} (-1)^m \binom{k}{m} \mu_{n+k-m} \qquad (k = 0, 1, 2, \cdots).$$

**DEFINITION 2b.**
$$\lambda_{k,m}(x) = \binom{k}{m} x^m (1-x)^{k-m} \qquad (k, m = 0, 1, 2, \cdots).$$

**DEFINITION 2c.**
$$\lambda_{k,m} = \binom{k}{m} (-1)^{k-m} \Delta^{k-m} \mu_m \qquad (k, m = 0, 1, 2, \cdots).$$

**DEFINITION 2d.** *A Bernstein polynomial $B_k[f(x)]$ for a function $f(x)$, defined on the interval $(0, 1)$, is*
$$B_k[f(x)] = \sum_{m=0}^{k} f\left(\frac{m}{k}\right) \lambda_{k,m}(x).$$

The degree of the polynomial is $k$ unless

$$\sum_{m=0}^{k} (-1)^{k-m} \binom{k}{m} f\left(\frac{m}{k}\right) = 0,$$

when it is of lower degree or identically zero. For example,

$$B_k[1] = \sum_{m=0}^{k} \lambda_{k,m}(x) = 1.$$

**DEFINITION 2e.** *The sequence $\{\mu_n\}_0^\infty$ satisfies Condition A if a constant $L$ exists such that*

$$\sum_{m=0}^{k} |\lambda_{k,m}| < L \qquad (k = 0, 1, 2, \cdots).$$

For example, if

(1) $$\mu_n = \int_0^1 t^n \, d\alpha(t) \qquad (n = 0, 1, 2, \cdots)$$

with $\alpha(t)$ of bounded variation in $(0, 1)$, then

$$\sum_{m=0}^{k} \lambda_{k,m} = \sum_{m=0}^{k} \binom{k}{m} \int_0^1 t^m (1-t)^{k-m} \, d\alpha(t) = \int_0^1 d\alpha(t)$$

$$\sum_{m=0}^{k} |\lambda_{k,m}| \leq \int_0^1 |d\alpha(t)| = V[\alpha(t)]_0^1.$$

That is, Condition $A$ is necessary that the sequence $\{\mu_n\}_0^\infty$ should have the form (1). In particular the sequences

$$\left\{\frac{1}{n+1}\right\}_0^\infty, \qquad \{c^n\}_0^\infty \qquad (0 < c \leqq 1)$$

satisfy Condition $A$.

**DEFINITION 2f.** *If $P_n(x)$ is the polynomial*

$$P_n(x) = \sum_{m=0}^n a_m x^m,$$

*an operator $M[P_n(x)]$, called the moment of $P_n(x)$ with respect to the sequence $\{\mu_n\}$, is defined as*

$$M[P_n(x)] = \sum_{m=0}^n a_m \mu_m.$$

For example,

$$M[x^n] = \mu_n, \qquad \sum_{m=0}^k M[\lambda_{k,m}(x)] = \sum_{m=0}^k \lambda_{k,m} = \mu_0.$$

If $\mu_n$ has the representation (1) then

$$M[P_n(x)] = \int_0^1 P_n(t)\, d\alpha(t).$$

Note that the operator is applicable only to polynomials.

We shall now prove that Condition $A$ is also sufficient that the sequence $\{\mu_n\}_0^\infty$ should have the representation (1). We need a preliminary result.

**LEMMA 2.** *If $n$ is a positive integer, then*

$$\lim_{k\to\infty} \prod_{i=0}^{n-1} \frac{kx - i}{k - i} = x^n$$

*uniformly for $0 \leqq x \leqq 1$.*

This is clear since each factor of the product approaches $x$ uniformly in the interval $0 \leqq x \leqq 1$.

We now prove:

**THEOREM 2a.** *If the sequence $\{\mu_n\}_0^\infty$ satisfies Condition $A$, then*

$$\mu_n = \lim_{k\to\infty} M[B_k[x^n]] \qquad (n = 0, 1, 2, \cdots).$$

For, by the binomial theorem we have for $k > n > 0$

$$x^n = x^n[(1-x) + x]^{k-n} = \sum_{m=0}^{k-n} \binom{k-n}{m} x^{m+n}(1-x)^{k-m-n}$$

$$= \sum_{m=n}^{k} \frac{m(m-1)\cdots(m-n+1)}{k(k-1)\cdots(k-n+1)} \lambda_{k,m}(x).$$

$$\mu_n = M[x^n] = \sum_{m=n}^{k} \frac{m(m-1)\cdots(m-n+1)}{k(k-1)\cdots(k-n+1)} \lambda_{k,m}.$$

$$\mu_n - M[B_k[x^n]] = \sum_{m=n}^{k} \left\{ \frac{m(m-1)\cdots(m-n+1)}{k(k-1)\cdots(k-n+1)} - \left(\frac{m}{k}\right)^n \right\} \lambda_{k,m}$$

$$- \sum_{m=0}^{n-1} \left(\frac{m}{k}\right)^n \lambda_{k,m}$$

$$= \sum_{m=n}^{k} \left\{ \frac{ky(ky-1)\cdots(ky-n+1)}{k(k-1)\cdots(k-n+1)} - y^n \right\} \lambda_{k,m}$$

$$- \sum_{m=0}^{n-1} \left(\frac{m}{k}\right)^n \lambda_{k,m},$$

where $y = m/k$. Let $\epsilon$ be an arbitrary positive number. By Lemma 2 we see that we can determine $k_0$ such that for $k > k_0$

$$\left| \frac{ky(ky-1)\cdots(ky-n+1)}{k(k-1)\cdots(k-n+1)} - y^n \right| < \epsilon$$

$$\left( y = \frac{m}{k},\ m = n, n+1, \cdots k \right),$$

and such that

$$\left| \sum_{m=0}^{n-1} \left(\frac{m}{k}\right)^n \lambda_{k,m} \right| < \left(\frac{n}{k}\right)^n L < \epsilon \qquad (k > k_0).$$

Hence

$$|\mu_n - M[B_k[x^n]]| < \epsilon L + \epsilon \qquad (k > k_0).$$

This gives us the desired result if $n = 1, 2, \cdots$. If $n = 0$

$$\mu_0 = M[B_k[1]].$$

By use of this result we can prove easily the main result of this section

THEOREM 2b. *A necessary and sufficient condition that $\{\mu_n\}_0^\infty$ should be a moment sequence is that it should satisfy Condition A.*

We have already seen that the condition is necessary. To prove it sufficient define a step-function $\alpha_k(t)$ which is normalized and has jumps $\lambda_{k,m}$ at points $m/k$,

$$\alpha_k\left(\frac{m}{k}+\right) - \alpha_k\left(\frac{m}{k}-\right) = \lambda_{k,m} \qquad (m = 1, 2, \cdots, k-1),$$

$$\alpha_k(0+) = \lambda_{k,0}, \qquad \alpha_k(0) = 0,$$

$$\alpha_k(1-) = \sum_{m=0}^{k-1} \lambda_{k,m},$$

$$\alpha_k(1) = \sum_{m=0}^{k} \lambda_{k,m} = \mu_0.$$

Then

$$M[B_k[x^n]] = \int_0^1 t^n \, d\alpha_k(t),$$

and by Theorem 2a,

$$\mu_n = \lim_{k \to \infty} \int_0^1 t^n \, d\alpha_k(t).$$

The total variation of $\alpha_k(t)$ is clearly

$$\sum_{m=0}^{k} |\lambda_{k,m}|,$$

which has an upper bound $L$, independent of $k$. Hence by Helly's theorem, i.e., Theorem 16.3 of Chapter I, there exists a subsequence $\{\alpha_{k_j}(t)\}_{j=0}^{\infty}$ of the sequence $\{\alpha_k(t)\}_0^{\infty}$ which approaches a limit $\alpha^*(t)$, of bounded variation in $0 \leq t \leq 1$. But

$$\mu_n = \lim_{j \to \infty} \int_0^1 t^n \, d\alpha_{k_j}(t)$$

$$= \lim_{j \to \infty} n \int_0^1 t^{n-1}[\alpha_{k_j}(1) - \alpha_{k_j}(t)] \, dt \qquad (n = 1, 2, \cdots)$$

$$\mu_0 = \lim_{j \to \infty} \alpha_{k_j}(1) = \alpha^*(1) = \int_0^1 d\alpha^*(t).$$

Since

$$|\alpha_{k_j}(1) - \alpha_{k_j}(t)| < 2L,$$

we may employ the Lebesgue limit theorem and obtain*

$$\mu_n = n \int_0^1 t^{n-1}[\alpha^*(1) - \alpha^*(t)] \, dt \qquad (n = 1, 2, \cdots)$$

$$= \int_0^1 t^n \, d\alpha^*(t) \qquad (n = 0, 1, 2, \cdots),$$

which is what we were to prove.

* One could avoid the integration by parts if Theorem 16.4 of Chapter I were used. The Lebesgue limit theorem is perhaps more familiar.

If $\alpha^*(t)$ is not normalized, we normalize it and denote the resulting function by $\alpha(t)$. Then by the uniqueness theorem, i.e., Theorem 6.1 of Chapter II, $\alpha^*(t) = \alpha(t)$ in the set $E$ of points of continuity of $\alpha(t)$. Hence at these points

$$\lim_{j \to \infty} \alpha_{k_j}(t) = \alpha(t)$$

Since every subsequence of $\{\alpha_k(t)\}_0^\infty$ has in it a subsequence which approaches $\alpha(t)$ at points of $E$ we have

(2) $$\lim_{k \to \infty} \alpha_k(t) = \alpha(t) \qquad (t \, \varepsilon \, E)$$

It can be shown in fact that (2) holds throughout the interval (0, 1)

We may use Theorem 2b to prove an important result of F. Riesz [1909] concerning linear functionals.

DEFINITION 2g. *To each function $f(x)$ continuous on $0 \leq x \leq 1$ let there correspond a number $L[f(x)]$. This correspondence is said to define a linear functional if*

(a) $$L[c_1 f_1(x) + c_2 f_2(x)] = c_1 L[f_1(x)] + c_2 L[f_2(x)]$$

*for every pair of constants $c_1$, $c_2$ and every pair of continuous functions $f_1(x), f_2(x)$;*

(b) $$|L[f(x)]| \leq M \, \|f(x)\|,$$

*where $M$ is some positive constant and $\|f(x)\|$ is the maximum value of $|f(x)|$ on $0 \leq x \leq 1$.*

For example, if

$$L[f(x)] = f(\tfrac{1}{2}),$$

or if

$$L[f(x)] = \int_0^1 f(x) \, dx,$$

we see easily that $L[f(x)]$ is a linear functional. In fact by reference to Chapter I we see that if

(3) $$L[f(x)] = \int_0^1 f(x) \, d\alpha(x)$$

with $\alpha(x)$ of bounded variation on $0 \leq x \leq 1$, then conditions (a) and (b) are satisfied with $M$ equal to the total variation of $\alpha(x)$ on $0 \leq x \leq 1$. Riesz's result is that (3) defines the most general linear functional defined on the set of continuous functions. We give a proof due to T. H. Hildebrandt and I. J. Schoenberg [1933].

THEOREM 2c. *Every linear functional $L[f(x)]$ defined on the set of functions continuous in $0 \leq x \leq 1$ has the form (3) with $\alpha(x)$ of bounded variation on $0 \leq x \leq 1$.*

106        THE MOMENT PROBLEM        [Ch. III

To prove this set
$$L[x^n] = \mu_n \qquad (n = 0, 1, 2, \cdots).$$

We show first that the sequence $\{\mu_n\}_0^\infty$ satisfies Condition A. We must determine a constant $N$ such that

(4) $\qquad \sum_{m=0}^{k} |\lambda_{k,m}| = \sum_{m=0}^{k} \binom{k}{m} |\Delta^{k-m} \mu_m| < N \qquad (k = 0, 1, 2, \cdots).$

But by choosing $\epsilon_m = \pm 1$ suitably we have

$$\sum_{m=0}^{k} |\lambda_{k,m}| = \sum_{m=0}^{k} \epsilon_m \binom{k}{m} (-1)^{k-m} \Delta^{k-m} \mu_m$$

$$= L\left[ \sum_{m=0}^{k} \epsilon_m \binom{k}{m} x^m (1-x)^{k-m} \right].$$

Here we have used property (a) of the functional $L$ and observed that
$$L[x^m(1-x)^{k-m}] = (-1)^{k-m} \Delta^{k-m} \mu_m.$$
Since
$$\left\| \sum_{m=0}^{k} \epsilon_m \binom{k}{m} x^m(1-x)^{k-m} \right\| \leq \left\| \sum_{m=0}^{k} \binom{k}{m} x^m(1-x)^{k-m} \right\| = 1,$$
we see by use of (b) that
$$\sum_{m=0}^{k} |\lambda_{k,m}| \leq M,$$
so that (4) holds with $M = N$. Hence by Theorem 2b

(5) $\qquad L[x^n] = \int_0^1 x^n \, d\alpha(x) \qquad (n = 0, 1, 2, \cdots)$

for some function $\alpha(x)$ of bounded variation on $0 \leq x \leq 1$.

Now let $f(x)$ be any function continuous on $0 \leq x \leq 1$ and let $\epsilon$ be an arbitrary positive number. By Weierstrass's theorem we can determine a polynomial $P(x)$ such that
$$|f(x) - P(x)| \leq \epsilon \qquad (0 \leq x \leq 1).$$
By (5), (a) and (b) it is clear that
$$L[f(x)] = L[f(x) - P(x)] + L[P(x)] = L[f(x) - P(x)] + \int_0^1 P(x) \, d\alpha(x)$$

$$\left| L[f(x)] - \int_0^1 f(x) \, d\alpha(x) \right| \leq M \, \|f(x) - P(x)\| + \left| \int_0^1 [P(x) - f(x)] \, d\alpha(x) \right|$$

$$\leq M\epsilon + \epsilon \int_0^1 |d\alpha(x)|.$$

Hence (3) follows, and our theorem is proved.

It can also be shown that Theorem 2b follows from Theorem 2c. Hence the problem of determining the general linear functional on the set of continuous functions is equivalent to that of determining the set of all moment sequences.

## 3. An Inversion Operator

Let us now introduce a new operator on the sequence $\{\mu_n\}_0^\infty$ by the following definition.

**DEFINITION 3.** *An operator $L_{k,t}\{\mu_n\}$ is defined by the relation*

$$L_{k,t}\{\mu\} = L_{k,t}\{\mu_n\} = (k+1)\lambda_{k,[kt]} \qquad (k=1,2,\cdots, 0 \leq t \leq 1).$$

The notation $[kt]$ means the largest integer contained in $kt$. By means of this operator we can prove:

**THEOREM 3.** *If $\{\mu_n\}_0^\infty$ satisfies Condition A then*

$$\mu_n - \mu_\infty = \lim_{k\to\infty} \int_0^1 t^n L_{k,t}\{\mu\}\, dt \qquad (n=0,1,2,\cdots).$$

For, by the law of the mean

$$\int_0^1 t^n L_{k,t}\{\mu\}\, dt = \frac{(k+1)}{k} \sum_{m=0}^{k-1} \lambda_{k,m} \left(\frac{m+\theta_m}{k}\right)^n,$$

where

$$0 < \theta_m < 1 \qquad (m=0,1,\cdots,k-1).$$

But we saw in section 2 that

$$\mu_n = \lim_{k\to\infty} \sum_{m=0}^{k} \left(\frac{m}{k}\right)^n \lambda_{k,m} \qquad (n=0,1,\cdots)$$

$$= \lim_{k\to\infty} \left[\lambda_{k,k} + \sum_{m=0}^{k-1} \left(\frac{m}{k}\right)^n \lambda_{k,m}\right].$$

To evaluate the first term we have

$$\lambda_{k,k} = \mu_k = \int_0^1 t^k\, d\alpha(t)$$

$$\mu_k = \{\alpha(1) - \alpha(1-)\} + k\int_0^1 t^{k-1}\{\alpha(1-) - \alpha(t)\}\, dt \qquad (k=1,2,\cdots)$$

$$\varlimsup_{k\to\infty} |\mu_k - \alpha(1) + \alpha(1-)| \leq \varlimsup_{t\to 1-} |\alpha(1-) - \alpha(t)| = 0$$

$$\mu_\infty = \alpha(1) - \alpha(1-).$$

Thus $\mu_\infty$ must exist under Conditions $A$, and it remains only to show that

$$\lim_{k \to \infty} \sum_{m=0}^{k-1} \left\{ \left( \frac{m+\theta_m}{k} \right)^n - \left( \frac{m}{k} \right)^n \right\} \lambda_{k,m} = 0$$

By the law of the mean we have

$$\left( \frac{m+\theta_m}{k} \right)^n - \left( \frac{m}{k} \right)^n = \theta_m \frac{n}{k} \left( \frac{m+\theta'_m}{k} \right)^{n-1}$$

where

$$0 < \theta'_m < \theta_m < 1 \qquad (m = 0, 1, \cdots, k-1).$$

Hence

$$\left| \sum_{m=0}^{k-1} \left\{ \left( \frac{m+\theta_m}{k} \right)^n - \left( \frac{m}{k} \right)^n \right\} \lambda_{k,m} \right| \leq \frac{n}{k} \sum_{m=0}^{k-1} |\lambda_{k,m}| < \frac{nL}{k},$$

so that the theorem is established.

## 4. Completely Monotonic Sequences

We now introduce the notion of a completely monotonic sequence.

DEFINITION 4. *The sequence $\{\mu_n\}_0^\infty$ is completely monotonic if its elements are non-negative and its successive differences are alternately non-positive and non-negative*

$$(-1)^k \Delta^k \mu_n \geq 0 \qquad (n, k = 0, 1, 2, \cdots)$$

An equivalent form for the definition is

$$\lambda_{k,m} \geq 0 \qquad (m, k = 0, 1, 2, \cdots).$$

For example, the sequences

(1) $$\left\{ \frac{1}{n+1} \right\}_0^\infty \qquad \{c^n\}_0^\infty \qquad (0 < c \leq 1)$$

are completely monotonic. Note that this class is included in the class of sequences which satisfy Condition A. For

$$\sum_{m=0}^k |\lambda_{k,m}| = \sum_{m=0}^k \lambda_{k,m} = \mu_0 = L.$$

We can now prove:

THEOREM 4a. *A necessary and sufficient condition that the sequence $\{\mu_n\}_0^\infty$ should have the expression*

(2) $$\mu_n = \int_0^1 t^n \, d\alpha(t) \qquad (n = 0, 1, 2, \cdots),$$

*where $\alpha(t)$ is non-decreasing and bounded for $0 \leq t \leq 1$, is that it should be completely monotonic.*

For the necessity of the condition we have

$$(-1)^k \Delta^k \mu_n = \int_0^1 t^n(1-t)^k \, d\alpha(t) \geqq 0 \qquad (n, k = 0, 1, 2, \cdots).$$

For the sufficiency we see at once that the given sequence must have the form (2) with $\alpha(t)$ of bounded variation on (0, 1) by Theorem 2b. But we showed in section 2 that if $\alpha(t)$ is normalized

$$\lim_{k \to \infty} \alpha_k(t) = \alpha(t)$$

at all points of continuity of $\alpha(t)$. But $\alpha_k(t)$ is non-decreasing since its jumps, $\lambda_{k,m}$, are non-negative. It follows that $\alpha(t)$ is non-decreasing if properly defined at its points of discontinuity. This completes the proof of the theorem.

THEOREM 4b. *A necessary and sufficient condition that the sequence $\{\mu_n\}$ should satisfy Condition A is that it should be the difference of two completely monotonic sequences.*

This is obvious since $\alpha(t)$ is of bounded variation if and only if it is the difference of two bounded non-decreasing functions.

It is easily seen directly that the sequences (1) have the form (2). In the first case $\alpha(t)$ is the non-decreasing function $t$; in the second it is a step-function with jump unity at $t = c$.

## 5. Function of $L^p$

In this section we discuss sequences §2 (1) where $\alpha(t)$ is the integral of a function of class $L^p$ $(p > 1)$. That is,

$$\mu_n = \int_0^1 t^n \varphi(t) \, dt \qquad (n = 0, 1, 2, \cdots)$$

$$\int_0^1 |\varphi(t)|^p \, dt < \infty.$$

We introduce a condition which will guarantee that a sequence will have this form.

DEFINITION 5. *The sequence $\{\mu_n\}_0^\infty$ satisfies Condition B for a given number $p > 1$ if there exists a constant $L$ such that*

$$(k+1)^{p-1} \sum_{m=0}^k |\lambda_{k,m}|^p < L \qquad (k = 0, 1, 2, \cdots).$$

For example, the sequence

$$\left\{\frac{1}{n+1-\theta}\right\}_0^\infty \qquad (0 < \theta < 1)$$

satisfies Condition $B$ with $p$ any number less than $1/\theta$ and greater than unity. Any sequence satisfying this condition also satisfies Condition $A$. For, by Hölder's inequality

$$\sum_{m=0}^{k} |\lambda_{k,m}| \leq (k+1)^{(p-1)/p} \left\{ \sum_{m=0}^{k} |\lambda_{k,m}|^p \right\}^{1/p}$$

$$< L^{1/p} \qquad (k = 0, 1, 2, \cdots).$$

The result to be established is contained in:

**THEOREM 5.** *Condition $B$ is necessary and sufficient that*

(1) $$\mu_n = \int_0^1 t^n \varphi(t)\, dt \qquad (n = 0, 1, \cdots),$$

*where $\varphi(t)$ belongs to $L^p$ on the interval $(0, 1)$.*

Suppose first that the given sequence has the representation (1). Then by Hölder's inequality

$$|\lambda_{k,m}| \leq \int_0^1 \lambda_{k,m}(t) |\varphi(t)|\, dt$$

$$|\lambda_{k,m}|^p \leq \left\{ \int_0^1 \lambda_{k,m}(t)\, dt \right\}^{p-1} \int_0^1 \lambda_{k,m}(t) |\varphi(t)|^p\, dt.$$

Since

$$\int_0^1 \lambda_{k,m}(t)\, dt = \binom{k}{m} \int_0^1 t^m (1-t)^{k-m}\, dt = \frac{1}{k+1},$$

we have

$$(k+1)^{p-1} \sum_{m=0}^{k} |\lambda_{k,m}|^p \leq \int_0^1 |\varphi(t)|^p\, dt,$$

so that Condition $B$ is satisfied.

Conversely, let $\{\mu_n\}$ satisfy Condition $B$. Then the sequence of functions $L_{k,t}\{\mu\}$ defined by Definition 3 satisfies the inequality

(2) $$\int_0^1 |L_{k,t}\{\mu\}|^p\, dt = \frac{(k+1)^p}{k} \sum_{m=0}^{k-1} |\lambda_{k,m}|^p < 2L \qquad (k = 1, 2, \cdots)$$

$$= |\lambda_{0,0}|^p < L \qquad (k = 0).$$

By Theorem 3

$$\mu_n - \mu_\infty = \lim_{k \to \infty} \int_0^1 t^n L_{k,t}\{\mu\}\, dt \qquad (n = 0, 1, 2, \cdots).$$

But by (2) and Theorem 17a of Chapter I it is possible to pick from the sequence $L_{k,t}\{\mu\}$ a subset $L_{k_j,t}\{\mu\}$ and to find a function $\varphi(t)$ of $L^p$ such that for every function $\psi(t)$ of $L^{p/(p-1)}$

$$\lim_{j\to\infty} \int_0^1 L_{k_j,t}\{\mu\}\psi(t)\,dt = \int_0^1 \varphi(t)\psi(t)\,dt.$$

In particular, if $\psi(t) = t^n$, then

$$\mu_n - \mu_\infty = \int_0^1 t^n \varphi(t)\,dt.$$

But from Condition B we see that

$$|\mu_k| = |\lambda_{k,k}| < \frac{L^{1/p}}{(k+1)^{(p-1)/p}},$$

so that

$$\mu_\infty = 0.$$

Hence our theorem is completely established.

## 6. Bounded Functions

We treat next the case in which the sequence $\{\mu_n\}$ has the form §5 (1) with $\varphi(t)$ bounded.

DEFINITION 6. *The sequence $\{\mu_n\}_0^\infty$ satisfies Condition C if there exists a constant $L$ such that*

(1) $\qquad (k+1)|\lambda_{k,m}| < L \qquad (k, m = 0, 1, 2, \cdots, m \leq k).$

For example, if

$$\mu_n = \frac{1}{n+1} \qquad (n = 0, 1, 2, \cdots),$$

then

$$(k+1)\lambda_{k,m} = 1.$$

Sequences satisfying Condition $C$ also satisfy $A$. For (1) implies

$$\sum_{m=0}^k |\lambda_{k,m}| < \sum_{m=0}^k \frac{L}{k+1} = L.$$

THEOREM 6. *Condition C is necessary and sufficient that the sequence $\{\mu_n\}_0^\infty$ should have the form*

(2) $\qquad \mu_n = \int_0^1 t^n \varphi(t)\,dt \qquad (n = 0, 1, 2, \cdots)$

*with $\varphi(t)$ bounded in the interval $0 \leq t \leq 1$.*

If $\{\mu_n\}_0^\infty$ is defined by (2), then

$$(k+1)\,|\lambda_{k,m}| \leq (k+1)\binom{k}{m}\int_0^1 t^m(1-t)^{k-m}|\varphi(t)|\,dt$$

$$\leq \underset{0\leq t\leq 1}{\text{u.b.}}|\varphi(t)|,$$

so that Condition $C$ is satisfied.

Conversely, if $C$ is satisfied, the sequence of functions $L_{k,t}\{\mu_n\}$ of Definition 3 satisfies the inequality

(3) $\quad |L_{k,t}\{\mu\}| = (k+1)\,|\lambda_{k,[kt]}| < L \quad (0 \leq t \leq 1, k = 0, 1, \cdots).$

Since Condition $A$ must also be satisfied, Theorem 3 gives

$$\mu_n - \mu_\infty = \lim_{k\to\infty}\int_0^1 t^n L_{k,t}\{\mu\}\,dt \quad (n = 0, 1, 2, \cdots).$$

But by (3) and Theorem 17b, Chapter I, it is possible to pick from the sequence $L_{k,t}\{\mu\}$ a subset $L_{k_j,t}\{\mu\}$ and to find a bounded function $\varphi(t)$ such that for every function $\psi(t)$ of class $L$

$$\lim_{j\to\infty}\int_0^1 L_{k_j,t}\{\mu\}\psi(t)\,dt = \int_0^1 \varphi(t)\psi(t)\,dt.$$

As in section 5

$$\mu_n - \mu_\infty = \int_0^1 t^n \varphi(t)\,dt \quad (n = 0, 1, \cdots).$$

From Condition $C$

$$|\mu_k| = |\lambda_{k,k}| < \frac{L}{k+1} = o(1) \quad (k\to\infty),$$

so that $\mu_\infty$ is zero, and our theorem is proved.

We observe that Theorem 6 is the limiting case $p = \infty$ of Theorem 5. The limiting case $p = 1$ is not obtained by setting $p = 1$ in Theorem 5. For, Condition $B$ with $p = 1$ becomes Condition $A$, which is sufficient to make $\alpha(t)$ of bounded variation but not sufficient to make it the integral of a function of $L$. We could show that (2) holds with $\varphi(t)$ belonging to $L$ if and only if

$$\lim_{j,k\to\infty}\int_0^1 |L_{k,t}\{\mu\} - L_{j,t}\{\mu\}|\,dt = 0.$$

We shall have no use for this result, and since it is not easily interpreted, independently of the operator $L_{k,t}\{\mu\}$, in terms of the sequence $\{\mu_n\}$ alone, we omit the proof here. We shall give the proof for the corresponding continuous case when we come to the study of the Laplace transform.

## 7. Hausdorff Summability

One of the important applications of the little moment problem occurs in the theory of the summability of divergent series. It was in fact the study of the latter theory which led Hausdorff [1921a] to investigate the moment problem. We give here a brief outline of the theory.*

We shall need to use the elements of matrix theory. The notations employed are

$$\alpha = ||a_{m,n}||_0^k = \begin{Vmatrix} a_{00} & a_{0i} & \cdots & a_{0k} \\ a_{10} & a_{11} & \cdots & a_{1k} \\ \cdot & \cdot & \cdots & \cdot \\ \cdot & \cdot & \cdots & \cdot \\ a_{k0} & a_{k1} & \cdots & a_{kk} \end{Vmatrix}.$$

$$\begin{cases} \alpha \cdot \beta = \alpha\beta = \gamma = ||c_{m,n}||_0^k \\ c_{mn} = \sum_{j=0}^{k} a_{mj} b_{jn}, \end{cases}$$

$$\beta = \alpha^{-1} \quad \text{if} \quad \alpha\beta = \beta\alpha = 1$$

We shall also use infinite matrices

$$\alpha = ||a_{m,n}||_0^\infty = \begin{Vmatrix} a_{00} & a_{01} & \cdots \\ a_{10} & a_{11} & \cdots \\ \cdot & \cdot & \cdots \\ \cdot & \cdot & \cdots \end{Vmatrix}$$

In particular we reserve the letters $\sigma$, $\tau$, $\mu$ and $\rho$ for the following matrices,

$$\sigma = \begin{Vmatrix} s_0 & 0 & 0 & \cdots \\ s_1 & 0 & 0 & \cdots \\ s_2 & 0 & 0 & \cdots \\ \cdot & \cdot & \cdot & \cdots \\ \cdot & \cdot & \cdot & \cdots \end{Vmatrix} \quad \tau = \begin{Vmatrix} t_0 & 0 & 0 & \cdots \\ t_1 & 0 & 0 & \cdots \\ t_2 & 0 & 0 & \cdots \\ \cdot & \cdot & \cdot & \cdots \\ \cdot & \cdot & \cdot & \cdots \end{Vmatrix}$$

$$\mu = \begin{Vmatrix} \mu_0 & 0 & 0 & \cdots \\ 0 & \mu_1 & 0 & \cdots \\ 0 & 0 & \mu_2 & \cdots \\ \cdot & \cdot & \cdot & \cdots \end{Vmatrix} \quad \rho = \begin{Vmatrix} 1 & 0 & 0 & 0 & 0 & \cdots \\ 1 & -1 & 0 & 0 & 0 & \cdots \\ 1 & -2 & 1 & 0 & 0 & \cdots \\ 1 & -3 & 3 & -1 & 0 & \cdots \\ \cdot & \cdot & \cdot & \cdot & \cdot & \cdots \\ \cdot & \cdot & \cdot & \cdot & \cdot & \cdots \end{Vmatrix}$$

* Compare E. Hille and J. D. Tamarkin [1933a], [1933b], [1933c] and [1934].

114                THE MOMENT PROBLEM                [CH. III

$\mu$ being called a *diagonal* matrix, and $\rho$ the *difference* matrix. Then

(1) $$\tau = \alpha\sigma$$

transforms the matrix $\sigma$ into $\tau$ or, if we consider only the first columns of $\sigma$ and $\tau$, the sequence $\{s_n\}_0^\infty$ into $\{t_n\}_0^\infty$.

We now introduce the notion of summability.

**DEFINITION 7.** *The sequence* $\{s_n\}_{n=0}^\infty$ *is summable by the matrix* $\alpha$ *to the sum* $s$ *if the sequence* $\{t_n\}_{n=0}^\infty$ *is defined by* (1) *and if*

(2) $$\lim_{n\to\infty} t_n = s.$$

More explicitly, this means that the series

$$t_m = \sum_{n=0}^\infty a_{mn} s_n \qquad (m = 0, 1, 2, \cdots)$$

all converge and that (2) holds.

For example, if

(3) $$\alpha = \begin{Vmatrix} 1 & 0 & 0 & 0 & \cdots \\ \frac{1}{2} & \frac{1}{2} & 0 & 0 & \cdots \\ \frac{1}{3} & \frac{1}{3} & \frac{1}{3} & 0 & \cdots \\ \cdot & \cdot & \cdot & \cdot & \cdots \end{Vmatrix}$$

then

$$t_m = \frac{s_0 + s_1 + \cdots + s_m}{m+1} \qquad (m = 0, 1, \cdots),$$

and we have Hölder or Cesàro summability of order one.

We show that the difference matrix $\rho$, whose elements are

$$\gamma_{m,n} = (-1)^n \binom{m}{n} \qquad (n = 0, 1, \cdots, m)$$

$$= 0 \qquad (n = m+1, m+2, \cdots)$$

is self-reciprocal,

$$\rho = \rho^{-1}.$$

It will be sufficient to prove that

(4) $$c_{m,n} = \sum_{j=0}^\infty \gamma_{mj}\gamma_{jn} = 1 \qquad m = n$$
$$= 0 \qquad m \neq n.$$

But

(5) $$\sum_{n=0}^\infty c_{m,n} x^n = \sum_{j=0}^\infty \gamma_{mj} \sum_{n=0}^\infty \gamma_{jn} x^n = \sum_{j=0}^\infty \gamma_{mj}(1-x)^j = x^m,$$

and comparing coefficients we have (4). The reason for the term "difference matrix" is clear since

$$\sum_{n=0}^{\infty} \gamma_{m,n} s_n = \sum_{n=0}^{m} (-1)^n \binom{m}{n} s_n = (-1)^m \Delta^m s_0$$

7.1. We are now ready to define Hausdorff summability.

DEFINITION 7.1. *The matrix* $\lambda$ *is a Hausdorff matrix corresponding to the sequence* $\{\mu_n\}_0^\infty$ *if* $\lambda = \rho\mu\rho^{-1}$. *The sequence* $\{s_n\}_0^\infty$ *is summable to* $s$ *in the Hausdorff sense corresponding to the sequence* $\{\mu_n\}_0^\infty$, *if the sequence* $\{t_n\}_0^\infty$,

$$\tau = \lambda\sigma,$$

*approaches* $s$ *as* $n$ *becomes infinite.*

It is easily seen that multiplication of Hausdorff matrices is commutative.

As an example, we show that the matrix §7 (3) is a Hausdorff matrix. If $\lambda = \| l_{m,n} \|$, and $\mu_n = 1/(n + 1)$, then

$$\sum_{n=0}^{\infty} l_{m,n} x^n = \sum_{n=0}^{\infty} x^n \sum_{j=0}^{\infty} \gamma_{mj} \mu_j \gamma_{jn} = \sum_{j=0}^{\infty} \gamma_{mj} \mu_j (1 - x)^j$$

$$= \sum_{j=0}^{\infty} \gamma_{mj} \int_0^1 t^j (1 - x)^j \, dt = \int_0^1 (1 - t + tx)^m \, dt$$

$$l_{m,n} = \binom{m}{n} \int_0^1 (1 - t)^{m-n} t^n \, dt = \frac{1}{m + 1} \qquad (n \leq m)$$

$$= 0 \qquad (n > m).$$

7.2. A method of summability is said to be *consistent** if every convergent sequence is summable by it to the actual limit of the sequence. We wish to determine what sequences $\{\mu_n\}$ lead to consistent Hausdorff summability. We base this study on a well known theorem of O. Toeplitz [1911], which we now prove.

THEOREM 7.2. *Summability by the matrix* $\| c_{m,n} \|$ *is consistent if and only if a constant* $K$ *exists such that*

(A)  $\qquad \sum_{n=0}^{\infty} | c_{m,n} | < K \qquad (m = 0, 1, 2, \cdots)$

(B)  $\qquad \lim_{m \to \infty} c_{m,n} = 0 \qquad (n = 0, 1, 2, \cdots)$

(C)  $\qquad \lim_{m \to \infty} \sum_{n=0}^{\infty} c_{m,n} = 1.$

* Other terms sometimes used are "regular" and "permanent".

We first prove the sufficiency of these conditions. If $(A)$, $(B)$, and $(C)$ hold, and if the limit of $s_m$ is $s$, then

$$|t_m - s| \leq \left|t_m - s \sum_{n=0}^{\infty} c_{m,n}\right| + |s|\left|1 - \sum_{n=0}^{\infty} c_{m,n}\right|$$

$$\varlimsup_{m \to \infty} |t_m - s| \leq \varlimsup_{m \to \infty} \left\{ \sum_{n=0}^{N} |c_{m,n}||s_n - s| \right.$$

$$\left. + \operatorname*{u.b.}_{N+1 \leq n \leq \infty} |s_n - s| \sum_{n=N+1}^{\infty} |c_{m,n}| \right\},$$

where the numbers $t_m$ are defined as in section 7. Here $N$ is an arbitrary positive integer. Hence by $(A)$ and $(B)$

$$\varlimsup_{m \to \infty} |t_m - s| \leq K \operatorname*{u.b.}_{N+1 \leq n < \infty} |s_n - s|$$

Letting $N$ become infinite we have

$$\varlimsup_{m \to \infty} |t_m - s| \leq K \varlimsup_{n \to \infty} |s_n - s| = 0,$$

from which it follows that $t_m$ approaches $s$.

Next suppose that $t_m$ approaches $s$ whenever $s_m$ does. Then that $(C)$ is necessary one sees by taking $s_n = 1$ for all $n$. Then

$$t_m = \sum_{n=0}^{\infty} c_{m,n} \qquad (m = 0, 1, \cdots),$$

and since $s_m$ and $t_m$ must both approach unity we have $(C)$.

To show that $(B)$ is necessary take $s_n = 1$ when $n = k$ and take all other terms zero. Then $s_n$ and $t_n$ approach zero. But

$$t_m = c_{m,k} \qquad (m = 0, 1, 2, \cdots),$$

so that $(B)$ holds.

Finally, to show that $(A)$ holds we first prove that all the series $(A)$ converge. For, suppose

$$\sum_{n=0}^{\infty} |c_{h,n}| = \infty$$

for some integer $h$. Then by a familiar theorem of Abel [2], one can determine a sequence $\{\epsilon_n\}_0^{\infty}$ tending to zero such that

$$\sum_{n=0}^{\infty} |c_{h,n}|\epsilon_n = \infty$$

Choose
$$s_n = \epsilon_n \operatorname{sgn} c_{hn},$$
so that
$$t_h = \sum_{n=0}^{\infty} \epsilon_n |c_{hn}| = \infty.$$

Since consistency implies that all terms of $\{t_n\}$ are defined we have a contradiction.

Set
$$\sum_{n=0}^{\infty} |c_{m,n}| = T_m < \infty \qquad (m = 0, 1, \cdots).$$

Then if $(A)$ is false,

(1) $$\underset{0 \leq m < \infty}{\text{u.b.}} T_m = \infty.$$

Let $n_1$ be an arbitrary positive integer and determine $m_1$ such that
$$\sum_{n=0}^{n_1-1} |c_{m_1,n}| < 1 \qquad T_{m_1} > 1^2 + 2.$$

This is possible by $(B)$ and (1). Now determine $n_2 > n_1$ such that
$$\sum_{n=n_2}^{\infty} |c_{m_1,n}| < 1, \qquad \sum_{n=n_1}^{n_2-1} |c_{m_1,n}| > 1^2$$

This is possible, since every series $(A)$ converges. Now determine $m_2$ so that
$$\sum_{n=0}^{n_2-1} |c_{m_2,n}| < 1 \qquad T_{m_2} > 2^2 + 2.$$

Continue the process. We shall have for the integers $n_r$, $m_r$, $n_{r+1}$,
$$\sum_{n=0}^{n_r-1} |c_{m_r,n}| < 1, \qquad \sum_{n=n_r}^{n_{r+1}-1} |c_{m_r,n}| > r^2, \qquad \sum_{n=n_{r+1}}^{\infty} |c_{m_r,n}| < 1.$$

Now define $\{s_n\}$ as follows:
$$\begin{aligned} s_n &= 0 & n < n_1 \\ &= \frac{1}{r} \operatorname{sgn} c_{m_r,n} & (n_r \leq n < n_{r+1}); \end{aligned}$$
so that
$$\lim_{n \to \infty} s_n = 0.$$

Also
$$t_{m_r} = \sum_{n=0}^{\infty} c_{m_r,n} s_n > \sum_{n_r}^{n_{r+1}-1} \frac{1}{r} |c_{m_r,n}| - \sum_{0}^{n_r-1} |c_{m_r,n}|$$
$$- \sum_{n_{r+1}}^{\infty} |c_{m_r,n}| > r - 2.$$

Hence
$$\lim_{r \to \infty} t_{m_r} = \infty,$$

contradicting the assumption that
$$\lim_{n \to \infty} t_n = \lim_{n \to \infty} s_n = 0.$$

Thus the theorem is proved. As an illustration we see that the matrix §7 (3), which leads to Hölder or Cesàro's summability of order one, corresponds to a consistent method of summability.

**7.3.** Before applying Theorem 7.2 to the Hausdorff method of summability we must first obtain the Hausdorff matrix in terms of the given sequence $\{\mu_n\}_0^\infty$. By definition of the Hausdorff transformation

(1) $$\tau = \rho\mu\rho^{-1}\sigma.$$

We wish to determine the elements $l_{m,n}$ so that this will have the form
$$t_m = \sum_{n=0}^{\infty} l_{m,n} s_n \qquad (m = 0, 1, 2, \cdots).$$

Equation (1) means
$$t_m = \sum_{j=0}^{m} (-1)^j \binom{m}{j} \mu_j \sum_{n=0}^{j} (-1)^n \binom{j}{n} s_n$$
$$= \sum_{n=0}^{m} s_n \sum_{j=n}^{m} (-1)^{j+n} \binom{m}{j}\binom{j}{n} \mu_j. \qquad (m = 0, 1, 2, \cdots)$$

Now employ the familiar identity
$$\binom{m}{j}\binom{j}{n} = \binom{m}{n}\binom{m-n}{j-n} \qquad (n \leq j \leq m),$$

and obtain
$$t_m = \sum_{n=0}^{m} \binom{m}{n} s_n \sum_{j=0}^{m-n} (-1)^j \binom{m-n}{j} \mu_{j+n}$$
$$= \sum_{n=0}^{m} (-1)^{m-n} \binom{m}{n} s_n \Delta^{m-n} \mu_n$$
$$= \sum_{n=0}^{m} \lambda_{m,n} s_n,$$

where $\lambda_{m,n}$ is the number introduced in Definition 2c. We have thus proved that $l_{m,n} = \lambda_{m,n}$ for $n = 0, 1, \cdots, m$ and $l_{m,n} = 0$ for $n = m + 1, m + 2, \cdots$.

THEOREM 7.3. *A matrix $\lambda$ is a Hausdorff matrix corresponding to the sequence $\{\mu_n\}$ if and only if it has the form*

$$\lambda = \|\lambda_{m,n}\|,$$

$$\lambda_{m,n} = \binom{m}{n}(-1)^{m-n}\Delta^{m-n}\mu_n \qquad (n = 0, 1, \cdots, m)$$

$$= 0 \qquad (n = m+1, m+2, \cdots).$$

7.4. We are now able to apply Toeplitz's theorem to the Hausdorff method of summability.

THEOREM 7.4. *The Hausdorff method of summability corresponding to the sequence $\{\mu_n\}_0^\infty$ is consistent if and only if*

$$(1) \qquad \mu_n = \int_0^1 t^n\, d\alpha(t) \qquad (n = 0, 1, \cdots),$$

*where $\alpha(t)$ is of bounded variation in $(0, 1)$ and*

$$(2) \qquad \alpha(0) = \alpha(0+) = 0, \quad \alpha(1) = 1.$$

To prove this we have only to apply Theorem 7.2 to the matrix $\lambda$. Condition $(A)$ of that theorem coincides with Condition $A$ of Theorem 2.1. Hence the sequence $\{\mu_n\}$ must have the form (1) with $\alpha(t)$ of bounded variation in $(0, 1)$. Conditions $(B)$ and $(C)$ of Theorem 7.2 become

$$(B) \qquad \lim_{m\to\infty} \binom{m}{n}\int_0^1 t^n(1-t)^{m-n}\, d\alpha(t) = 0 \qquad (n = 0, 1, 2, \cdots)$$

$$(C) \qquad \lim_{m\to\infty}\int_0^1 d\alpha(t) = 1.$$

The latter shows that if we take $\alpha(0) = 0$ then $\alpha(1) = 1$. By a familiar Abelian argument it is easily seen that

$$\lim_{m\to\infty}\binom{m}{n}\int_0^1 t^n(1-t)^{m-n}\, d\alpha(t) = 0 \qquad (n = 1, 2, \cdots)$$

$$\lim_{m\to\infty}\int_0^1 (1-t)^m\, d\alpha(t) = \alpha(0+),$$

from which we see at once that $\alpha(0+) = 0$.

The example of Hölder or Cesàro summability of order one serves to illustrate the theorem. We have seen that this method is Hausdorff's method with

$$\mu_n = \frac{1}{n+1} = \int_0^1 t^n \, dt \qquad (n = 0, 1, \cdots).$$

Here $\alpha(t) = t$, a function which satisfies all conditions of the theorem.

7.5. As another example we show that Hölder's method of summation of higher order is also a Hausdorff method. Set

$$H_n^1 = \frac{s_0 + s_1 + \cdots + s_n}{n+1}$$

$$H_n^p = \frac{H_0^{p-1} + H_1^{p-1} + \cdots + H_n^{p-1}}{n+1} \qquad (p = 2, 3, \cdots).$$

DEFINITION 7.5. *The sequence $\{s_n\}_0^\infty$ is summable by Hölder's method of order $p$, $(H, p)$, to $s$ if*

$$\lim_{n \to \infty} H_n^p = s.$$

We have already seen that the sequence $\{H_n^1\}_0^\infty$ is the first column of the matrix

$$\lambda \sigma = \rho \mu \rho^{-1} \sigma$$

$$\mu_n = \frac{1}{n+1} \qquad (n = 0, 1, 2, \cdots).$$

Since $\{H_n^2\}$ is found from $\{H_n^1\}$ precisely as $\{H_n^1\}$ is found from $\{s_n\}$, it is clear that $\{H_n^2\}_0^\infty$ is the first column of the matrix

$$\lambda^2 \sigma = \rho \mu^2 \rho^{-1} \sigma$$

and in general that $\{H_n^p\}$ is the first column of the matrix

$$\lambda^p \sigma = \rho \mu^p \rho^{-1} \sigma$$

We have thus proved:

THEOREM 7.5. *The method of summability $(H, p)$ is a Hausdorff method corresponding to the sequence*

$$\mu_n = \frac{1}{(n+1)^p} \qquad (n = 0, 1, 2, \cdots).$$

Clearly $(H, p)$ is consistent since

$$\frac{1}{(n+1)^p} = \frac{1}{(p-1)!} \int_0^1 t^n \left[ \log \frac{1}{t} \right]^{p-1} dt.$$

That is,
$$\alpha(t) = \frac{1}{(p-1)!} \int_0^t \left[\log \frac{1}{u}\right]^{p-1} du \qquad (0 \leq t \leq 1),$$
$$\alpha(0+) = 0, \qquad \alpha(1) = 1,$$
and all conditions of Theorem 7.4 are satisfied.

7.6. We treat next Cesàro's method of higher order.

DEFINITION 7.6. *The sequence $\{s_n\}$ is summable by Cesàro's method of order $p$, $(C, p)$, to $s$ if*
$$\lim_{m \to \infty} \frac{1}{\binom{m+p}{m}} \sum_{n=0}^{m} \binom{m-n+p-1}{m-n} s_n = s.$$

THEOREM 7.6. *The method $(C, p)$ is a Hausdorff method corresponding to the sequence*
$$\mu_n = \frac{1}{\binom{n+p}{n}} = p \int_0^1 t^n (1-t)^{p-1} dt \qquad (n = 0, 1, 2, \cdots).$$

For, by Theorem 7.3

(1)
$$\lambda_{m,n} = \binom{m}{n} p \int_0^1 t^n (1-t)^{m-n+p-1} dt \qquad (n = 0, 1, \cdots, m)$$
$$= 0 \qquad (n = m+1, m+2, \cdots).$$

Evaluating the integral (1) we obtain
$$\lambda_{m,n} = \frac{\binom{m-n+p-1}{m-n}}{\binom{m+p}{m}}$$

Again it is clear that $(C, p)$ is consistent since in this case
$$\alpha(t) = 1 - (1-t)^p$$
$$\alpha(0) = 0, \qquad \alpha(1) = 1.$$

7.7. We say that the method of summability by the matrix $\lambda$ is *stronger*\* than that by the matrix $\lambda'$ if whenever the latter sums a sequence to a sum $s$ the former does also. Two methods are *equivalent* if each is stronger than the other. If we are dealing with Hausdorff matrices we can obtain a very simple criterion for the comparative

---
\* We are using the terms "stronger" and "not weaker" interchangeably here.

strength of two matrices in terms of the sequences $\{\mu_n\}_0^\infty$ which generate them.

Let $\{\mu_n\}$ and $\{\mu_n'\}$ be two sequences giving rise to the matrices $\lambda$ and $\lambda'$ respectively:

$$\lambda = \rho\mu\rho^{-1} \qquad \lambda' = \rho\mu'\rho^{-1}.$$

If $\{s_n\}$ is the sequence to be summed and $\sigma$ its matrix (that is, the matrix with the elements $s_n$ in the first column and with zeros elsewhere), we set

$$\tau = \lambda\sigma \qquad \tau' = \lambda'\sigma.$$

Then it is very easy to express $\tau$ in terms of $\tau'$. In fact

$$\sigma = \rho(\mu')^{-1}\rho^{-1}\tau',$$

where $(\mu')^{-1}$ is the matrix with elements $1/\mu_n$ $(n = 0, 1, \cdots)$ in the principal diagonal and with zeros elsewhere.

Then

$$\tau = \rho\mu\rho^{-1}\rho(\mu')^{-1}\rho^{-1}\tau' = \rho\mu(\mu')^{-1}\rho^{-1}\tau'.$$

To say that the method generated by $\mu$ is stronger than that generated by $\mu'$ is to say that whenever the sequence in the first column of $\tau'$ converges to a sum $t$ that in the first column of $\tau$ does also. That is, the Hausdorff matrix

$$\rho\mu(\mu')^{-1}\rho^{-1}$$

is consistent. By Theorem 7.4 this is true if and only if

(1) $$\frac{\mu_n}{\mu_n'} = \int_0^1 t^n \, d\alpha(t) \qquad (n = 0, 1, 2, \cdots).$$

where $\alpha(t)$ is of bounded variation in $(0, 1)$ and

(2) $$\alpha(0) = \alpha(0+) = 0, \qquad \alpha(1) = 1.$$

We have thus proved:

THEOREM 7.7. *The Hausdorff method of summability corresponding to the sequence $\{\mu_n\}_0^\infty$ is stronger than that corresponding to $\{\mu_n'\}_0^\infty$ if and only if (1) and (2) hold.*

From this we see easily that $(C, q)$ is stronger than $(C, p)$ if $q > p$. For, by Theorem 7.6 the quotient (1) becomes in this case

(3) $$\frac{q!\,(n+p)!}{p!\,(n+q)!} \qquad (n = 0, 1, \cdots).$$

But this is

$$\frac{q!}{p!\,(q-p-1)!}\int_0^1 t^{n+p}(1-t)^{q-p-1}\,dt \qquad (q > p;\ n = 0, 1, \cdots),$$

so that (1) is satisfied. To show that (2) holds we have

$$\frac{1}{(q-p-1)!}\frac{q!}{p!}\int_0^1 t^p(1-t)^{q-p-1}\,dt = 1.$$

Also $(H, q)$ is stronger than $(H, p)$ if $q > p$. For, in this case the quotient (1) becomes

$$\frac{1}{(n+1)^{q-p}} = \frac{1}{(q-p-1)!}\int_0^1 t^n\left[\log\frac{1}{t}\right]^{q-p-1}dt$$

and

$$\frac{1}{(q-p-1)!}\int_0^1 \left[\log\frac{1}{t}\right]^{q-p-1}dt = 1.$$

7.8. Finally, we show that the Hölder and Cesàro methods $(C, p)$ and $(H, p)$ are equivalent. We have only to show that the two sequences

(1) $$\frac{(n+1)^p}{\binom{n+p}{n}}, \quad \frac{\binom{n+p}{n}}{(n+1)^p} \qquad (n = 0, 1, 2, \cdots)$$

have the expression §7.7 (1), (2). For the first sequence we have

$$\frac{(n+1)^p\,p!}{(n+1)(n+2)\cdots(n+p)} = \int_0^\infty e^{-nt}\,d\psi(t) \qquad (n = 0, 1, \cdots),$$

where

$$\psi(0) = 0$$

$$\psi(t) = e^{-t}\frac{d^p}{dt^p}[e^t(1 - e^{-t})^p] \qquad (t > 0)$$

$$= \sum_{j=0}^p \binom{p}{j}(-1)^{j+p}(j-1)^p e^{-jt}.$$

Since $\psi(\infty) = 1$, the result is clearly true for $n = 0$. For $n > 0$, we have by successive integration by parts

$$\int_0^\infty e^{-nt}\,d\psi(t) = n(n+1)^p \int_0^\infty e^{-nt}[1 - e^{-t}]^p\,dt$$

$$= n(n+1)^p(-1)^p \Delta^p n^{-1}$$

$$= \frac{(n+1)^p\,p!}{(n+1)(n+2)\cdots(n+p)}.$$

In each integration by parts the integrated part vanishes. Note that
$$\psi(0+) = p!$$
$$\lim_{t \to 0+} \frac{d^j}{dt^j}[e^t(1 - e^{-t})^p] = 0 \qquad (j = 0, 1, \ldots, p-1).$$

Since

(2)
$$\int_0^\infty e^{-nt} d\psi(t) = \int_0^1 t^n d\alpha(t)$$
$$\alpha(t) = 1 - \psi\left(\log \frac{1}{t}\right) \qquad (0 < t \leq 1)$$
$$\alpha(0) = 0,$$

we have the representation desired ($\alpha(t)$ of bounded variation, $\alpha(0+) = 0$, $\alpha(1) = 1$) for the first of the sequences (1). For the second we have
$$\frac{(n+1)(n+2)\cdots(n+p)}{(n+1)^p p!} = \int_0^\infty e^{-nt} d\psi(t) \qquad (n = 0, 1, 2, \ldots),$$
where
$$\psi(0) = 0, \qquad \psi(0+) = \frac{1}{p!},$$
$$\psi'(t) = \frac{1}{p!(p-1)!} \overbrace{e^{-t}\frac{d}{dt} e^{-t}\frac{d}{dt} \cdots e^{-t}\frac{d}{dt}}^{p} [e^{(p-1)t} t^{p-1}] \qquad (t > 0).$$

This last differential operator is of order $p$. To show this, we have first
$$\int_0^\infty e^{-nt} d\psi(t) = \frac{1}{p!} + \int_0^\infty e^{-nt}\psi'(t)\, dt.$$

Since
$$\lim_{t \to 0+} \overbrace{e^{-t}\frac{d}{dt} \cdots e^{-t}\frac{d}{dt}}^{p-1} [e^{(p-1)t} t^{p-1}] = (p-1)!,$$
we have by means of a second integration by parts
$$\int_0^\infty e^{-nt} d\psi(t) = \frac{(n+1)}{p!(p-1)!} \int_0^\infty e^{-(n+1)t} \overbrace{e^{-t}\frac{d}{dt} \cdots e^{-t}\frac{d}{dt}}^{p-1} [e^{(p-1)t} t^{p-1}]\, dt$$
$$(n = 0, 1, 2, \ldots).$$

# §8] FURTHER MOMENT PROBLEMS

Continuing the process we obtain

$$\int_0^\infty e^{-nt} d\psi(t) = \frac{(n+1)(n+2)\cdots(n+p)}{p!(p-1)!} \int_0^\infty e^{-(n+1)t} t^{p-1} dt$$

$$= \frac{(n+1)(n+2)\cdots(n+p)}{p!(n+1)^p}$$

Since the above computation holds for $n = 0$ we have in particular

$$\psi(\infty) = \int_0^\infty d\psi(t) = 1.$$

Now defining $\alpha(t)$ by equations (2), our proof is complete.

We note that the proof of the equivalence of $(C, p)$ and $(H, p)$ could have been given more simply by observing that the sequences

$$\left\{\frac{n}{n+a}\right\}_{n=0}^\infty \qquad \left\{\frac{n+a}{n}\right\}_{n=0}^\infty$$

are both moment sequences for any positive $a$, and that the product of two moment sequences is again a moment sequence. We have preferred to obtain the explicit Stieltjes integral representation for the sequences in question.

## 8. Statement of Further Moment Problems

We treat next the moment problem of Hamburger, of which that of Hausdorff already discussed and that of Stieltjes to follow are special cases. The problem of Hamburger [1920a] is to determine for what sequences $\{\mu_n\}_0^\infty$ there will exist a non-decreasing function $\alpha(t)$ such that

$$(1) \qquad \mu_n = \int_{-\infty}^\infty t^n d\alpha(t) \qquad (n = 0, 1, 2, \cdots).$$

The problem of Stieltjes [1894] differs from this only in the limits of integration. The lower limit is to be replaced by zero. Alternatively, we may say that in the Stieltjes problem we add the additional restriction on $\alpha(t)$ that it should be constant for negative values of the independent variable. If we further restrict this problem by the demand that $\alpha(t)$ should also be constant in the interval $(1, \infty)$, we have again the Hausdorff problem.

It is important to observe that neither the Hamburger nor the Stieltjes problem will in general have *unique* non-decreasing solutions $\alpha(t)$ For, compute the following integrals:

$$(2) \qquad \int_{-\infty}^\infty t^n \varphi(t) dt \qquad (n = 0, 1, 2, \cdots),$$

where

$$\varphi(t) = e^{-t^{1/4}} \sin(t^{1/4}) \quad (0 \leq t < \infty)$$
$$= 0 \quad (-\infty < t < 0).$$

Making the change of variable $t^{1/4} = u$, the integrals (2) become

$$4\int_0^\infty e^{-u} u^{4n+3} \sin u \, du = -2i \int_0^\infty [e^{(i-1)u} - e^{-(i+1)u}] u^{4n+3} du$$

$$= -2i \left[ \frac{(4n+3)!}{(1-i)^{4n+4}} - \frac{(4n+3)!}{(1+i)^{4n+4}} \right]$$

$$= -2i 4^{-(n+1)} (4n+3)! [e^{i\pi(n+1)} - e^{-i\pi(n+1)}]$$

$$= 4^{-n}(4n+3)! \sin(n+1)\pi = 0 \quad (n = 0, 1, 2, \cdots).$$

Thus, all the moments of the function $\varphi(t)$ are zero.

Now suppose

$$\mu_n = 4(4n+3)! \quad (n = 0, 1, 2, \cdots).$$

Simple computations show that the function

$$\alpha(t) = 0 \quad (-\infty < t \leq 0)$$

$$\alpha(t) = \int_0^t e^{-u^{1/4}} du \quad (0 \leq t < \infty)$$

is then a non-decreasing solution of equations (1). But clearly the function

$$\alpha^*(t) = 0 \quad (-\infty < t \leq 0)$$

$$\alpha^*(t) = \int_0^t e^{-u^{1/4}} [1 - \sin(u^{1/4})] du \quad (0 \leq t < \infty)$$

is a distinct non-decreasing solution. This specific example shows that a non-decreasing solution of equations (1), if it exists, need not be unique. Many authors have studied conditions under which the solution is unique, but since we shall have no use for the results in our study of the Laplace integral we do not include them here.* The above example is due to T. J. Stieltjes.†

## 9. The Moment Operator

We seek next to obtain conditions on the sequence $\{\mu_n\}_0^\infty$ which will insure that the corresponding Hamburger moment problem shall have

* See, however, section 16 of the present chapter.
† See Stieltjes [1894] p. J 105.

at least one non-decreasing solution. We follow the method of M. Riesz [1922]. We begin with several definitions.

DEFINITION 9a. *The moment* $M[P(t)]$ *of a polynomial*

$$P(t) = \sum_{k=0}^{n} \alpha_k t^k$$

*with respect to the sequence* $\{\mu_n\}_0^\infty$ *is*

$$M[P(t)] = \sum_{k=0}^{n} \alpha_k \mu_k.$$

We regard the definition as defining an operator $M$ which applies to any polynomial $P(t)$. This operator is clearly distributive:

$$M[c_1 P_1(t) + c_2 P_2(t)] = c_1 M[P_1(t)] + c_2 M[P_2(t)],$$

where $c_1$ and $c_2$ are any constants, $P_1(t)$ and $P_2(t)$ are any polynomials.

DEFINITION 9b. *The sequence* $\{\mu_n\}_0^\infty$ *is positive if the moment of every non-negative polynomial is non-negative.*

If the sequence $\{\mu_n\}_0^\infty$ is positive we say that the corresponding operator $M$ is positive when applied to polynomials. Under these conditions we wish to show that the definition of $M$ can be extended so as to apply to a larger class $E$ of functions and so as to remain positive and distributive. We first define the class $E$.

Let $\{\xi_n\}_1^\infty$ be the set of all rational numbers arranged in some order. The set is of course dense on the interval $(-\infty, \infty)$. Let

$$h_m(t) = 1 \qquad (t \leq \xi_m)$$
$$= 0 \qquad (t > \xi_m).$$

DEFINITION 9c. *The set of functions* $E$ *is the set of all linear combinations (with real constants of combination) of a finite number of the functions*

$$\cdots, t^2, t, 1, h_1(t), h_2(t), \cdots.$$

The result to be established can now be stated as follows:

THEOREM 9. *If* $\{\mu_n\}_0^\infty$ *is a positive sequence, there exists an operator* $M$ *which is applicable to the class of functions* $E$, *is positive and distributive, and reduces to the moment of a polynomial when applied to a polynomial.*

We define the set of functions $E_1$ as the set of all linear combinations (with real constants of combination) of a finite number of the functions

$$h_1(t), 1, t, t^2, \cdots.$$

We further define two constants $\underline{h}_1$ and $\bar{h}_1$ as follows:

$$\underline{h}_1 = \underset{p(t)<h_1(t)}{\text{u.b.}} M[p(t)]$$

$$\bar{h}_1 = \underset{h_1(t)<P(t)}{\text{l.b.}} M[P(t)].$$

In the definition of $\underline{h}_1$, for example, the notation means the least upper bound of all the numbers $M[p(t)]$ for which $p(t)$ is a polynomial less than $h_1(t)$. We observe that $\underline{h}_1 \leq \bar{h}_1$. For, if

$$p(t) < h_1(t) < P(t),$$

then

$$M[p(t)] \leq M[P(t)],$$

since $M$ is a positive distributive operator when applied to polynomials. Since polynomials $p(t)$ and $P(t)$ certainly exist less than and greater than $h_1(t)$ respectively, it follows that $\underline{h}_1$ and $\bar{h}_1$ are finite numbers, and that $\underline{h}_1 \leq \bar{h}_1$.

Now define $M[h_1(t)]$ by the equation

$$M[h_1(t)] = h_1 = \frac{\underline{h}_1 + \bar{h}_1}{2},$$

and define $M[P(t) + c_1 h_1(t)]$ for any polynomial $P(t)$ and any real constant $c_1$ by the equation

$$M[P(t) + c_1 h_1(t)] = M[P(t)] + c_1 h_1.$$

It is then clear that $M$ remains a distributive operator when applied to functions of $E_1$. To show that it also remains positive, suppose that $p(t)$ is a polynomial and $c_1$ a constant such that

$$p(t) + c_1 h_1(t) \geq 0 \qquad (c_1 \neq 0)$$

If $c_1 > 0$, we have

$$-\frac{p(t)}{c_1} \leq h_1(t).$$

By definition of $\underline{h}_1$ and $h_1$, it follows that

$$-\frac{M[p(t)]}{c_1} = M\left[-\frac{p(t)}{c_1}\right] \leq \underline{h}_1 \leq h_1$$

$$M[p(t) + c_1 h_1(t)] \geq 0.$$

A similar proof, using the definition of $\bar{h}_1$, holds when $c_1 < 0$.

We next define the set $E_2$ of all functions which are linear combinations of a finite number of functions in $E_1$ and $h_2(t)$. We set

$$\underline{h}_2 = \underset{f_1(t) < h_2(t)}{\text{u.b.}} M[f_1(t)]$$

$$\overline{h}_2 = \underset{h_2(t) < F_1(t)}{\text{l.b.}} M[F_1(t)],$$

where $f_1(t)$ and $F_1(t)$ are functions of $E_1$ satisfying the indicated inequalities. Set

$$M[h_2(t)] = h_2 = \frac{\underline{h}_2 + \overline{h}_2}{2}$$

$$M[F_1(t) + c_2 h_2(t)] = M[F_1(t)] + c_2 h_2$$

for any function $F_1(t)$ of $E_1$ and any real constant $c_2$. Just as in the preceding case it may be shown that the operator $M$ as now extended to apply to functions of $E_2$ remains positive and distributive.

It is easily seen that we may continue the process started, thus making $M$ a positive distributive operator applicable to the whole set $E$. This completes the proof of the theorem. It should be noted that the only use made of the specific properties of the functions $\{h_n(t)\}_0^\infty$ was in noting that it was possible to find a polynomial greater and a polynomial less than each.

## 10. The Hamburger Moment Problem

We are now in a position to solve completely the Hamburger problem.

THEOREM 10. *A necessary and sufficient condition that there should exist at least one non-decreasing function $\alpha(t)$ such that*

(1) $$\mu_n = \int_{-\infty}^{\infty} t^n \, d\alpha(t) \qquad (n = 0, 1, 2, \cdots),$$

*all the integrals converging, is that the sequence $\{\mu_n\}_0^\infty$ should be positive.*

First suppose that $\alpha(t)$ is a non-decreasing solution of equations (1), and that $P(t)$ is an arbitrary non-negative polynomial,

$$P(t) = \sum_{k=0}^{n} \alpha_k t^k.$$

Then

$$M[P(t)] = \sum_{k=0}^{n} \alpha_k \mu_k = \int_{-\infty}^{\infty} P(t) \, d\alpha(t) \geqq 0,$$

so that the sequence $\{\mu_n\}$ is positive. That is, the condition is necessary.

Conversely, suppose that the sequence $\{\mu_n\}_0^\infty$ is positive. We shall

130    THE MOMENT PROBLEM    [Ch. III

exhibit a non-decreasing solution $\alpha(t)$ of (1). It is defined at the rational points by the equations

$$\alpha(\xi_m) = M[h_m(t)] \qquad (m = 0, 1, 2, \cdots).$$

Then if $\xi_i < \xi_j$, we have from the positive character of $M$

$$h_i(t) \leqq h_j(t)$$
$$\alpha(\xi_i) \leqq \alpha(\xi_j).$$

That is $\alpha(t)$ is non-decreasing in so far as it has been defined. To complete the definition, let $\eta$ be an arbitrary irrational number, and let

$$\bar{\zeta} = \underset{\xi_m > \eta}{\text{l.b.}}\ \alpha(\xi_m)$$

$$\underline{\zeta} = \underset{\xi_m < \eta}{\text{u.b.}}\ \alpha(\xi_m)$$

$$\alpha(\eta) = \frac{\bar{\zeta} + \underline{\zeta}}{2}.$$

Since $\alpha(t)$ is non-decreasing on the rational points it is clear that $\underline{\zeta}$ is not greater than $\bar{\zeta}$ and that $\alpha(t)$ as now completely defined is non-decreasing.

Let $n$ be an arbitrary positive integer. We wish to prove that for this $n$

$$\mu_n = \int_{-\infty}^{\infty} t^n\, d\alpha(t).$$

Since $\alpha(t)$ is non-decreasing, it will be sufficient to show that to an arbitrary positive $\epsilon$ there corresponds a positive $T_0$ such that for every pair of *rational* numbers $T_1$ and $T_2$ greater than $T_0$

(2)    $$\left| \int_{-T_1}^{T_2} t^n\, d\alpha(t) - \mu_n \right| < \epsilon.$$

Let $m$ be an integer such that $2m > n$. Set*

$$\epsilon' = \frac{\epsilon}{2(\mu_0 + \mu_{2m})}$$

* Since $t^{2m} \geqq 0$ we have $\mu_{2m} \geqq 0$. We may assume that $\mu_0 > 0$, for if $\mu_0 = 0$ all the $\mu_n$ are zero and $\alpha(t) \equiv 0$ is a solution of our problem. To show that $\mu_1 = 0$ we have from the fact that $(x + c)^2 \geqq 0$ for any $c$ that $2c\mu_1 + \mu_2 \geqq 0$. This is only possible if $\mu_1 = 0$. Again since $(x^2 + c)^2 \geqq 0$ we have $2c\mu_2 + \mu_4 \geqq 0$ whence $\mu_2 = 0$. By considering successively the polynomials $(x^2 + cx)^2$, $(x^3 + cx)^2$, $(x^3 + cx^2)^2$, $(x^4 + cx^2)^2$, $\cdots$ we see that all the $\mu_n$ are zero.

and determine $T_0$ such that

(3) $$|t^n| < \epsilon' t^{2m} \qquad (|t| > T_0).$$

Choose $T_1$ and $T_2$ any two rational numbers greater than $T_0$. Divide up the interval $(-T_1, T_2)$ into $p$ sub-intervals by rational points $t_i = \xi_{k_i}$:

$$t_0 = -T_1 < t_1 < t_2 < \cdots < t_p = T_2,$$

choosing $p$ so large and the sub-intervals so small that the oscillation of $t^n$ in each sub-interval is less than $\epsilon'$ and such that

(4) $$\left| \sum_{i=0}^{p-1} t_{i+1}^n [\alpha(t_{i+1}) - \alpha(t_i)] - \int_{-T_1}^{T_2} t^n \, d\alpha(t) \right| < \frac{\epsilon}{2}.$$

This is possible by the uniform continuity of $t^n$ and by the definition of the Stieltjes integral.

Next define a function $V(t)$ as follows:

$$V(t) = 0 \qquad (t \leq -T_1, t > T_2)$$
$$= t_{i+1}^n \qquad (t_i < t \leq t_{i+1}; i = 0, 1, \cdots, p-1).$$

Then

$$V(t) = \sum_{i=0}^{p-1} t_{i+1}^n [h_{k_{i+1}}(t) - h_{k_i}(t)]$$

and

$$|V(t) - t^n| \leq \epsilon' t^{2m} \qquad (t \leq -T_1, t > T_2)$$

by (3). Also

$$|V(t) - t^n| \leq \sum_{i=0}^{p-1} \epsilon'[h_{k_{i+1}}(t) - h_{k_i}(t)] = \epsilon' \qquad (-T_1 < t \leq T_2),$$

so that

$$|V(t) - t^n| \leq \epsilon' + \epsilon' t^{2m} \qquad (-\infty < t < \infty).$$

Now $V(t)$ certainly belongs to one of the sets $E_j$, so that the operator $M$ is positive and distributive as applied to $V(t)$. That is,

(5) $$|M[V(t)] - \mu_n| \leq \epsilon'(\mu_0 + \mu_{2m}) = \frac{\epsilon}{2}.$$

But

$$M[V(t)] = \sum_{i=0}^{p-1} t_{i+1}^n [\alpha(t_{i+1}) - \alpha(t_i)].$$

Combining (4) and (5) we obtain (2), which is what we set out to establish.

## 11. Positive Definite Sequences

We further classify positive sequences into *positive definite* and *positive semidefinite sequences*.

DEFINITION 11a. *A sequence $\{\mu_n\}_0^\infty$ is positive definite if the moment of every non-negative polynomial which is not identically zero is greater than zero.*

DEFINITION 11b. *A sequence $\{\mu_n\}_0^\infty$ is positive semidefinite if it is positive and if there exists a non-negative polynomial not identically zero whose moment is zero.*

For example, the sequence $\{n!\}_0^\infty$ is positive definite since the relations

$$P(t) \geq 0 \qquad P(t) \not\equiv 0$$

imply

$$M[P(t)] = \int_0^\infty P(t) e^{-t} \, dt > 0.$$

The sequence

$$1, 0, 0, \cdots$$

is positive semidefinite since in this case

$$M[P(t)] = P(0) \geq 0$$

$$M[t^2] = 0$$

Observe that a positive sequence is either definite or semidefinite.

We wish to characterize these two cases by conditions on the sequence $\{\mu_n\}_0^\infty$. We shall need:

LEMMA 11. *Every real non-negative polynomial is the sum of the squares of two real polynomials.*

Since the polynomial is real its imaginary roots, if any, occur in conjugate imaginary pairs. The degree of the polynomial must be even or zero. Hence its factored form must be

$$\prod_{i=0}^{n} [(x - \alpha_i)^2 + \beta_i^2] \prod_{i=0}^{m} (x - \gamma_i)^2,$$

where $\alpha_i$, $\beta_i$, $\gamma_i$ are real numbers. By use of the trivial identity

$$|x_1 + iy_1|^2 |x_2 + iy_2|^2 = |(x_1 + iy_1)(x_2 + iy_2)|^2,$$

we have

$$[(x - \alpha_i)^2 + \beta_i^2][(x - \alpha_j)^2 + \beta_j^2] = [(x - \alpha_i)(x - \alpha_j) - \beta_i\beta_j]^2$$
$$+ [\beta_j(x - \alpha_i) + \beta_i(x - \alpha_j)]^2.$$

That is, the product of the sum of two squares by the sum of two squares is the sum of two squares. Repeated application of this result proves the lemma.

THEOREM 11. *A necessary and sufficient condition that the sequence $\{\mu_n\}_0^\infty$ should be positive definite (semidefinite) is that the quadratic forms*

(1) $$\sum_{i=0}^{n}\sum_{j=0}^{n} \mu_{i+j}\xi_i\xi_j \qquad (n = 0, 1, 2, \cdots)$$

*should be positive definite (semidefinite\*).*

First suppose that the quadratic forms (1) are all positive (definite or semidefinite). Let $P(t)$ be an arbitrary non-negative polynomial. By the lemma it is the sum of the squares of two other polynomials $P_1(t)$ and $P_2(t)$:

$$P_1(t) = \sum_{i=0}^{n} \alpha_i t^i, \qquad P_2(t) = \sum_{i=0}^{m} \beta_i t^i,$$

so that

$$M[P(t)] = \sum_{i=0}^{n}\sum_{j=0}^{n} \alpha_i\alpha_j\mu_{i+j} + \sum_{i=0}^{m}\sum_{j=0}^{m} \beta_i\beta_j\mu_{i+j},$$

which is non-negative by hypothesis. Hence $\{\mu_n\}_0^\infty$ is positive. Next suppose that the forms (1) are positive definite. That is, no form can vanish unless all of its variables vanish. Then if the polynomial $P(t)$ above is not identically zero the $\alpha_i$ and $\beta_i$ are not all zero from which we see that $M[P(t)]$ is actually greater than zero, and the sequence $\{\mu_n\}$ is positive definite.

Conversely, suppose that $\{\mu_n\}_0^\infty$ is a positive definite sequence. Let $n$ be an arbitrary positive integer and $\xi_0, \xi_1, \cdots, \xi_n$ arbitrary constants not all zero. We wish to show that

$$\sum_{i=0}^{n}\sum_{j=0}^{n} \mu_{i+j}\xi_i\xi_j > 0.$$

This follows from the definition of a positive definite sequence, since the polynomial

$$\left[\sum_{i=0}^{n} \xi_i t^i\right]^2$$

---

\* By this we mean that all of the forms are positive and at least one of them is semidefinite.

is non-negative and not identically zero. If the sequence $\{\mu_n\}_0^\infty$ is only known to be positive then we have

$$M[(\xi_0 + \xi_1 t + \cdots + \xi_n t^n)^2] = \sum_{i=0}^{n} \sum_{j=0}^{n} \mu_{i+j} \xi_i \xi_j \geq 0,$$

so that the forms (1) are at least positive (definite or semidefinite). Since a positive sequence or form is either definite or semidefinite, this completes the proof of the theorem.

COROLLARY 11. *Any completely monotonic sequence is a positive sequence.*

For by Theorem 4a the completely monotonic sequence $\{\mu_n\}_0^\infty$ has the integral representation

$$\mu_n = \int_0^1 t^n \, d\alpha(t) \qquad (n = 0, 1, 2, \cdots),$$

where $\alpha(t)$ is non-decreasing. Hence the quadratic forms

$$\sum_{i=0}^{n} \sum_{j=0}^{n} \mu_{i+j} \xi_i \xi_j = \int_0^1 \left( \sum_{i=0}^{n} t^i \xi_i \right)^2 d\alpha(t) \qquad (n = 0, 1, 2, \cdots)$$

are clearly positive.

## 12. Determinant Criteria

We next investigate the effect of the definite or semidefinite character of a moment sequence on its integral representation.

THEOREM 12a. *A necessary and sufficient condition that there should exist a non-decreasing function $\alpha(t)$ with infinitely many points of increase (with a finite number of points of increase) such that*

(1) $$\mu_n = \int_{-\infty}^{\infty} t^n \, d\alpha(t) \qquad (n = 0, 1, 2, \cdots)$$

*is that the sequence $\{\mu_n\}_0^\infty$ should be positive definite (semidefinite).*

We show first that if there exists a non-decreasing solution $\alpha(t)$ of (1) with but a finite number of points of increase $t_0, t_1, \cdots, t_m$, then the sequence $\{\mu_n\}_0^\infty$ is positive semidefinite. This follows since

$$M[(t - t_0)^2 (t - t_1)^2 \cdots (t - t_m)^2]$$
$$= \int_{-\infty}^{\infty} (t - t_0)^2 (t - t_1)^2 \cdots (t - t_m)^2 \, d\alpha(t) = 0.$$

Conversely, if the sequence $\{\mu_n\}_0^\infty$ is positive semidefinite it has a representation (1) with non-decreasing $\alpha(t)$ by Theorem 10. If $\alpha(t)$

had infinitely many points of increase, then for every non-negative polynomial $P(t)$ not identically zero we should have by Theorem 7c of Chapter I

$$M[P(t)] = \int_{-\infty}^{\infty} P(t)\, d\alpha(t) > 0,$$

contrary to hypothesis. Hence $\alpha(t)$ has but a finite number of points of increase. By virtue of Theorem 10 and of the fact that every positive sequence is either definite or semidefinite the proof is complete.

Theorem 12a a gives us another proof of Corollary 11. For, a completely monotonic sequence has, by Theorem 4a, the representation (1) with $\alpha(t)$ non-decreasing in (0, 1) and constant elsewhere.

Another result of the same nature, but less useful since it has no counterpart for the positive semidefinite case is contained in:

THEOREM 12b. *A necessary and sufficient condition that equations* (1) *should have a non-decreasing solution $\alpha(t)$ with infinitely many points of increase is that*\*

(2) $\quad \mu_0 > 0, \quad \begin{vmatrix} \mu_0 & \mu_1 \\ \mu_1 & \mu_2 \end{vmatrix} > 0, \quad \begin{vmatrix} \mu_0 & \mu_1 & \mu_2 \\ \mu_1 & \mu_2 & \mu_3 \\ \mu_2 & \mu_3 & \mu_4 \end{vmatrix} > 0 \cdots.$

This follows at once from Theorem 11 and a familiar necessary and sufficient condition from algebra that a quadratic form should be positive definite.†

It is clear further that if equations (1) have a non-decreasing solution $\alpha(t)$ with a finite number of points of increase then the determinants (2) are positive or zero. For this is true if the forms §11 (1) are all positive (definite or semidefinite). But it is not true conversely that equations (1) have a non-negative solution whenever determinants (2) are non-negative. For example, if

$$\mu_n = \begin{cases} 1 & (n = 0, 1, 2, 3) \\ 0 & (n = 4, 5, \cdots), \end{cases}$$

---

\* These determinants are called Hankel determinants, having been introduced by H. Hankel [1861] in his thesis.

† For a simple proof of the theorem in question see L. M. Blumenthal [1928].

we have

$$\mu_0 > 0, \qquad \begin{vmatrix} \mu_0 & \mu_1 \\ \mu_1 & \mu_2 \end{vmatrix} = 0,$$

$$\begin{vmatrix} \mu_0 & \mu_1 & \mu_2 \\ \mu_1 & \mu_2 & \mu_3 \\ \mu_2 & \mu_3 & \mu_4 \end{vmatrix} = 0, \qquad \begin{vmatrix} \mu_0 & \mu_1 & \mu_2 & \mu_3 \\ \mu_1 & \mu_2 & \mu_3 & \mu_4 \\ \mu_2 & \mu_3 & \mu_4 & \mu_5 \\ \mu_3 & \mu_4 & \mu_5 & \mu_6 \end{vmatrix} > 0,$$

with all successive determinants zero. But the quadratic form

$$\sum_{i=0}^{2} \sum_{j=0}^{2} \xi_i \xi_j \mu_{i+j}$$

is neither positive definite nor positive semidefinite. For it reduces to $-1$ when

$$\xi_0 = -1, \qquad \xi_1 = 0, \qquad \xi_2 = 1,$$

Hence by Theorems 10 and 11 equations (1) can have no non-decreasing solution $\alpha(t)$.

### 13. The Stieltjes Moment Problem

The Stieltjes problem may be treated as a special case of the Hamburger problem.

THEOREM 13a. *A necessary and sufficient condition that there should exist a non-decreasing function $\alpha(t)$ such that*

(1) $$\mu_n = \int_0^\infty t^n \, d\alpha(t) \qquad (n = 0, 1, 2, \cdots),$$

*the integrals all converging, is that the sequences $\{\mu_n\}_0^\infty$ and $\{\mu_n\}_1^\infty$ should be positive, or that the quadratic forms*

(2) $$\sum_{i=0}^{n} \sum_{j=0}^{n} \mu_{i+j} \xi_i \xi_j \qquad (n = 0, 1, 2, \cdots)$$

(3) $$\sum_{i=0}^{n} \sum_{j=0}^{n} \mu_{i+j+1} \xi_i \xi_j \qquad (n = 0, 1, 2, \cdots)$$

*should be positive (definite or semidefinite\*).*

The equivalence of the two forms of the condition is apparent by Theorem 11. We prove the result in the latter form involving quadratic

---
\* See the footnote to Theorem 11.

forms. For the necessity, let the sequence $\{\mu_n\}_0^\infty$ have the form (1) By Theorems 10 and 11 the forms (2) are positive, since we may regard $\alpha(t)$ as constant in the interval $(-\infty, 0)$. Furthermore,

$$\mu_{n+1} = \int_0^\infty t^n \, d\beta(t) \qquad (n = 0, 1, \cdots)$$

$$\beta(t) = \int_0^t u \, d\alpha(u) \qquad (t \geqq 0).$$

Since $\beta(t)$ is also non-decreasing, Theorems 10 and 11 show that the forms (3) are also positive.

Conversely, let the forms (2) and (3) be positive. Consider the new sequence $\{\nu_n\}_0^\infty$ where

$$\nu_{2n} = \mu_n \qquad (n = 0, 1, \cdots)$$

$$\nu_{2n+1} = 0 \qquad (n = 0, 1, \cdots).$$

Now if $n$ is odd

$$\sum_{i=0}^n \sum_{j=0}^n \nu_{i+j} \xi_i \xi_j = \sum_{i=0}^{\frac{n-1}{2}} \sum_{j=0}^{\frac{n-1}{2}} \mu_{i+j} \xi_{2i} \xi_{2j} + \sum_{i=0}^{\frac{n-1}{2}} \sum_{j=0}^{\frac{n-1}{2}} \mu_{i+j+1} \xi_{2i+1} \xi_{2j+1},$$

and if $n$ is even

$$\sum_{i=0}^n \sum_{j=0}^n \nu_{i+j} \xi_i \xi_j = \sum_{i=0}^{\frac{n}{2}} \sum_{j=0}^{\frac{n}{2}} \mu_{i+j} \xi_{2i} \xi_{2j} + \sum_{i=0}^{\frac{n}{2}-1} \sum_{j=0}^{\frac{n}{2}-1} \mu_{i+j+1} \xi_{2i+1} \xi_{2j+1}.$$

This shows that the sequence $\{\nu_n\}_0^\infty$ is positive, and hence by Theorem 10 that there exists a non-decreasing function $\beta(t)$ such that

$$\nu_n = \int_{-\infty}^\infty t^n \, d\beta(t) \qquad (n = 0, 1, 2, \cdots),$$

or that

(4)
$$\mu_n = \int_{-\infty}^\infty t^{2n} \, d\beta(t) \qquad (n = 0, 1, 2, \cdots)$$

$$0 = \int_{-\infty}^\infty t^{2n+1} \, d\beta(t) \qquad (n = 0, 1, 2, \cdots).$$

Set

$$\gamma(t) = \frac{\beta(t) - \beta(-t)}{2} \qquad (-\infty < t < \infty).$$

This function is odd and also satisfies equations (4) It is non-decreasing. Set $\alpha(t) = 2\gamma(t^{1/2})$ $(t \geqq 0)$ Then

$$\mu_n = \int_{-\infty}^\infty t^{2n} \, d\gamma(t) = \int_0^\infty t^{2n} \, d\gamma(t) - \int_0^\infty t^{2n} \, d\gamma(-t)$$

by an obvious change of variable. But since $\gamma(t)$ is odd, this gives

$$\mu_n = 2 \int_0^\infty t^{2n} \, d\gamma(t) \qquad (n = 0, 1, 2, \cdots)$$

$$= \int_0^\infty t^n \, d[2\gamma(t^{1/2})]$$

$$= \int_0^\infty t^n \, d\alpha(t).$$

Since $\alpha(t)$ is non-decreasing in the interval $0 \leq t < \infty$, we have the desired result.

It is now clear that we also have the following results.

THEOREM 13b. *A necessary and sufficient condition that equations* (1) *should have a non-decreasing solution* $\alpha(t)$ *with infinitely many points of increase is that the forms* (2) *and* (3) *should all be positive definite or that the determinants*

(5)
$$\mu_0, \quad \begin{vmatrix} \mu_0 & \mu_1 \\ \mu_1 & \mu_2 \end{vmatrix}, \quad \begin{vmatrix} \mu_0 & \mu_1 & \mu_2 \\ \mu_1 & \mu_2 & \mu_3 \\ \mu_2 & \mu_3 & \mu_4 \end{vmatrix}, \cdots$$

$$\mu_1, \quad \begin{vmatrix} \mu_1 & \mu_2 \\ \mu_2 & \mu_3 \end{vmatrix}, \quad \begin{vmatrix} \mu_1 & \mu_2 & \mu_3 \\ \mu_2 & \mu_3 & \mu_4 \\ \mu_3 & \mu_4 & \mu_5 \end{vmatrix}, \cdots$$

should all be greater than zero.

THEOREM 13c. *A necessary and sufficient condition that equations* (1) *should have a non-decreasing solution* $\alpha(t)$ *with a finite number of points of increase is that the forms* (2) *and* (3) *should all be positive, at least one of them being positive semidefinite.*

We observe that it is not sufficient that the determinants (4) should be all non-negative.

## 14. Moments of Functions of Bounded Variation

From analogy with the Hausdorff problem one might expect that it would be desirable to consider the Hamburger and Stieltjes problems for the case in which $\alpha(t)$ is of bounded variation on the appropriate infinite interval. R. P. Boas* [1939] has observed that in this case there is no problem, that every sequence leads to a soluble Stieltjes or Ham-

* For references to this unpublished result see J. Shohat [1938] and G. Pólya [1938].

§14]  FUNCTIONS OF BOUNDED VARIATION  139

burger problem if we regard any function of bounded variation as a solution. We give the proof of Boas. It will of course be sufficient to consider the Stieltjes case.

THEOREM 14. *The equations*

$$\mu_n = \int_0^\infty t^n \, d\alpha(t) \qquad (n = 0, 1, 2, \cdots)$$

*always have a solution* $\alpha(t)$ *of bounded variation for which*

$$\int_0^\infty |\, d\alpha(t)\,| < \infty.$$

We set up two other sequences $\{\lambda_n\}_0^\infty$, $\{\nu_n\}_0^\infty$, such that

(1) $\qquad \mu_n = \lambda_n - \nu_n$

(2) $\qquad \lambda_n = \int_0^\infty t^n \, d\beta(t)$

(3) $\qquad \nu_n = \int_0^\infty t^n \, d\gamma(t) \qquad (n = 0, 1, 2, \cdots),$

where $\beta(t)$ and $\gamma(t)$ are bounded non-decreasing functions. First choose $\lambda_0$, $\lambda_1$, $\nu_0$, $\nu_1$ as any positive numbers satisfying (1). Now proceed by induction. Suppose we have already determined $\lambda_k$, $\nu_k$ for $k = 0, 1, 2, \cdots, 2n - 1$ so that (1) holds and so that the determinants

(4)
$$[\lambda_0, \lambda_1, \cdots, \lambda_{2k}] = \begin{vmatrix} \lambda_0 & \lambda_1 & \cdots & \lambda_k \\ \lambda_1 & \lambda_2 & \cdots & \lambda_{k+1} \\ \cdot & \cdot & \cdots & \cdot \\ \lambda_k & \lambda_{k+1} & \cdots & \lambda_{2k} \end{vmatrix}$$

$$[\lambda_1, \lambda_2, \cdots, \lambda_{2k+1}] = \begin{vmatrix} \lambda_1 & \lambda_2 & \cdots & \lambda_{k+1} \\ \lambda_2 & \lambda_3 & \cdots & \lambda_{k+2} \\ \cdot & \cdot & \cdots & \cdot \\ \lambda_{k+1} & \lambda_{k+2} & \cdots & \lambda_{2k+1} \end{vmatrix},$$

(5) $\qquad [\nu_0, \nu_1, \cdots, \nu_{2k}], \quad [\nu_1, \nu_2, \cdots, \nu_{2k+1}]$

are positive for $k = 0, 1, \cdots, n - 1$. We now define $\lambda_{2n}$, $\nu_{2n}$, $\lambda_{2n+1}$, $\nu_{2n+1}$. We have with undetermined $\lambda_{2n}$

(6) $\qquad [\lambda_0, \lambda_1, \cdots, \lambda_{2n}] = \lambda_{2n}[\lambda_0, \lambda_1, \cdots, \lambda_{2n-2}] + P,$

where $P$ is a polynomial in $\lambda_0, \lambda_1, \cdots, \lambda_{2n-1}$; and similarly for $[\nu_0, \nu_1, \cdots, \nu_{2n}]$. Since $[\lambda_0, \lambda_1, \cdots, \lambda_{2n-2}]$ and $[\nu_0, \nu_1, \cdots, \nu_{2n-2}]$ are both greater than zero we can choose $\lambda_{2n}$ and $\nu_{2n}$ positive and so large that $\lambda_{2n} - \nu_{2n} = \mu_{2n}$ and that

$$[\lambda_0, \lambda_1, \cdots, \lambda_{2n}] > 0, \qquad [\nu_0, \nu_1, \cdots, \nu_{2n}] > 0.$$

Now observe that (6) holds with all subscripts increased by unity, $P$ now being a polynomial in $\lambda_1, \lambda_2, \cdots, \lambda_{2n}$, a similar equation holding for the $\nu_k$. With $\lambda_{2n}$ and $\nu_{2n}$ now determined we proceed exactly as above to determine $\lambda_{2n+1}$ and $\nu_{2n+1}$. This completes the induction. But by Theorem 13b if the determinants (4) and (5) are positive for $k = 0, 1, 2, \cdots$ equations (2) and (3) have bounded non-decreasing solutions $\beta(t)$ and $\gamma(t)$ respectively, so that when $\alpha(t)$ is defined as $\beta(t) - \gamma(t)$ our proof is complete.

In section 8 we gave the example of Stieltjes to show the existence of a function, not a constant, all the moments of which are zero. This also follows from Theorem 14. For, by this result, there exists a non-constant function $\alpha(t)$ such that

$$\int_0^\infty t^n \, d\alpha(t) = 1 \qquad (n = 1)$$
$$= 0 \qquad (n = 0, 2, 3, 4, \cdots)$$
$$\int_0^\infty |d\alpha(t)| < \infty.$$

Setting

$$\beta(t) = \alpha(t^{1/2}),$$

we have

$$\int_0^\infty t^n \, d\beta(t) = \int_0^\infty t^{2n} \, d\alpha(t) = 0 \qquad (n = 0, 1, 2, \cdots).$$

The function $\beta(t)$ is the example required.

## 15. A Sufficient Condition for the Solubility of the Stieltjes Problem

By a slight modification of the method employed in Section 14 Boas showed that any sequence which increases sufficiently rapidly leads to a soluble Stieltjes problem [with non-decreasing $\alpha(t)$]. More precisely, the result is:

THEOREM 15. *If*

(1) $$\mu_0 \geq 1, \qquad \mu_n \geq (n\mu_{n-1})^n \qquad (n = 1, 2, \cdots),$$

*then the equations*

$$\mu_n = \int_0^\infty t^n \, d\alpha(t) \qquad (n = 0, 1, 2, \cdots)$$

*have a non-decreasing solution $\alpha(t)$*

§15]  SOLUBILITY OF STIELTJES PROBLEM  141

An example of a sequence satisfying (1) is $\mu_0 = 1$, $\mu_n = n^{n^n}$ for $n = 1, 2, \cdots$. As in the previous section

(2) $\quad [\mu_0, \mu_1, \cdots, \mu_{2n}] = \mu_{2n}[\mu_0, \mu_1, \cdots, \mu_{2n-2}] + \sum_{k=n}^{2n-1} \pm \mu_k D_k,$

where the $D_k$ are $n$-rowed minors of $[\mu_0, \mu_1, \cdots, \mu_{2n}]$ not containing $\mu_{2n}$. Similarly

(3) $\quad [\mu_1, \mu_2, \cdots, \mu_{2n+1}] = \mu_{2n+1}[\mu_1, \mu_2, \cdots, \mu_{2n-1}] + \sum_{k=n+1}^{2n} \pm \mu_k D'_k,$

where the $D'_k$ are $n$-rowed minors of $[\mu_1, \mu_2, \cdots, \mu_{2n+1}]$ not containing $\mu_{2n+1}$.

Suppose that for $k \leq m - 1$ we have showed that

(4) $\quad [\mu_0, \mu_1, \cdots, \mu_{2k}] \geq 1, \quad [\mu_1, \mu_2, \cdots, \mu_{2k+1}] \geq 1.$

We will prove the same inequalities for $k = m$. By (1) we see that the sequence $\{\mu_n\}_0^\infty$ is non-decreasing and hence that no element of the sequence is less than unity. Hence

$$\mu_n > 2\left(\frac{n}{2}\right)^{\frac{n+4}{4}} (\mu_{n-1})^{\frac{n+2}{2}} \qquad (n = 2, 3, \cdots).$$

In particular

(5) $\quad \begin{aligned} \mu_{2n} &> 2n^{\frac{n+2}{2}}(\mu_{2n-1})^{n+1} \geq 1 + n^{\frac{n+2}{2}}(\mu_{2n-1})^{n+1} & (n = 1, 2, \cdots) \\ \mu_{2n+1} &> 2n^{\frac{n+2}{2}}(\mu_{2n})^{n+1} \geq 1 + n^{\frac{n+2}{2}}(\mu_{2n})^{n+1} & (n = 1, 2, \cdots). \end{aligned}$

The elements of $D_k$ are not greater than $\mu_{2m-1}$ and those of $D'_k$ not greater than $\mu_{2m}$ when $k$ ranges over the integer indicated in the summations (2) and (3). By Hadamard's upper bound for a determinant

$$|D_k| \leq m^{\frac{m}{2}}(\mu_{2m-1})^m \qquad (k = m, m+1, \cdots, 2m-1)$$

$$|D'_k| \leq m^{\frac{m}{2}}(\mu_{2m})^m \qquad (k = m+1, m+2, \cdots, 2m).$$

Hence

$$\left|\sum_{k=m}^{2m-1} \pm \mu_k D_k\right| \leq m(\mu_{2m-1})m^{\frac{m}{2}}(\mu_{2m-1})^m$$

$$\left|\sum_{k=m+1}^{2m} \pm \mu_k D'_k\right| \leq m(\mu_{2m})m^{\frac{m}{2}}(\mu_{2m})^m,$$

so that by (2) and (3)

$$[\mu_0, \mu_1, \cdots, \mu_{2m}] \geqq \mu_{2m} - m^{\frac{m+2}{2}}(\mu_{2m-1})^{m+1}$$

$$[\mu_1, \mu_2, \cdots, \mu_{2m+1}] \geqq \mu_{2m+1} - m^{\frac{m+2}{2}}(\mu_{2m})^{m+1}$$

But by (5) we see that (4) is established for $k = m$. By induction (4) now holds for all $k$, and by Theorem 13b the Stieltjes moment problem corresponding to the moments (1) has a non-decreasing solution $\alpha(t)$. The theorem is thus established.

## 16. Indeterminacy of Solution

Making use of the previous result Boas showed further that any sequence of sufficiently rapid growth leads to a Stieltjes problem which has more than one non-decreasing solution.

THEOREM 16. *If*

$$\lambda_0 \geqq 1$$

$$\lambda_2 \geqq (2\lambda_1 + 2)^2$$

$$\lambda_n \geqq (n\lambda_{n-1})^n \qquad (n = 1, 3, 4, 5, \cdots)$$

$$\mu_n = \lambda_{2n} \qquad (n = 0, 1, 2, \cdots),$$

*then there are at least two essentially distinct non-decreasing functions $\alpha(t)$ such that*

(1) $$\mu_n = \int_0^\infty t^n \, d\alpha(t) \qquad (n = 0, 1, 2, \cdots).$$

For, by Theorem 15 there exists a function $\beta(t)$ which is positive, non-decreasing, and such that

$$\lambda_n = \int_0^\infty t^n \, d\beta(t) \qquad (n = 0, 1, \cdots),$$

whence

$$\mu_n = \int_0^\infty t^n \, d\beta(t^{1/2}) \qquad (n = 0, 1, 2, \cdots).$$

Next let $\{\nu_n\}_0^\infty$ be a sequence which is identical with the sequence $\{\lambda_n\}_0^\infty$ except that $\nu_1 = \lambda_1 + 1$. Then

$$\nu_n \geqq (n\nu_{n-1})^n \qquad (n = 1, 2, 3, \cdots).$$

This is obvious if $n$ is neither 1 nor 2. But

$$\lambda_1 + 1 = \nu_1 \geqq \nu_0 = \lambda_0$$

$$\nu_2 = \lambda_2 \geqq (2\nu_1)^2 = (2\lambda_1 + 2)^2.$$

# INDETERMINACY OF SOLUTION

Hence by Theorem 15 there is a positive non-decreasing function $\gamma(t)$ such that

$$\nu_n = \int_0^\infty t^n \, d\gamma(t) \qquad (n = 0, 1, 2, \cdots)$$

$$\mu_n = \int_0^\infty t^n \, d\gamma(t^{1/2}) \qquad (n = 0, 1, 2, \cdots).$$

Clearly $\beta(t^{1/2})$ is essentially distinct from $\gamma(t^{1/2})$, for otherwise we should have

$$\int_0^\infty t \, d\beta(t) = \int_0^\infty t \, d\gamma(t),$$

which is impossible since $\nu_1$ and $\lambda_1$ are not equal.

# CHAPTER IV

# ABSOLUTELY AND COMPLETELY MONOTONIC FUNCTIONS

## 1. Introduction

Absolutely monotonic functions were first introduced by S. Bernstein [1914] as functions which are non-negative with non-negative derivatives of all orders. He proved that such functions are necessarily analytic. He showed later* that if a function is absolutely monotonic on the negative real axis then it can be represented there by a Laplace-Stieltjes integral with non-decreasing determining function, and conversely. Somewhat earlier F. Hausdorff [1921a] had proved a similar result for completely monotonic sequences (see Chapter III), which essentially contained the Bernstein result.† Bernstein was evidently unaware of Hausdorff's result, and his proof followed entirely independent lines. The author [1931] later gave an independent proof of the theorem without knowing of Bernstein's work.

## 2. Elementary Properties of Absolutely Monotonic Functions

We give first Bernstein's original definition.

DEFINITION 2a. *A function $f(x)$ is absolutely monotonic in the interval $a < x < b$ if it has non-negative derivatives of all orders there:*

(1) $\qquad f^{(k)}(x) \geqq 0 \qquad (a < x < b; k = 0, 1, 2, \cdots )$.

DEFINITION 2b. *A function is absolutely monotonic in $a \leqq x \leqq b$ ($a \leqq x < b$, or $a < x \leqq b$) if it is continuous there and satisfies* (1).

It is clear that if $f(x)$ is absolutely monotonic in $a < x < b$ that $f(a+)$ exists. If $f(x)$ is defined to be $f(a+)$ when $x = a$ then it becomes absolutely monotonic in $a \leqq x < b$. But a similar situation does not obtain when $x = b$. The function $f(x) = -x^{-1}$ is absolutely monotonic in the interval $-\infty < x < 0$ but not in $-\infty < x \leqq 0$.

We list a number of examples, giving the regions in which the functions are absolutely monotonic.

1. $f(x) = c \geqq 0$ $\hfill (-\infty < x < \infty)$

2. $f(x) = \sum_{k=0}^{n} a_k x^k \quad (a_k \geqq 0)$ $\hfill (0 \leqq x < \infty)$

---

\* S. Bernstein [1928] p. 56.

† Compare also F. Hausdorff [1921b], Theorem III, p. 287, from which the theorem could easily be derived.

## §2] ELEMENTARY PROPERTIES 145

3. $f(x) = \sum_{k=0}^{\infty} a_k x^k < \infty \quad (0 \leqq x \leqq \rho, a_k \geqq 0) \qquad (0 \leqq x \leqq \rho)$

4. $f(x) = \sum_{k=0}^{\infty} a_k(-1)^k x^{-k} < \infty \quad (-\infty < x < -c < 0, a_k \geqq 0)$

$\qquad\qquad\qquad\qquad\qquad\qquad\qquad\qquad\qquad (-\infty < x < -c)$

5. $f(x) = \int_0^{\infty} e^{xt} d\alpha(t) < \infty \quad (\alpha(t)\!\uparrow, -\infty < x < -c)$

$\qquad\qquad\qquad\qquad\qquad\qquad\qquad\qquad\qquad (-\infty < x < -c)$

6. $f(x) = -\log(-x) \qquad\qquad\qquad (-1 \leqq x < 0)$

7. $f(x) = \sin^{-1} x \qquad\qquad\qquad\qquad (0 \leqq x \leqq 1)$.

The last example shows that an absolutely monotonic function in $(a \leqq x \leqq b)$ need not have a left-hand derivative at $b$. The following elementary theorem is easily established.

**THEOREM 2a.** *If $f_1(x)$ and $f_2(x)$ are absolutely monotonic in $a < x < b$ then the following functions are also:*

$$a_1 f_1(x) + a_2 f_2(x) \qquad (a_1 \geqq 0, a_2 \geqq 0)$$

$$f_1(x) f_2(x)$$

$$f_1^{(k)}(x) \qquad (k = 0, 1, 2, \cdots)$$

*If, in addition, $a < f_2(x) < b$, then $f_1(f_2(x))$ is absolutely monotonic in $a < x < b$.*

As examples of the last part of this theorem we see that the functions $e^{e^x}$ and $e^{-1/x}$ are absolutely monotonic for all $x$ and for all negative $x$ respectively.

We next introduce completely monotonic functions as the continuous analogues of Hausdorff's completely monotonic sequences.

**DEFINITION 2c.** *The function $f(x)$ is completely monotonic in $(a, b)$ if and only if $f(-x)$ is absolutely monotonic in $(-b, -a)$.*

It is obvious that such a function satisfies the inequalities

$$(-1)^k f^{(k)}(x) \geqq 0 \qquad (a < x < b).$$

**THEOREM 2b.** *If $f_1(x)$ is absolutely monotonic in $a < x < b$, if $f_2(x)$ is completely monotonic there, and if $a < f_2(x) < b$, then $f_1(f_2(x))$ is completely monotonic there.*

It must not be supposed that a completely monotonic function of such a function has the same property. Thus if $f_1(x) = x^{-1}$ and $f_2(x) = x^{-2}$, both functions being completely monotonic for $0 < x < \infty$ we have

146 ABSOLUTELY MONOTONIC FUNCTIONS [CH. IV

$f_1(f_2(x)) = x^2$ and $f_1(f_1(x)) = 1$, of which the second function is completely monotonic in $0 < x < \infty$, the first is not.

### 3. Analyticity of Absolutely Monotonic Functions

We now show that any completely or absolutely monotonic function is necessarily analytic. It will be sufficient to treat one of these classes of functions.

**THEOREM 3a.** *If $f(x)$ is absolutely monotonic in $a \leqq x < b$, then it can be extended analytically into the complex z-plane ($z = x + iy$), and the function $f(z)$ will be analytic in the circle*

$$|z - a| < b - a.$$

It is clear that at $x = a$ the function $f(x)$ has right-hand derivatives* of all orders at $x = a$, which we denote by $f^{(k)}(a)$ for $k = 1, 2, 3, \cdots$. Then by Taylor's formula with exact remainder

$$(1) \quad f(x) = f(a) + f'(a)(x - a) + \cdots + f^{(n)}(a)\frac{(x-a)^n}{n!} + R_n(x)$$

$$R_n(x) = \int_a^x \frac{(x-t)^n}{n!} f^{(n+1)}(t) \, dt$$

$$= \frac{(x-a)^{n+1}}{n!} \int_0^1 (1-t)^n f^{(n+1)}(a + [x-a]t) \, dt.$$

Since $f^{(n+2)}(x)$ is non-negative by hypothesis, it is clear that $f^{(n+1)}(a + [x-a]t)$ is a non-decreasing function of $x$ when $t$ is fixed, so that if $a \leqq x \leqq c < b$, we have

$$0 \leqq R_n(x) \leqq \frac{(x-a)^{n+1}}{n!} \int_0^1 (1-t)^n f^{(n+1)}(a + [c-a]t) \, dt$$

$$= \frac{(x-a)^{n+1}}{(c-a)^{n+1}} \left[ f(c) - f(a) - f'(a)(c-a) - \cdots - f^{(n)}(a)\frac{(c-a)^n}{n!} \right]$$

$$\leqq f(c) \left(\frac{x-a}{c-a}\right)^{n+1}.$$

Hence

$$\lim_{n \to \infty} R_n(x) = 0 \qquad (a \leqq x < c < b),$$

---

* This of course involves a definition of successive derivatives at $x = a$. For example, we know at once that $f(a) = \lim f(a + \delta)$, that $\lim f'(a + \delta)$ exists. If we define $f'(a)$ as this limit it is clear by the law of the mean that $f'(a)$ is the right-hand derivative of $f(x)$ at $x = a$ and that $f'(x)$ is continuous in the interval $a \leqq x < b$. In this way we can proceed step by step to the definition of any derivative at $x = a$.

and since $c$ is arbitrary

$$f(x) = \sum_{n=0}^{\infty} f^{(n)}(a) \frac{(x-a)^n}{n!} \qquad (a \leq x < b).$$

We have only to define $f(z) = f(x + iy)$ as

$$f(z) = \sum_{n=0}^{\infty} f^{(n)}(a) \frac{(z-a)^n}{n!}$$

to complete the proof of the theorem.

COROLLARY 3a. *If $f(x)$ is absolutely monotonic in $(a, b)$ and is zero at $x = c > a$, then it is identically zero.*

As a special case of the theorem we may have $a = -\infty$. Then $f(z)$ is analytic in the half-plane $x < b$. We observe that although Theorem 3a shows the possibility of the analytic extension of $f(x)$ leftward on the $x$-axis to the point $x = 2a - b$ there is no guarantee that $f(x)$ will be absolutely monotonic in the enlarged interval. For example $\sin^{-1} x$ is absolutely monotonic in $(0, 1)$ but not in $(-1, 0)$.

THEOREM 3b. *A necessary and sufficient condition that it should be possible to expand the function $f(x)$ in a series of powers of $(x - a)$ converging for $a \leq x < b$ is that $f(x)$ should be the difference of two functions absolutely monotonic in $a \leq x < b$.*

That the condition is sufficient is evident from Theorem 3a. That it is necessary is evident from the equations

$$f(x) = \sum_{n=0}^{\infty} a_n (x-a)^n = \sum_{n=0}^{\infty} |a_n| (x-a)^n - \sum_{n=0}^{\infty} (|a_n| - a_n)(x-a)^n.$$

## 4. Bernstein's Second Definition

We may introduce an equivalent definition of absolutely monotonic functions making less continuity requirements. This is also due to S. Bernstein [1914].

DEFINITION 4. *A function $f(x)$ is absolutely monotonic in $a \leq x < b$ if and only if*

(1) $$\Delta_h^n f(x) = \sum_{k=0}^{n} (-1)^{n-k} \binom{n}{k} f(x + kh) \geq 0$$

*for all non-negative integers $n$ and for all $x$ and $h$ such that*

$$a \leq x < x + h < \cdots < x + nh < b.$$

Until we shall have established the equivalence of the two definitions we shall say that a function satisfying the conditions of Section 2 is absolutely monotonic $(D)$, while the function of Definition 4 will be called absolutely monotonic (Δ). We begin with certain preliminary results.

**THEOREM 4.** *If $f(x)$ is absolutely monotonic (Δ) in $a \leqq x < b$ it is non-negative, non-decreasing, convex, and continuous there.*

The first two conclusions follow from (1) with $n = 0$ and $n = 1$ respectively. The function $f(x)$ is convex in $a \leqq x < b$ if for any points $x$ and $y$ of that interval

$$f\left(\frac{x+y}{2}\right) \leqq \frac{f(x)}{2} + \frac{f(y)}{2}.$$

But this follows from (1) with $n = 2$. Finally, let $x_0$ be an arbitrary point of $a < x < b$. Since $f(x)$ is non-decreasing it has limit values on the right and left and

$$f(x_0-) \leqq f(x_0) \leqq f(x_0+)$$

But for $h$ sufficiently small

$$2f(x_0 + h) \leqq f(x_0) + f(x_0 + 2h)$$

$$2f(x_0) \leqq f(x_0 + h) + f(x_0 - h),$$

whence

$$f(x_0+) \leqq f(x_0)$$

$$2f(x_0) \leqq f(x_0+) + f(x_0-)$$

$$f(x_0+) \leqq f(x_0-).$$

From these inequalities the continuity of $f(x)$ at $x_0$ is apparent. A similar proof holds if $x_0 = a$.

**COROLLARY 4.** *If $a \leqq \beta_1 < \beta_2 < \beta_3 < b$, then*

(2)
$$\begin{vmatrix} f(\beta_1) & \beta_1 & 1 \\ f(\beta_2) & \beta_2 & 1 \\ f(\beta_3) & \beta_3 & 1 \end{vmatrix} \leqq 0.$$

This is a familiar consequence of the convexity of $f(x)$ For completeness we include the proof. Set $\beta_3 - \beta_1 = h$ and

$$x_k = \beta_1 + \frac{k}{n} h \qquad (k = 0, 1, \cdots, n)$$

for an arbitrary positive integer $n$. Then

(3) $\quad 0 \leqq f(x_1) - f(x_0) \leqq f(x_2) - f(x_1) \leqq \cdots \leqq f(x_n) - f(x_{n-1})$

by the definition of convexity. The average of the first $m$ terms ($m < n$) of this sequence is clearly not greater than the average of the next $n - m$ terms. That is

$$\frac{f(x_m) - f(x_0)}{m} \leq \frac{f(x_n) - f(x_m)}{n - m},$$

or

$$\frac{f\left(\beta_1 + \frac{mh}{n}\right) - f(\beta_1)}{mh/n} \leq \frac{f(\beta_3) - f\left(\beta_1 + \frac{mh}{n}\right)}{(n - m)h/n}.$$

Now let $n$ and $m$ become infinite in such a manner that

$$\beta_1 + \frac{m}{n} h \to \beta_2.$$

Then by the continuity of $f(x)$

$$\frac{f(\beta_2) - f(\beta_1)}{\beta_2 - \beta_1} \leq \frac{f(\beta_3) - f(\beta_2)}{\beta_3 - \beta_2},$$

which is equivalent to (2).

## 5. Existence of One-sided Derivatives

We can now show that $f(x)$ has a right-hand derivative and a left-hand derivative, which we denote by $f'_r(x)$ and $f'_l(x)$ respectively.

THEOREM 5. *If $f(x)$ is absolutely monotonic $(\Delta)$ in $a \leq x < b$ and if $a < x < y < b$, then right- and left-hand derivatives of $f(x)$ exist and*

$$f'_l(x) \leq f'_r(x) \leq f'_l(y).$$

To prove this we have from Theorem 4 and Corollary 4 that

(1) $$0 \leq \frac{f(x + \delta_1) - f(x)}{\delta_1} \leq \frac{f(x + \delta_2) - f(x)}{\delta_2} \qquad (0 < \delta_1 < \delta_2),$$

(2) $$\frac{f(x - \delta_2) - f(x)}{-\delta_2} \leq \frac{f(x - \delta_1) - f(x)}{-\delta_1}$$
$$\leq \frac{f(x + \delta_1) - f(x)}{\delta_1} \qquad (0 < \delta_1 < \delta_2).$$

Inequalities (1) show the existence of $f'_r(x)$. Inequalities (2) show the existence of $f'_l(x)$ and that

$$f'_l(x) \leq f'_r(x).$$

The second of the required inequalities is proved by another application of Corollary 4. Since (1) holds also for $x = a$, the existence of $f'_r(a)$ is also assured.

## 6. Higher Differences of Absolutely Monotonic Functions

We next show that a difference of $f(x)$ of any order, $\Delta^k f(x)$, is a nondecreasing function of $x$.

THEOREM 6. *If $f(x)$ is absolutely monotonic ($\Delta$) in $a \leq x < b$, and if $a < x < y < b$, then*

$$\Delta_h^k f(x) \leq \Delta_h^k f(y),$$

*where for a given integer $k$, the number $h$ must be chosen so small that*

$$y + kh < b.$$

We first prove the identity

$$\Delta_h f(x) = \sum_{i=0}^{n-1} \Delta_{h/n} f\left(x + \frac{ih}{n}\right).$$

This follows from the equations

$$\sum_{i=0}^{n-1} \Delta_{h/n} f\left(x + \frac{ih}{n}\right) = \sum_{i=0}^{n-1} \left[ f\left(x + \frac{i+1}{n} h\right) - f\left(x + \frac{ih}{n}\right) \right]$$

$$= \sum_{i=1}^{n} f\left(x + \frac{ih}{n}\right) - \sum_{i=0}^{n-1} f\left(x + \frac{ih}{n}\right)$$

$$= f(x + h) - f(x).$$

Repeated application of this identity gives

$$\Delta_h^k f(x) = \sum_{i_1=0}^{n-1} \sum_{i_2=0}^{n-1} \cdots \sum_{i_k=0}^{n-1} \Delta_{h/n}^k f\left(x + [i_1 + i_2 + \cdots + i_k] \frac{h}{n}\right).$$

Now apply the difference operator $\Delta_{h/n}$ to both sides of this equation and obtain

$$\Delta_{h/n} \Delta_h^k f(x) = \sum_{i_1=0}^{n-1} \sum_{i_2=0}^{n-1} \cdots \sum_{i_k=0}^{n-1} \Delta_{h/n}^{k+1} f\left(x + [i_1 + i_2 + \cdots + i_k] \frac{h}{n}\right).$$

The right-hand side is positive since $f(x)$ is absolutely monotonic. Hence

$$\Delta_{h/n} \Delta_h^k f(x) = \Delta_h^k f\left(x + \frac{h}{n}\right) - \Delta_h^k f(x) \geq 0.$$

Applying this result successively we obtain the inequalities

$$\Delta_h^k f(x) \leq \Delta_h^k f\left(x + \frac{h}{n}\right) \leq \Delta_h^k f\left(x + \frac{2h}{n}\right) \leq \cdots.$$

Hence for any positive integer $m$ we have

$$\Delta_h^k f(x) \leq \Delta_h^k f\left(x + \frac{mh}{n}\right).$$

Now allowing $m$ and $n$ to become infinite in such a way that $x + mn^{-1}h$ tends to $y$, we have the desired result.

## 7. Equivalence of Bernstein's Two Definitions

We are now able to show the equivalence of Definitions 2b and 4.

THEOREM 7. *A function $f(x)$ is absolutely monotonic $(D)$ in $a \leq x < b$ if and only if it is absolutely monotonic $(\Delta)$ there.*

That $(D)$ implies $(\Delta)$ is obvious by use of the law of the mean.

Conversely, if $f(x)$ is absolutely monotonic $(\Delta)$, then by Theorem 5 the left-handed derivative $f_l'(x)$ exists in $a < x < b$. It is itself absolutely monotonic $(\Delta)$ in $a < x < b$ since by Theorem 6

$$\Delta_h^k f(x - \delta) \leq \Delta_h^k f(x) \qquad (a < x - \delta < x < b),$$

$$\Delta_h^k \frac{f(x - \delta) - f(x)}{-\delta} \geq 0$$

(1) $$\Delta_h^k f_l'(x) \geq 0.$$

Hence by Theorem 4 $f_l'(x)$ is continuous. By Theorem 5

$$f_l'(x) \leq f_r'(x) \leq f_l'(y) \qquad (x < y).$$

Allowing $y$ to approach $x$, we obtain

$$f_r'(x) = f_l'(x),$$

so that $f'(x)$ exists in $a < x < b$ and by (1) is absolutely monotonic $(\Delta)$ there. Applying this result to $f'(x)$ we see that $f''(x)$ exists and is absolutely monotonic. In this way we see that all derivatives of $f(x)$ exist in $a < x < b$ and are all non-negative there. By a remark following Definition 2b it is clear that $f(x)$ is absolutely monotonic $(D)$ in $a \leq x < b$.

COROLLARY 7. *If the functions*

$$u_n(x) \qquad (n = 0, 1, 2, \cdots)$$

*are absolutely monotonic in $a \leq x < b$ and if*

$$f(x) = \sum_{n=0}^{\infty} u_n(x),$$

*the series converging there, then $f(x)$ is absolutely monotonic there.*

For we clearly have

$$\Delta^k f(x) = \sum_{n=0}^{\infty} \Delta^k u_n(x) \geq 0 \qquad (k = 0, 1, \cdots, a \leq x < b),$$

provided only the points of subdivision of $(a, b)$ involved in the computation of $\Delta^k f(x)$ should all lie in $(a, b)$.

## 8. Bernstein Polynomials

Before taking up a third equivalent definition of absolutely monotonic functions, we must first discuss in further detail the polynomials of Bernstein defined earlier:

$$B_p[f] = \sum_{m=0}^{p} f\left(\frac{m}{p}\right) \lambda_{p,m}(x)$$

$$\lambda_{p,m}(x) = \binom{p}{m} x^m (1-x)^{p-m}.$$

LEMMA 8. *If $p$ is any positive integer, then*

$$px(1-x) = \sum_{m=0}^{p} (px - m)^2 \lambda_{p,m}(x).$$

To prove this differentiate the identity

$$[e^y + (1-x)]^p = \sum_{m=0}^{p} \binom{p}{m} e^{my}(1-x)^{p-m}$$

twice with respect to $y$ and set $y = \log x$. We thus obtain

$$1 = \sum_{m=0}^{p} \lambda_{p,m}(x)$$

$$px = \sum_{m=0}^{p} m \lambda_{p,m}(x)$$

$$px + p(p-1)x^2 = \sum_{m=0}^{p} m^2 \lambda_{p,m}(x).$$

Multiplying the first of these three equations by $p^2 x^2$, the second by $-2px$, and adding their sum to the third gives the desired identity.

THEOREM 8. *If $f(x)$ is continuous in $0 \leq x \leq 1$, then*

$$\lim_{p \to \infty} B_p[f] = f(x)$$

*uniformly in that interval.*

## BERNSTEIN POLYNOMIALS

For a fixed $x$ of $(0, 1)$ and a fixed positive integer $p$ set

$$\epsilon_p(x) = \max \left| f(x) - f\left(\frac{m}{p}\right) \right|$$

where the maximum is taken for all those values of $m$ for which

$$\left| \frac{m}{p} - x \right| < p^{-1/4}$$

By the uniform continuity of $f(x)$ it is clear that numbers $\epsilon_p$ exist such that

$$\epsilon_p(x) < \epsilon_p \qquad (0 \leq x \leq 1)$$

$$\lim_{p \to \infty} \epsilon_p = 0.$$

Now

$$f(x) - B_p[f] = \sum{}' \left[ f(x) - f\left(\frac{m}{p}\right) \right] \lambda_{p,m}(x)$$

$$+ \sum{}'' \left[ f(x) - f\left(\frac{m}{p}\right) \right] \lambda_{p,m}(x),$$

where the first summation is over all integers from 0 to $p$ inclusive for which $|m - px| \leq p^{3/4}$ and the second summation is over the remainder of those integers. If

$$M = \max_{0 \leq x \leq 1} |f(x)|,$$

we have

$$|f(x) - B_p[f]| < \sum{}' \epsilon_p \lambda_{p,m}(x) + 2M \sum{}'' \lambda_{p,m}(x).$$

The first sum on the right is not greater than $\epsilon_p$. For the second we have by Lemma 8

$$p^{3/2} \sum{}'' \lambda_{p,m}(x) < \sum_{m=0}^{p} (m - px)^2 \lambda_{p,m}(x) = px(1 - x) \leq \frac{p}{4}$$

$$\sum{}'' \lambda_{p,m}(x) < \frac{p^{-1/2}}{4}.$$

That is $|f(x) - B_p[f]|$ is less than a function of $p$ which tends to zero as $p$ becomes infinite. This proves the theorem.

We note in passing that the Weierstrass approximation theorem for continuous functions is a corollary to the present result.

## 9. Definition of Grüss

We come next to a definition introduced by G. Grüss [1935].

**DEFINITION 9.** *A function $f(x)$ is absolutely monotonic in $0 \leq x \leq 1$ if it is continuous there and if all the entries in the following table are non-negative for each positive integer $n$:*

$$\begin{array}{cccccc}
f(0) & & & & & \\
& \Delta f(0) & & & & \\
f\left(\dfrac{1}{n}\right) & & \Delta^2 f(0) & & & \\
& \Delta f\left(\dfrac{1}{n}\right) & & & & \\
f\left(\dfrac{2}{n}\right) & & \vdots & & \cdots & \Delta^n f(0), \\
\vdots & \vdots & & & & \\
f\left(\dfrac{n-1}{n}\right) & & \Delta^2 f\left(\dfrac{n-2}{n}\right) & & & \\
& \Delta f\left(\dfrac{n-1}{n}\right) & & & & \\
f(1) & & & & & \\
\end{array}$$

where

$$\Delta f(x) = f\left(x + \frac{1}{n}\right) - f(x).$$

It is easy to see how the definition should be modified to apply to an arbitrary finite interval. For the present we shall say that a function is absolutely monotonic $(G)$ when it satisfies Definition 9.

**THEOREM 9a.** *A function is absolutely monotonic $(G)$ in $(0 \leq x \leq 1)$ if and only if it is continuous there and for each positive integer $n$*

$$f(0) \geq 0, \quad \Delta f(0) \geq 0, \cdots \Delta^n f(0) \geq 0.$$

That is, it is sufficient (and clearly necessary) that the elements in the top side of the above triangular array should all be non-negative. This follows since

$$\Delta^{k-1} f(x) = \Delta^k f\left(x - \frac{1}{n}\right) + \Delta^{k-1} f\left(x - \frac{1}{n}\right).$$

**THEOREM 9b.** *A function is absolutely monotonic $(G)$ in $0 \leq x \leq 1$ if and only if it is the uniform limit of a sequence of polynomials with non-negative coefficients.*

First suppose that $f(x)$ is absolutely monotonic. Then

$$B_p[f] = \sum_{m=0}^{p} f\left(\frac{m}{p}\right) \lambda_{p,m}(x) = \sum_{m=0}^{p} \binom{p}{m} f\left(\frac{m}{p}\right) \sum_{k=0}^{p-m} \binom{p-m}{k} (-1)^k x^{k+m}$$

$$= \sum_{m=0}^{p} f\left(\frac{m}{p}\right) \binom{p}{m} \sum_{k=m}^{p} (-1)^{k-m} \binom{p-m}{k-m} x^k$$

$$= \sum_{k=0}^{p} \binom{p}{k} x^k \sum_{m=0}^{k} (-1)^{k-m} \binom{k}{m} f\left(\frac{m}{p}\right) = \sum_{k=0}^{p} \binom{p}{k} \Delta^k f(0) x^k.$$

That is, the Bernstein polynomials of an absolutely monotonic function have non-negative coefficients. By Theorem 8 the first half of our theorem is proved.

Next suppose that to an arbitrary positive $\epsilon$ there corresponds a polynomial

$$g(x) = \sum_{k=0}^{N} c_k x^k \qquad (c_k \geq 0,\ k = 0, 1, \cdots, N),$$

such that

$$|f(x) - g(x)| < \epsilon \qquad (0 \leq x \leq 1).$$

But from this inequality follows

$$|\Delta_\delta^n f(x) - \Delta_\delta^n g(x)| < 2^n \epsilon \qquad (n = 0, 1, 2, \cdots)$$

provided only that the points $x, x + \delta, \cdots, x + n\delta$ all lie in $(0, 1)$. Suppose $f(x)$ were not absolutely monotonic. Then we could find an integer $\lambda$ such that

$$\Delta^\lambda f(0) = -l^2 < 0.$$

Since $\Delta^\lambda g(0) \geq 0$ we have

(1) $$|\Delta^\lambda f(0) - \Delta^\lambda g(0)| \geq l^2$$

for any $g(x)$ with non-negative coefficients. Now choose $\epsilon < l^2 2^{-(\lambda+1)}$. Then

$$|\Delta^\lambda f(0) - \Delta^\lambda g(0)| < 2^\lambda l^2 2^{-(\lambda+1)} = \frac{l^2}{2}.$$

But this contradicts (1) and our theorem is proved.

## 10. Equivalence of Bernstein and Grüss Definitions

We can now show the equivalence of the Grüss definition with that of Bernstein.

THEOREM 10. *A function $f(x)$ is absolutely monotonic $(G)$ in $0 \leq x \leq 1$ if and only if it is absolutely monotonic $(\Delta)$ or $(D)$ there.*

That $(\Delta)$ or $(D)$ implies $(G)$ is trivial.

Conversely, we need only show that if $f(x)$ is absolutely monotonic $(G)$ that all differences $\Delta_h^k f(x)$ are non-negative. Suppose the contrary; that is, for some positive integer $k$, some $x_1$ in $0 \leq x < 1$, and some $h$ for which

$$0 \leq x_1 < x_1 + h < x_1 + 2h < \cdots < x_1 + kh \leq 1$$

we have

$$\Delta_h^k f(x_1) = -l^2 < 0$$

Then by Theorem 9b there exists a polynomial $g(x)$ with non-negative coefficients such that

$$|f(x) - g(x)| < \frac{l^2}{2^{k+1}} \qquad (0 \leq x \leq 1),$$

whence

$$|\Delta_h^k f(x) - \Delta_h^k g(x)| < \frac{l^2}{2}.$$

But since $\Delta_h^k g(x)$ is non-negative the left-hand side of this inequality is not less than $l^2$ when $x = x_1$. The contradiction is evident, and the theorem is proved.

## 11. Additional Properties of Absolutely Monotonic Functions

We prove here several additional properties of absolutely monotonic functions.

THEOREM 11a. *If $f(x)$ is absolutely monotonic in $0 \leq x \leq a$, and if*

(1) $$f(x) = \sum_{n=0}^{\infty} A_n x^n, \qquad (0 \leq x \leq a),$$

*the series converging for $|x| < \rho$ ($\rho > a$), then $f(x)$ is absolutely monotonic for $0 \leq x < \rho$. If any coefficient $A_k$ of the series (1) is zero, then $f(x)$ cannot be absolutely monotonic in any two sided neighborhood of $x = 0$ unless it is a polynomial.*

For by the proof of Theorem 3a all the coefficients $A_n$ of (1) are non-negative. Hence $f(x)$ will be absolutely monotonic as far to the right of $x = 0$ as it can be extended analytically. On the other hand if $A_k = 0$, then $f^{(k)}(0) = 0$; and if $f(x)$ were absolutely monotonic in $-\delta < x < \delta$, ($\delta > 0$), then $f^{(k)}(x)$ would be non-negative, non-decreasing there, and hence identically zero in $-\delta < x < 0$. That is, $f(x)$ is a polynomial of degree at most $k - 1$. This completes the proof of the theorem.

It should be observed that the converse of the last statement of the theorem is not true. Thus the function

$$f(x) = (1 + x) \cosh x = \sum_{n=0}^{\infty} \frac{x^{2n}}{2n!} + \frac{x^{2n+1}}{2n!},$$

has all its derivatives positive at the origin yet ceases to be absolutely monotonic to the left of $x = 0$. But

$$f^{(2n)}(x) = (1 + x) \cosh x + 2n \sinh x,$$

and for $x < 0$

$$\lim_{n \to \infty} f^{(2n)}(x) = -\infty.$$

If $x$ is negative, however near to the origin, $f^{(2n)}(x)$ will be negative for $n$ sufficiently large.

We also observe that if the function $f(x)$ defined by (1) has a derivative of order $k$ which becomes infinite as $x$ approaches $\rho$ then each derivative of higher order will have the same property. For, if $f^{(k+1)}(x)$ approached a finite limit as $x$ approaches $\rho$ the integral

$$\int_0^x f^{(k+1)}(t)\, dt = f^{(k)}(x) - f^{(k)}(0),$$

and hence also $f^{(k)}(x)$, would approach a finite limit.

An example of a function which is absolutely monotonic in $0 \leq x \leq 1$ but in no larger interval is the function of Fredholm

$$f(x) = \sum_{n=0}^{\infty} e^{-n} x^{n^2}.$$

This function clearly has a singularity at $x = 1$ and hence cannot be absolutely monotonic for $x > 1$. Yet it has continuous derivatives of all orders at $x = 1$. Since $f''(0) = 0$, $f(x)$ is not absolutely monotonic in any negative interval by Theorem 11a.

THEOREM 11b. *If $f(x)$ is absolutely monotonic in $-c \leq x \leq 0$, then $f(e^x - c)$ has the same property in $-\infty < x \leq \log c$.*

For one has by successive differentiation

$$\frac{d^n}{dx^n} f(e^x - c) = \sum_{k=1}^{n} B_k^n f^{(k)}(e^x - c) e^{kx}$$

where the $B_k^n$ are positive integers.

Note that the inverse change of variable does not conserve the property. Thus it is not true that if $f(x)$ is absolutely monotonic in $-\infty <$

$x \leq \log c$ then $f(\log(x + c))$ is absolutely monotonic in $-c \leq x \leq 0$. For example, $e^{x/2}$ has the property in $-\infty < x < \infty$; yet $\sqrt{x + 1}$ fails to have it in $(-1 \leq x \leq 0)$.

**THEOREM 11c.** *If $f(x)$ is absolutely monotonic for $-\infty < x < c$, the constant $c$ being positive, negative, or zero, then*

$$|f(x + iy)| \leq f(x) \qquad (-\infty < x < c, -\infty < y < \infty).$$

For let $x_0 + iy_0$ be an arbitrary point for which $x_0 < c$. Then if $-a < c$, the equation

$$(2) \qquad f(x + iy) = \sum_{k=0}^{\infty} \frac{f^{(k)}(-a)}{k!}(x + iy + a)^k$$

is valid in the circle

$$|x + iy + a| < c + a.$$

Clearly we may choose $a$ so large that the point $x_0 + iy_0$ will be inside this circle. Since the coefficients of the series (2) are non-negative it follows that $|f(x_0 + iy_0)|$ is not greater than the value of $f(x)$ at the point $x$ where the circle with center at $-a$ and through $x_0 + iy_0$ cuts the axis of reals to the right of $-a$. That is,

$$|f(x_0 + iy_0)| \leq f\left(-a + \sqrt{(x_0 + a)^2 + y_0^2}\right).$$

Since the left-hand side is independent of $a$ we may make the most of the inequality by allowing $a$ to become positively infinite. The function $f(x)$ being continuous, we obtain

$$|f(x_0 + iy_0)| \leq f(x_0).$$

This completes the proof of the theorem. It is obvious that a similar result holds for completely monotonic functions on $c < x < \infty$.

**THEOREM 11d.** *If $f(x)$ is completely monotonic in $a \leq x < \infty$ and if $\delta$ is any positive number, then the sequence $\{f(a + n\delta)\}_{n=0}^{\infty}$ is completely monotonic.*

For, if $k$ is any positive integer

$$\Delta^k f(a + n\delta) = \Delta^{k-1} f(a + n\delta + \delta) - \Delta^{k-1} f(a + n\delta),$$

and by the law of the mean

$$\Delta^k f(a + n\delta) = \Delta^{k-1} f'(\xi_1)\delta \qquad (a + n\delta < \xi_1 < a + n\delta + \delta).$$

A second application of the law of the mean gives

$$\Delta^k f(a + n\delta) = \Delta^{k-2} f''(\xi_2)\delta^2 \qquad (\xi_1 < \xi_2 < \xi_1 + \delta).$$

§11]  ADDITIONAL PROPERTIES  159

Finally,
$$\Delta^k f(a + n\delta) = f^{(k)}(\xi_k)\delta^k \qquad (\xi_{k-1} < \xi_k < \xi_{k-1} + \delta).$$

Clearly $\xi_k$ is a number in the interval $(a + n\delta, a + n\delta + k\delta)$. Since $f(x)$ is completely monotonic,
$$(-1)^k \Delta^k f(a + n\delta) \geqq 0 \qquad (k, n = 0, 1, 2, \cdots),$$
and this is what we wished to prove.

THEOREM 11e. *If $f(x)$ is completely monotonic in $c < x < \infty$, then for any number $a > c$ and any positive $\delta$ the sequence $\{(-1)^n \Delta_\delta^n f(a)\}_{n=0}^\infty$ is completely monotonic.*

For, by Theorem 11d the sequence $\{f(a + n\delta)\}_{n=0}^\infty$ is completely monotonic. Hence by Theorem 4a of Chapter III there exists a function $\alpha(t)$ which is non-decreasing and bounded in $(0, 1)$ such that
$$f(a + n\delta) = \int_0^1 t^n \, d\alpha(t) \qquad (n = 0, 1, 2, \cdots).$$

Hence by direct computation of the successive differences of $f(a)$ we have
$$(-1)^n \Delta_\delta^n f(a) = \int_0^1 (1 - t)^n \, d\alpha(t).$$

By an obvious change of variable this becomes
$$(-1)^n \Delta_\delta^n f(a) = \int_0^1 t^n \, d\beta(t),$$
where
$$\beta(t) = -\alpha(1 - t).$$

Since $\beta(t)$ is itself non-decreasing and bounded, we have our result at once by a second application of Theorem 4a of Chapter III.

THEOREM 11f. *If $f(x)$ is completely monotonic in $c < x < \infty$, then for any number $a > c$ the sequences*
$$\{(-1)^n f^{(n)}(a)\}_{n=0}^\infty, \qquad \{(-1)^n f^{(n)}(a)\}_{n=1}^\infty$$
*are positive.*[*]

Let $\delta$ be an arbitrary positive number, $n$ an arbitrary non-negative integer and $\{\xi_i\}_0^\infty$ an arbitrary real sequence. Then by Theorem 11e

(1)
$$\sum_{i=0}^n \sum_{j=0}^n \frac{(-1)^{i+j}}{\delta^{i+j}} \Delta^{i+j} f(a) \xi_i \xi_j = \sum_{i=0}^n \sum_{j=0}^n \xi_i \xi_j \int_0^1 \left(\frac{t}{\delta}\right)^{i+j} d\beta(t)$$
$$= \int_0^1 \left[\sum_{i=0}^n \left(\frac{t}{\delta}\right)^i \xi_i\right]^2 d\beta(t) \geqq 0.$$

* For the meaning of a "positive" sequence see Definition 9b of Chapter III.

Here $\beta(t)$ is the non-decreasing function employed in the proof of Theorem 11e.

In a similar way we have

(2) $$\sum_{i=0}^{n} \sum_{j=0}^{n} \frac{(-1)^{i+j+1}}{\delta^{i+j+1}} \Delta_\delta^{i+j+1} f(a) \xi_i \xi_j = \int_0^1 \frac{t}{\delta} \left[ \sum_{i=0}^{n} \left(\frac{t}{\delta}\right)^i \xi_i \right]^2 d\beta(t) \geq 0.$$

Allowing $\delta$ to approach zero in (1) and (2) we obtain

$$\sum_{i=0}^{n} \sum_{j=0}^{n} (-1)^{i+j} f^{(i+j)}(a) \xi_i \xi_j \geq 0$$

$$\sum_{i=0}^{n} \sum_{j=0}^{n} (-1)^{i+j+1} f^{(i+j+1)}(a) \xi_i \xi_j \geq 0 \qquad (n = 0, 1, 2, \cdots),$$

which is what we wished to prove.

## 12. Bernstein's Theorem

In this section we give a proof of Bernstein's theorem described in Section 1.

THEOREM 12a. *A necessary and sufficient condition that $f(x)$ should be completely monotonic in $0 \leq x < \infty$ is that*

(1) $$f(x) = \int_0^\infty e^{-xt} d\alpha(t),$$

*where $\alpha(t)$ is bounded and non-decreasing and the integral converges for $0 \leq x < \infty$.*

To prove that the condition is sufficient we have only to observe that if (1) is valid, then

$$(-1)^k f^{(k)}(x) = \int_0^\infty e^{-xt} t^k d\alpha(t) \geq 0 \qquad (x > 0; k = 0, 1, 2, \cdots)$$

$$f(0) = f(0+) = \alpha(\infty) - \alpha(0).$$

Conversely, let $f(x)$ be a completely monotonic function in $0 \leq x < \infty$. Then the sequence $\{f(n)\}_0^\infty$ is completely monotonic by Theorem 11d. Hence by Hausdorff's theorem, there exists a bounded non-decreasing function $\beta(t)$ defined in $(0 \leq t \leq 1)$ such that

$$f(n) = \int_0^1 t^n d\beta(t) = \int_{0+}^1 t^n d\beta(t) \qquad (n = 1, 2, \cdots)$$

$$= \int_0^\infty e^{-nt} d\alpha(t) = \lim_{R \to \infty} \int_0^R e^{-nt} d\alpha(t),$$

where
$$\alpha(t) = -\beta(e^{-t})$$
Set
$$g(x) = \int_0^\infty e^{-xt} d\alpha(t) \qquad (x \geq 0),$$

$G(x) = g(x+1), \qquad F(x) = f(x+1), \qquad \Phi(x) = G(x) - F(x).$

It is clear that
$$\Phi(n) = 0 \qquad (n = 0, 1, 2, \cdots).$$

Moreover $\Phi(x + iy)$ is certainly analytic for $x \geq 0$ Also

$$|G(x + iy)| \leq \int_0^\infty e^{-xt} e^{-t} d\alpha(t) = g(x+1) \leq g(1) \qquad (x \geq 0).$$

By Theorem 11c
$$|F(x + iy)| \leq F(x) \leq f(1) \qquad (x \geq 0).$$

Hence $|\Phi(x + iy)|$ is bounded in the half-plane $x \geq 0$.

We are now in a position to apply a theorem of F. Carlson [1921] to the effect that any function $f(x + iy)$ bounded and analytic in the half-plane $x \geq 0$ must vanish identically if it vanishes at the points $x = 0, 1, 2, \cdots$. That is,

$$f(x) = g(x) = \int_0^\infty e^{-xt} d\alpha(t) \qquad (x \geq 1).$$

From the definition of $\alpha(t)$ one sees that it is non-decreasing and bounded.* Since $f(x)$ and $g(x)$ are both analytic for $x > 0$ and continuous for $x \geq 0$, it follows that $f(x) = g(x)$ for $x \geq 0$.

**THEOREM 12b.** *A necessary and sufficient condition that $f(x)$ should be completely monotonic for $0 < x < \infty$ is that*

$$f(x) = \int_0^\infty e^{-xt} d\alpha(t),$$

*where $\alpha(t)$ is non-decreasing and the integral converges for $0 < x < \infty$.*

The proof of the sufficiency follows as before by successive differentiation.

To prove the necessity let $\delta$ be an arbitrary positive number and apply Theorem 12a to the function $f(x + \delta)$ Then

$$f(x + \delta) = \int_0^\infty e^{-xt} d\beta_\delta(t),$$

---
* The author is indebted to G. H. Hardy for suggesting this method of proof.

where $\beta_\delta(t)$ is a bounded non-decreasing function and the integral converges for $0 \leq x < \infty$. Then

(2) $$f(x) = \int_0^\infty e^{-xt}\, d\alpha(t)$$

$$\alpha(t) = \int_0^t e^{\delta u}\, d\beta_\delta(u).$$

The integral (2) converges for $x > \delta$. But $\alpha(t)$ is independent of $\delta$ by the uniqueness theorem. Hence (2) converges for $x > 0$. It is clear from its definition that $\alpha(t)$ is non-decreasing, and the proof is complete.

It is evident that a change of variable would make the theorems of this section applicable to an arbitrary interval of the form $(c, \infty)$.

THEOREM 12c. *A necessary and sufficient condition that $f(x)$ should be absolutely monotonic in $-\infty < x < 0$ is that*

$$f(x) = \int_0^\infty e^{xt}\, d\alpha(t),$$

*where $\alpha(t)$ is non-decreasing and the integral converges for $-\infty < x < 0$.*
This result follows from Theorem 12b by an obvious change of variable.

### 13. Alternative Proof of Bernstein's Theorem

We give here a second proof of the necessity of the condition of Theorem 12a. Under our hypothesis it is clear by Theorem 11d that for any positive integer $m$ the sequence $\left\{f\left(\dfrac{n}{m}\right)\right\}_{n=0}^\infty$ is completely monotonic. By Hausdorff's theorem

(1) $$f\left(\frac{n}{m}\right) = \int_0^1 t^n\, d\beta_m(t) \qquad (n = 0, 1, 2, \cdots),$$

where $\beta_m(t)$ is a non-decreasing bounded function, which we may assume to be normalized without loss of generality. Then

$$f(n) = \int_0^1 t^{mn}\, d\beta_m(t)$$

$$= \int_0^1 t^n\, d\beta_m(t^{1/m}).$$

By the uniqueness theorem

$$\beta_m(t^{1/m}) = \beta_1(t),$$

so that (1) becomes after a suitable change of variable*

(2)
$$f\left(\frac{n}{m}\right) = \int_0^1 t^{n/m}\, d\beta_1(t) = \int_{0+}^1 t^{n/m}\, d\beta_1(t)$$
$$= \int_0^\infty e^{-nt/m}\, d\alpha(t) \qquad (n, m = 1, 2, \cdots).$$

Here $\alpha(t)$ is the non-decreasing bounded function

$$\alpha(t) = -\beta_1(e^{-t}) \qquad (0 \leq t < \infty).$$

Since the functions $f(x)$ and

$$\int_0^\infty e^{-xt}\, d\alpha(t)$$

are both continuous for $0 \leq x < \infty$ it follows that

$$f(x) = \int_0^\infty e^{-xt}\, d\alpha(t)$$

in that interval. This completes the proof.†

## 14. Interpolation by Completely Monotonic Functions

We next discuss the relation between completely monotonic functions and completely monotonic sequences.‡ More explicitly, we wish to determine what sequences $\{a_n\}_0^\infty$ are of such a nature that there exists a completely monotonic function $f(x)$ taking on the values $a_n$ at the positive integral points:

$$f(n) = a_n \qquad (n = 0, 1, 2, \cdots).$$

In order to simplify the statement of the theorem we introduce

DEFINITION 14a. *A completely monotonic sequence* $\{a_n\}_0^\infty$ *is minimal if it ceases to be completely monotonic when $a_0$ is decreased.*

An example of such a sequence is $\{n^{-1}\}_1^\infty$. For if the first element of the sequence is replaced by $(1 - \epsilon)$ where $\epsilon > 0$, we have

$$(-1)^n \Delta^n a_0 = \frac{1}{n+1} - \epsilon \qquad (n = 0, 1, 2, \cdots)$$

---

\* It should be observed that we do not assume $\beta_1(t)$ to be continuous at $t = 0$, though this could be proved. Since $n$ and $m$ are both positive in (2) it is not necessary to compute $\beta_1(0+)$.

† The author is indebted to J. D. Tamarkin for pointing out that the above method of proof is essentially contained in Hausdorff [1921b].

‡ Compare D. V. Widder [1931] p. 882.

This ceases to be positive if $n$ is sufficiently large. On the other hand the sequence $2, 2^{-1}, 3^{-1}, 4^{-1}, \cdots$ is not minimal, for it remains completely monotonic when the first term is decreased to unity.

**THEOREM 14a.** *A completely monotonic sequence $\{a_n\}_0^\infty$ is minimal if and only if*

$$a_n = \int_0^1 t^n \, d\chi(t) \qquad (n = 0, 1, 2, \cdots),$$

*where $\chi(t)$ is a non-decreasing bounded function continuous at $t = 0$.*

We prove first the necessity of the condition. Let $\chi(0) = 0, \chi(0+) = b > 0$. We show that $\{a_n\}_0^\infty$ is not minimal. Set

$$\psi(t) = \chi(t) - \chi(0+) \qquad (0 < t \leq 1)$$

$$\psi(0) = 0$$

Then

$$\int_0^1 t^n \, d\psi(t) = \int_0^1 t^n \, d\chi(t) = a_n \qquad (n = 1, 2, 3, \cdots)$$

$$\int_{0+}^1 d\psi(t) = \chi(1) - \chi(0+) = a_0 - b.$$

Since $\psi(t)$ is itself non-decreasing, the set $a_0 - b, a_1, a_2, \cdots$ is itself completely monotonic, so that the given set cannot have been minimal.

Conversely, if $\chi(0) = \chi(0+) = 0$, then $\{a_n\}_0^\infty$ is minimal. For, it is easily seen that when $\chi(0) = \chi(0+) = 0$

$$\lim_{n \to \infty} (-1)^n \Delta^n a_0 = \int_0^1 (1 - t)^n \, d\chi(t) = 0.$$

If $b_0 = a_0 - \epsilon, b_1 = a_1, b_2 = a_2, \cdots$, then

$$(-1)^n \Delta^n b_0 = (-1)^n \Delta^n a_0 - \epsilon,$$

a number which is clearly negative for $n$ sufficiently large. This completes the proof of the theorem.

**THEOREM 14b** *A necessary and sufficient condition that there should exist a function $f(x)$ completely monotonic in $0 \leq x < \infty$ such that*

$$f(n) = a_n \qquad (n = 0, 1, 2, \cdots)$$

*is that $\{a_n\}_0^\infty$ should be a minimal completely monotonic sequence.*

The condition is sufficient. For, by Theorem 14a, if $\{a_n\}_0^\infty$ is minimal

$$a_n = \int_0^1 t^n \, d\chi(t) = \int_{0+}^1 t^n \, d\chi(t) \qquad (n = 0, 1, 2, \cdots),$$

where $\chi(t)$ is non-decreasing in $0 \leq t \leq 1$ and continuous at $t = 0$. If we set

(1) $$f(x) = \int_0^\infty e^{-xt}\, d\alpha(t)$$
$$\alpha(t) = -\chi(e^{-t}),$$

clearly $f(x)$ is completely monotonic in $0 \leq x < \infty$, and

$$f(n) = a_n \qquad (n = 1, 2, \cdots)$$
$$f(0) = \chi(1) - \chi(0+) = a_0.$$

Suppose, conversely, that $f(x)$ is completely monotonic for $x \geq 0$. Then it has the representation (1), the integral converging for $x \geq 0$ and $\alpha(t)$ being non-decreasing and bounded by Theorem 12a. Then

$$f(n) = a_n = \int_0^\infty e^{-nt}\, d\alpha(t) \qquad (n = 0, 1, 2, \cdots),$$

and $\{a_n\}_0^\infty$ is a completely monotonic sequence by Theorem 11d. To show that it is also minimal we have

$$a_n = \lim_{R \to \infty} \int_0^R e^{-nt}\, d\alpha(t) = \int_{0+}^1 t^n\, d\beta(t) \qquad (n = 0, 1, 2, \cdots)$$
$$\beta(t) = -\alpha(\log t^{-1}) \qquad (0 < t \leq 1).$$

The function $\beta(t)$ is undefined at $t = 0$. If $\beta(0)$ is defined as $-\alpha(\infty)$, we have $\beta(0) = \beta(0+)$ and

$$a_n = \int_0^1 t^n\, d\beta(t) \qquad (n = 0, 1, 2, \cdots).$$

The proof is now completed by an appeal to Theorem 14a.

## 15. Absolutely Monotonic Functions with Prescribed Derivatives at a Point

We treat in this section a related problem, the determination of an absolutely monotonic function with prescribed derivatives at a point.*

THEOREM 15. *A necessary and sufficient condition that there should exist at least one absolutely monotonic function $f(x)$ on the interval $-\infty < x \leq 0$ such that*

$$f^{(n)}(0-) = a_n \qquad (n = 0, 1, 2, \cdots)$$

---

* A proof of Theorem 15 independent of the Stieltjes moment problem was given by S. Bernstein [1928].

166    ABSOLUTELY MONOTONIC FUNCTIONS    [CH. IV

*is that the quadratic forms*

$$(1) \qquad \sum_{i=0}^{n}\sum_{j=0}^{n} a_{i+j}\xi_i\xi_j, \qquad \sum_{i=0}^{n}\sum_{j=0}^{n} a_{i+j+1}\xi_i\xi_j \qquad (n = 0, 1, 2, \cdots)$$

*should all be positive.*

The condition is necessary. For, if $f(x)$ is absolutely monotonic in $-\infty < x \leqq 0$, then

$$(2) \qquad f(x) = \int_0^\infty e^{xt}\, d\alpha(t) \qquad (-\infty < x \leqq 0),$$

where $\alpha(t)$ is non-decreasing. If

$$\int_0^\infty t^k\, d\alpha(t) = \infty$$

for some integer $k$, we should have

$$\lim_{x \to 0-} f^{(k)}(x) = \infty$$

contrary to assumption. Hence by Theorem 4.3 of Chapter II

$$(3) \qquad a_n = \int_0^\infty t^n\, d\alpha(t) \qquad (n = 0, 1, 2, \cdots),$$

and the forms (1) must be positive by the theorem of Stieltjes, i.e., Theorem 13a of Chapter III.

Conversely, if these forms are positive, then by the theorem of Stieltjes there exists a function $\alpha(t)$ which is non-decreasing and such that (3) holds. Define $f(x)$ by (2): it is clearly absolutely monotonic on $(-\infty, 0)$ and its successive derivatives take on the prescribed values.

COROLLARY 15. *A sufficient condition that there should exist at least one absolutely monotonic function $f(x)$ on the interval $-\infty < x \leqq 0$ such that*

$$(4) \qquad f^{(n)}(0) = a_n \qquad (n = 0, 1, 2, \cdots)$$

*is that*

$$a_0 > 0, \qquad \begin{vmatrix} a_0 & a_1 \\ a_1 & a_2 \end{vmatrix} > 0, \qquad \begin{vmatrix} a_0 & a_1 & a_2 \\ a_1 & a_2 & a_3 \\ a_2 & a_3 & a_4 \end{vmatrix} > 0, \cdots;$$

$$a_1 > 0, \qquad \begin{vmatrix} a_1 & a_2 \\ a_2 & a_3 \end{vmatrix} > 0, \qquad \begin{vmatrix} a_1 & a_2 & a_3 \\ a_2 & a_3 & a_4 \\ a_3 & a_4 & a_5 \end{vmatrix} > 0, \cdots.$$

## 16. Hankel Determinants whose Elements are the Derivatives of an Absolutely Monotonic Function

A result which is closely related to the previous one is the following.

**THEOREM 16.** *If $f(x)$ is absolutely monotonic in $-\infty < x < 0$, then for any negative $x$*

(1)
$$f(x) \geq 0, \quad \begin{vmatrix} f(x) & f'(x) \\ f'(x) & f''(x) \end{vmatrix} \geq 0, \quad \begin{vmatrix} f(x) & f'(x) & f''(x) \\ f'(x) & f''(x) & f'''(x) \\ f''(x) & f'''(x) & f^{(4)}(x) \end{vmatrix} \geq 0, \cdots;$$

$$f'(x) \geq 0, \quad \begin{vmatrix} f'(x) & f''(x) \\ f''(x) & f'''(x) \end{vmatrix} \geq 0, \quad \begin{vmatrix} f'(x) & f''(x) & f'''(x) \\ f''(x) & f'''(x) & f^{(4)}(x) \\ f'''(x) & f^{(4)}(x) & f^{(5)}(x) \end{vmatrix} \geq 0, \cdots.$$

For, by Bernstein's theorem

$$f(x) = \int_0^\infty e^{xt} d\alpha(t) \qquad (-\infty < x < 0)$$

with $\alpha(t)$ non-decreasing. Hence

(2)
$$\sum_{i=0}^n \sum_{j=0}^n f^{(i+j)}(x)\xi_i\xi_j = \int_0^\infty e^{xt}\left(\sum_{i=0}^n \xi_i t^i\right)^2 d\alpha(t) \geq 0$$

$$\sum_{i=0}^n \sum_{j=0}^n f^{(i+j+1)}(x)\xi_i\xi_j = \int_0^\infty e^{xt} t\left(\sum_{i=0}^n \xi_i t^i\right)^2 d\alpha(t) \geq 0.$$

Since the forms (2) are positive, the determinants (1) are non-negative.

If $f(x)$ is completely monotonic for $0 \leq x < \infty$ and if in the determinants (1) the function $f^{(k)}(x)$ is replaced by $(-1)^k f^{(k)}(x)$, the determinants are then non-negative for all positive $x$.

**COROLLARY 16.** *If $f(x)$ is absolutely monotonic in $-\infty < x \leq 0$ and not identically zero, then $f(x)$ is logarithmically convex for negative $x$; if $f(x)$ is not a constant, then*

$$\frac{f(x)}{f'(x)} \geq \frac{f'(x)}{f''(x)} \geq \frac{f''(x)}{f'''(x)} \geq \cdots.$$

For, to prove that $f(x)$ is logarithmically convex we have only to show that

(3)
$$\frac{d^2 \log f(x)}{dx^2} = \frac{f(x)f''(x) - f'(x)f'(x)}{f^2(x)} \geq 0.$$

But this follows from (1).

If $f(x)$ is not a constant then no derivative of $f(x)$ can vanish for negative $x$. For, if it did the analytic non-decreasing character of that derivative would then make it identically zero. That is, $f(x)$ would be a polynomial

$$f(x) = \sum_{k=0}^{n} a_k x^k$$

with non-negative coefficients. But such a function or its derivative would become negative in the neighborhood of $x = -\infty$ unless constant. Hence we may write (3) as

$$\frac{f(x)}{f''(x)} \geqq \frac{f'(x)}{f'''(x)}.$$

Applying the same argument successively to $f'(x), f''(x), \cdots$ we obtain the desired result.*

## 17. Laguerre Polynomials

We wish now to give another proof of Theorem 12a by use of Laguerre polynomials.† In this section we collect certain facts about these polynomials which we shall need. We take as our definition of the Laguerre polynomial $L_n(x)$ of degree $n$

$$(1) \qquad L_n(x) = \frac{e^x}{n!} \frac{d^n}{dx^n} [e^{-x} x^n]$$

$$(2) \qquad = \sum_{k=0}^{n} \binom{n}{k} \frac{(-x)^k}{k!}.$$

From this definition it is seen by use of integration by parts that the Laguerre polynomials form a normal orthogonal set with respect to the weight function $e^{-x}$:

$$(3) \qquad \int_0^\infty e^{-x} L_n(x) L_m(x)\, dx = 0 \qquad m \neq n$$

$$= 1 \qquad m = n.$$

From (1) it is clear, by use of Rolle's theorem that $L_n(x)$ has exactly $n$ real positive roots. We prove first:‡

THEOREM 17a. *For any non-negative value of $x$*

$$(4) \qquad |L_n(x)| \leqq e^{x/2} \qquad (n = 0, 1, 2, \cdots).$$

---

* Compare S. Bernstein [1928] and G. Doetsch [1937b] p. 270.
† See D. V. Widder [1935].
‡ Compare G. Szegö [1939] p. 159. The theorem was proved first by G. Szegö [1918].

§17] LAGUERRE POLYNOMIALS 169

This is a consequence of the fact that the weight function $e^{-x}$ is monotonic. Let the $n$ roots of $L_n(x)$ be $x_1, x_2, \cdots, x_n$ arranged in order of increase. Then integration by parts gives

$$\int_0^x e^{-t}[L_n(t)]^2 \, dt = 1 - e^{-x}[L_n(x)]^2 + 2 \int_0^x e^{-t} L_n(t) L_n'(t) \, dt.$$

We need only show that

$$F(x) = \int_0^x e^{-t} L_n(t) L_n'(t) \, dt = -\int_x^\infty e^{-t} L_n(t) L_n'(t) \, dt \leq 0 \quad (0 < x < \infty).$$

Since $L_n(t) L_n'(t)$ is clearly negative for $(0 \leq t < x_1)$ and positive for $(x_n < t < \infty)$ the result is obvious in these intervals. But if $2 \leq k \leq n$

$$F(x) = \int_0^x e^{-t} [Q(t)]^2 \left[\frac{L_n(t)}{Q(t)}\right] \left[\frac{L_n(t)}{Q(t)}\right]' dt$$
$$+ \int_0^x e^{-t} Q(t) Q'(t) \left[\frac{L_n(t)}{Q(t)}\right]^2 dt$$

(5) $\quad F(x) = -\int_x^\infty e^{-t} [Q(t)]^2 \left[\frac{L_n(t)}{Q(t)}\right] \left[\frac{L_n(t)}{Q(t)}\right]' dt$
$$+ \int_0^x e^{-t} Q(t) Q'(t) \left[\frac{L_n(t)}{Q(t)}\right]^2 dt$$

$$Q(t) = (t - x_k)(t - x_{k+1}) \cdots (t - x_n)$$

The integrand in the first integral (5) is positive for $t > x_{k-1}$, that in the second is negative for $t < x_k$, so that $F(x) \leq 0$ for $x_{k-1} \leq x \leq x_k$. Thus the result is completely established.

THEOREM 17b.* *The series*

$$K(x, y, t) = \sum_{k=0}^\infty L_n(x) L_n(y) t^n$$

*converges for* $(0 \leq x < \infty)$, $(0 \leq y < \infty)$ *and* $(0 \leq t < 1)$ *to the value*

$$K(x, y, t) = \frac{1}{1-t} e^{-(x+y)t/(1-t)} J_0\left(\frac{2\sqrt{-xyt}}{1-t}\right).$$

Here $J_0(x)$ is Bessel's function

$$J_0(x) = \sum_{n=0}^\infty \frac{(-1)^n}{n! n!} \left(\frac{x}{2}\right)^{2n}$$

* Compare G. Szegö [1939] p. 98, where references are given. The present proof follows G. H. Hardy.

We evaluate first the generating function for the Laguerre polynomials

(6) $$k(x, t) = \sum_{n=0}^{\infty} L_n(x) t^n.$$

By virtue of Theorem 17a the series converges for $(0 \leq x < \infty, |t| < 1)$. The Laplace transform of $L_n(x)$ is easily obtained from (2),

(7) $$\int_0^{\infty} e^{-sx} L_n(x)\, dx = \frac{(s-1)^n}{s^{n+1}} \qquad (\sigma > 0),$$

so that by integration of (6), easily justified, we have

$$\int_0^{\infty} e^{-sx} k(x, t)\, dx = \frac{1}{s - ts + t} \qquad (\sigma > 1/2).$$

But

$$\frac{1}{s - ts + t} = \frac{1}{1 - t} \int_0^{\infty} e^{-sx} e^{-tx/(1-t)}\, dx \qquad (\sigma > 0).$$

Hence by the uniqueness theorem

$$k(x, t) = \frac{1}{1-t} e^{-tx/(1-t)} \qquad (0 \leq x < \infty, |t| < 1).$$

We apply the same method to $K(x, y, t)$. By (7)

$$\int_0^{\infty} e^{-sx} K(x, y, t)\, dx = \frac{1}{s} \sum_{n=0}^{\infty} L_n(y) \left(\frac{ts - t}{s}\right)^n$$

$$= \frac{1}{s} k\left(y, \frac{ts - t}{s}\right)$$

(8) $$= \frac{1}{s - st + t} e^{-(s-1)ty/(s-st+t)} \qquad (\sigma > 1/2).$$

On the other hand

$$\int_0^{\infty} e^{-sx} J_0\left(\frac{2\sqrt{-xyt}}{1-t}\right) dx = \int_0^{\infty} e^{-sx} \sum_{n=0}^{\infty} \frac{1}{n!n!} \frac{(xyt)^n}{(1-t)^{2n}} dx$$

$$= \sum_{n=0}^{\infty} \frac{1}{n!} \frac{(yt)^n}{s^{n+1}(1-t)^{2n}}$$

$$= \frac{1}{s} e^{yt/s(1-t)^2} \qquad (\sigma > 0.)$$

Replacing $s$ by $s + t(1-t)^{-1}$ in this last equation and multiplying it by $(1-t)^{-1} e^{-yt(1-t)}$, we obtain

$$\text{(9)} \quad \frac{1}{1-t} \int_0^\infty e^{-sx} e^{-(x+y)t/(1-t)} J_0\left(\frac{2\sqrt{-xyt}}{1-t}\right) dx \\ = \frac{1}{s - st + t} e^{-(s-1)ty/(s-st+t)}$$

at least for $\sigma > 0$. But (8) and (9) give us two determining functions for the same generating function. Equating determining functions gives the desired results. There is no difficulty in justifying the term by term integration employed.

## 18. A Linear Functional

Let $f(x)$ be completely monotonic for $0 \leq x < \infty$,

$$(-1)^k f^{(k)}(x) \geq 0 \qquad (k = 0, 1, 2, \cdots)$$

$$f(0+) < \infty.$$

We define a linear functional $I$ operating on functions of the form

$$e^{-x} P(x) = e^{-x} \sum_{k=0}^n a_k x^k$$

by the equation

$$\text{(1)} \qquad I[e^{-x} P(x)] = \sum_{k=0}^n a_k (-1)^k f^{(k)}(1).$$

Observe that if we knew $f(x)$ to be a generating function

$$f(x) = \int_0^\infty e^{-xt} d\alpha(t),$$

a fact which we shall prove here independently of previous proofs, we should have

$$I[e^{-x} P(x)] = \int_0^\infty e^{-t} P(t) d\alpha(t).$$

This interpretation of the operator $I$ may serve as a guide to the discovery of the results of the present section, results which we must *prove*, however, directly from the definition (1). We need the following preliminary result.

LEMMA 18 *If $P(x)$ is a polynomial which is non-negative for $(0 \leq x < \infty)$ then*

(2) $$P(x) = A^2(x) + B^2(x) + xC^2(x) + xD^2(x)$$

*where $A(x)$, $B(x)$, $C(x)$, and $D(x)$ are real polynomials.*

This is well-known,* but we give a proof for completeness. Clearly the coefficients of $P(x)$ must be real, and hence any imaginary roots must occur in conjugate imaginary pairs. Any real roots of odd multiplicity must be negative Hence

$$P(x) = \prod_{i=1}^{N} [(x - \alpha_i)^2 + \beta_i^2] \prod_{i=1}^{M} (x + \gamma_i),$$

where the $\alpha_i$, $\beta_i$, $\gamma_i$ are real (not necessarily distinct) and the $\gamma_i$ are positive. The result is trivial if there is just one factor in the second product. For, by Lemma 11, Chapter III, the first product, being a positive polynomial, is the sum of squares of polynomials, $Q^2(x) + R^2(x)$. Hence we have

$$[Q^2(x) + R^2(x)][x + \gamma] = [\sqrt{\gamma}Q]^2 + [\sqrt{\gamma}R]^2 + x\{Q^2 + R^2\}.$$

It is now clear that we have only to show that if $P(x)$ has the form (2) then $(x + \gamma)P(x)$ has the same form. But, if $P(x)$ is defined by (2), then

$$(x + \gamma)P(x) = [\gamma A^2 + \gamma B^2 + x^2 C^2 + x^2 D^2] + x[A^2 + B^2 + \gamma C^2 + \gamma D^2].$$

Again applying Lemma 11, Chapter III, we have

$$(x + \gamma)P(x) = L^2 + M^2 + x[Q^2 + R^2],$$

where $L$, $M$, $Q$ and $R$ are real polynomials. This completes the proof.

THEOREM 18a. *If $P(x)$ is a non-negative polynomial for $(0 \leq x < \infty)$ then $I[e^{-x}P(x)]$ is non-negative*

By virtue of the lemma we have only to show that for any real polynomial

$$P(x) = \sum_{k=0}^{n} a_k x^k$$

we have

$$I[e^{-x}P^2(x)] = \sum_{i=0}^{n} \sum_{j=0}^{n} (-1)^{i+j} f^{(i+j)}(1) a_i a_j \geq 0$$

$$I[e^{-x}xP^2(x)] = \sum_{i=0}^{n} \sum_{j=0}^{n} (-1)^{i+j+1} f^{(i+j+1)}(1) a_i a_j \geq 0.$$

But this follows from Theorem 11$f$.

* Compare G. Pólya and G. Szegö [1925], vol. 2, p. 82.

THEOREM 18b. *If $P(x)$ is a polynomial such that*

(3) $$P(x) < e^x \qquad (0 \leq x < \infty),$$

*then there exists a polynomial $Q_m(x)$ of the form*

$$Q_m(x) = \sum_{k=0}^{m} \frac{x^k}{k!}$$

*which is greater than $P(x)$ for $0 \leq x < \infty$.*

For, if $n$ is the degree of $P(x)$, there exists a number $R$ such that

(4) $$P(x) < Q_{n+1}(x) \qquad (x \geq R).$$

Since $Q_n(x)$ approaches $e^x$ uniformly in $(0 \leq x \leq R)$, for any positive $\epsilon$ there is a positive integer $q$ such that

$$e^x - \epsilon < Q_{n+q}(x) \qquad (0 \leq x \leq R).$$

Suppose the $\epsilon$ chosen so that

$$P(x) < e^x - \epsilon \qquad (0 \leq x \leq R),$$

which is possible by (3). Then

(5) $$P(x) < Q_{n+q}(x) \qquad (0 \leq x \leq R).$$

Since for each positive $x$ the sequence $\{Q_n(x)\}_0^\infty$ is increasing, we see from (4) that (5) also holds for $x \geq R$, and our theorem is proved.

THEOREM 18c. *If $P(x)$ is a polynomial and if*

$$P(x)e^{-x} \leq M \qquad (0 \leq x < \infty),$$

*then*

$$I[P(x)e^{-x}] \leq Mf(0).$$

Observe first that $M$ is not negative since $P(x)e^{-x}$ tends to zero as $x$ becomes infinite. If $\eta$ is an arbitrary positive number, then

$$P(x) < Me^x + \eta \qquad (0 \leq x < \infty).$$

By Theorem 18b there is a polynomial $Q_m(x)$ such that

$$P(x) < MQ_m(x) + \eta.$$

By Theorem 18a

$$I[e^{-x}P(x)] \leq M \sum_{k=0}^{m} (-1)^k \frac{f^{(k)}(1)}{k!} + \eta f(1).$$

The inequalities are only strengthened if the finite sum is replaced by the corresponding infinite series, which by Theorem 3a converges to $f(0)$. That is,

$$I[e^{-x}P(x)] \leq Mf(0) + \eta f(1).$$

Since $\eta$ was arbitrary, our result is established.

COROLLARY 18c. *If*

$$\lambda_n = I[e^{-x}L_n(x)] = \sum_{k=0}^{n}\binom{n}{k}\frac{f^{(k)}(1)}{k!},$$

*then*

$$|\lambda_n| \leq f(0) \qquad (n = 0, 1, 2, \cdots).$$

This follows at once from Theorem 17a.

THEOREM 18d. *The series*

$$A_t(y) = \sum_{n=0}^{\infty} \lambda_n L_n(y) t^n$$

*converges for* $(0 \leq y < \infty, 0 \leq t < 1)$, *and*

(6) $$0 \leq A_t(y) \leq \frac{f(0)e^{y/2}}{1-t} \qquad (0 \leq y < \infty, 0 \leq t < 1)$$

By Theorem 17a and Corollary 18c

$$|A_t(y)| \leq \sum_{n=0}^{\infty} f(0)e^{y/2}t^n = \frac{f(0)e^{y/2}}{1-t}$$

Since the series

$$\sum_{n=0}^{\infty} e^{-x}L_n(x)L_n(y)t^n$$

converges uniformly in $(0 \leq x < \infty)$ for fixed $y$ and $t$, and since $K(x, y, t)e^{-x}$, its sum, is non-negative by Theorem 17b, it follows that for any positive $\eta$ there is an integer $m_0$ such that for all $m > m_0$

$$\eta + \sum_{n=0}^{m} e^{-x}L_n(x)L_n(y)t^n > 0 \qquad (0 \leq x < \infty)$$

By Theorem 18c

$$\eta f(0) + \sum_{n=0}^{m} \lambda_n L_n(y)t^n \geq 0 \qquad (m > m_0)$$

$$\eta f(0) + A_t(y) \geq 0.$$

Since $\eta$ was arbitrary, our result is established.

## 19. Bernstein's Theorem

We can now give a new proof of Bernstein's theorem. We assume that the functions $f(x)$ and $A_t(y)$ are defined as in section 18 and prove:

**THEOREM 19a.** *The integral*

$$\int_0^\infty e^{-xy} A_t(y)\, dy$$

*converges for $x > 1/2$ and*

$$f(x) = \lim_{t \to 1-} \int_0^\infty e^{-xy} A_t(y)\, dy \qquad (1 \leq x < 2).$$

By §17 (7) we have

$$\begin{aligned}
\int_0^\infty e^{-xy} A_t(y)\, dy &= \int_0^\infty e^{-xy} \sum_{n=0}^\infty \lambda_n L_n(y) t^n\, dy \\
&= \sum_{n=0}^\infty \lambda_n t^n \frac{(x-1)^n}{x^{n+1}} \qquad (1/2 < x).
\end{aligned} \tag{1}$$

To justify the term-by-term integration we have only to employ Theorem 17a and Corollary 18c and observe that the integral

$$\int_0^\infty e^{-xy} e^{y/2}\, dy$$

converges for $x > 1/2$.

The series (1) can be transformed as follows:

$$\begin{aligned}
\sum_{n=0}^\infty \lambda_n t^n \frac{(x-1)^n}{x^{n+1}} &= \frac{1}{x} \sum_{n=0}^\infty \left(t - \frac{t}{x}\right)^n \sum_{k=0}^n \binom{n}{k} \frac{f^{(k)}(1)}{k!} \\
&= \frac{1}{x} \sum_{k=0}^\infty \frac{f^{(k)}(1)}{k!} \sum_{n=k}^\infty \binom{n}{k} \left(t - \frac{t}{x}\right)^n \\
&= \sum_{k=0}^\infty \frac{f^{(k)}(1)}{k!} \frac{(tx - t)^k}{(x + t - tx)^{k+1}} \qquad (1 \leq x < 2).
\end{aligned} \tag{2}$$

To justify the interchange in the order of summation we need only show that the double series converges absolutely. This follows since

$$\sum_{n=0}^\infty \left(t - \frac{t}{x}\right)^n \sum_{k=0}^n \binom{n}{k} \frac{f^{(k)}(1)}{k!} \ll \sum_{n=0}^\infty \left(t - \frac{t}{x}\right)^n 2^n f(0). \tag{3}$$

Here we have used the obvious inequality

$$0 \leq (-1)^k \frac{f^{(k)}(1)}{k!} \leq f(0) \qquad (k = 0, 1, 2, \ldots).$$

But the dominant series (3) converges for $(1 \leq x < 2, 0 \leq t < 1)$. Finally, by letting $t$ approach unity in (2) we obtain

$$\lim_{t \to 1-} \int_0^\infty e^{-xy} A_t(y) \, dy = \sum_{k=0}^\infty \frac{f^{(k)}(1)}{k!} (x - 1)^k = f(x).$$

This step is justified since the series (2) converges uniformly in $(0 \leq t \leq 1)$, as one sees by the relation

$$\sum_{k=0}^\infty \frac{f^{(k)}(1)}{k!} \frac{(tx - t)^k}{(x + t - tx)^{k+1}} \ll \sum_{k=0}^\infty (-1)^k f^{(k)}(1) \frac{(x - 1)^k}{k!}$$

$$(1 \leq x < 2, 0 \leq t \leq 1).$$

This completes the proof.

THEOREM 19b. *If $f(x)$ is completely monotonic for $0 \leq x < \infty$, then*

$$f(x) = \int_0^\infty e^{-xt} \, d\alpha(t) \qquad (0 \leq x < \infty),$$

*where $\alpha(t)$ is bounded and non-decreasing in $(0 \leq t < \infty)$.*

For, set

$$\alpha_t(x) = \int_0^x e^{-y} A_t(y) \, dy \qquad (0 \leq x < \infty).$$

By Theorem 18d, $\alpha_t(x)$ is non-decreasing, and by (1)

$$0 \leq \alpha_t(x) \leq \lambda_0 = f(1).$$

By Helly's theorem we can choose a sequence of positive numbers $\{t_i\}_0^\infty$ tending to unity and a bounded non-decreasing function $\beta(x)$ such that

$$\lim_{i \to \infty} \alpha_{t_i}(x) = \beta(x) \qquad (0 \leq x < \infty).$$

Then by Theorem 19a

$$f(x) = \lim_{i \to \infty} \int_0^\infty e^{-(x-1)y} \, d\alpha_{t_i}(y) = \lim_{t \to 1-} \int_0^\infty e^{-xy} A_t(y) \, dy \quad (1 \leq x < 2).$$

In the usual way this gives

$$f(x) = \int_0^\infty e^{-(x-1)y} \, d\beta(y)$$

(4)
$$= \int_0^\infty e^{-xy} \, d\alpha(y) \qquad (1 \leq x < 2),$$

where

$$\alpha(y) = \int_0^y e^u \, d\beta(u) \qquad (y \geq 0).$$

But the integral (4) represents an analytic function at least for $x > 1$. Since $f(x)$ is completely monotonic for $0 < x < \infty$ it is analytic there Consequently the integral (4) must converge for $x > 0$ by Theorem 5b of Chapter II. Finally, we see that $\alpha(y)$ is bounded, for otherwise $\alpha(\infty) = \infty$. This would imply, since $\alpha(y)$ is monotonic, that $f(0+) = \infty$, contrary to hypothesis. This completes the proof.

## 20. Completely Convex Functions

By analogy with the completely monotonic functions of S. Bernstein we introduce the class of completely convex functions according to the following definition.

DEFINITION 20  *A function $f(x)$ is completely convex in an interval $(a, b)$ if it has derivatives of all orders there and if*

$$(-1)^k f^{(2k)}(x) \geqq 0 \qquad (k = 0, 1, 2, \cdots)$$

*in that interval.*

For example the functions $\sin x$ and $\cos x$ are completely convex in the intervals $(0, \pi)$ and $\left(-\dfrac{\pi}{2}, \dfrac{\pi}{2}\right)$ respectively. We shall show that any such function is necessarily entire.* We need two preliminary results. The first is a familiar result of J. Hadamard.†

LEMMA 20a.  *If $f(x)$ is of class $C^2$ in an interval $(a, b)$ and if*

$$M_k = \underset{a \leqq x \leqq b}{\text{u.b.}} |f^{(k)}(x)| \qquad (k = 0, 1, 2),$$

*then*

(1) $$M_1 \leqq \frac{2M_0}{b - a} + \frac{M_2}{2}(b - a).$$

For, by Taylor's theorem we have

$$f(b) - f(x) = (b - x)f'(x) + f''(\xi)\frac{(b - x)^2}{2} \qquad (x < \xi < b)$$

$$f(a) - f(x) = (a - x)f'(x) + f''(\eta)\frac{(a - x)^2}{2} \qquad (a < \eta < x).$$

Subtracting these two equations gives

$$f(b) - f(a) = (b - a)f'(x) + f''(\xi)\frac{(b - x)^2}{2} - f''(\eta)\frac{(a - x)^2}{2}.$$

---
* See D V Widder [1940]. The proof here presented was given later by R. P. Boas.
† See T. Carleman [1926] p. 11.

178    ABSOLUTELY MONOTONIC FUNCTIONS    [Ch. IV

From this, inequality (1) is immediate if we choose for $x$ a value at which $f'(x)$ attains its maximum $M_1$.

LEMMA 20b.  *If $f(x)$ is of class $C^2$ and if $f(x)$ and $-f''(x)$ are non-negative in $(a, b)$, then*

(2) $$f(x) \leq \frac{2}{b-a} \int_a^b f(x)\, dx \qquad (a \leq x \leq b).$$

For, suppose that

$$\underset{a \leq x \leq b}{\text{u.b.}}\, f(x) = f(c) \qquad (a \leq c \leq b)$$

Then by comparing the area under the curve $y = f(x)$ with the area of the triangle with vertices $(a, 0)$, $(b, 0)$, $(c, f(c))$ we obtain (2).

We are now in a position to prove the result stated above.

THEOREM 20.  *If $f(x)$ is completely convex in an interval $(a, b)$, it is entire.*

By consideration of the function $f(a + (b - a)x)$ one sees that there is no restriction in replacing the given interval by the interval $(0, 1)$. Integration by parts gives

$$\int_0^1 f(x) \sin \pi x\, dx = \frac{f(1) + f(0)}{\pi} - \frac{1}{\pi^2} \int_0^1 f''(x) \sin \pi x\, dx,$$

so that

$$-\frac{1}{\pi^2} \int_0^1 f''(x) \sin \pi x\, dx \leq \int_0^1 f(x) \sin \pi x\, dx.$$

Since the function $-f''(x)$ is also completely convex the same result gives

$$\frac{1}{\pi^4} \int_0^1 f^{(4)}(x) \sin \pi x\, dx \leq -\frac{1}{\pi^2} \int_0^1 f''(x) \sin \pi x\, dx.$$

$$\leq \int_0^1 f(x) \sin \pi x\, dx.$$

Repeating the process we obtain

$$\frac{(-1)^k}{\pi^{2k}} \int_0^1 f^{(2k)}(x) \sin \pi x\, dx \leq \int_0^1 f(x) \sin \pi x\, dx = A.$$

Now let $\delta$ be any positive number less than $\frac{1}{2}$. Then

$$0 \leq (-1)^k \int_\delta^{1-\delta} f^{(2k)}(x)\, dx \leq \frac{A \pi^{2k}}{\sin \pi \delta},$$

and by Lemma 20b

$$(-1)^k f^{(2k)}(x) \leq \frac{2A\pi^{2k}}{(1-2\delta)\sin \pi\delta} \qquad (\delta \leq x \leq 1-\delta).$$

By Lemma 20a

$$(-1)^{k+1} f^{(2k+1)}(x) \leq \frac{4A\pi^{2k}}{(1-2\delta)^2 \sin \pi\delta} + \frac{A\pi^{2k+2}}{\sin \pi\delta}.$$

That is,

$$f^{(k)}(x) = O(\pi^k) \qquad (k \to \infty)$$

uniformly in $(\delta, 1-\delta)$. By use of Taylor's formula with remainder we recognize at once that the power series expansion of $f(x)$ must converge for all $x$ and converge to $f(x)$ in $(0, 1)$. This proves* the theorem.

* We have in fact proved a little more: that $f(x)$ is at most of order unity, type $\pi$.

# CHAPTER V

# TAUBERIAN THEOREMS

## 1. Abelian Theorems for the Laplace Transform

The theorems which we shall treat first are called Abelian because they are generalizations of a familiar result of Abel. This states that if $f(x)$ is defined by the series

$$f(x) = \sum_{n=0}^{\infty} a_n x^n \qquad (|x| < 1),$$

the series converging for $x$ numerically less than one, then

$$\lim_{x \to 1-} f(x) = \sum_{n=0}^{\infty} a_n$$

whenever the series on the right converges.

An integral analogue of this result is that if $f(s)$ is defined for real $s > 0$ by the convergent integral

$$f(s) = \int_0^\infty e^{-st} a(t)\, dt,$$

then

$$\lim_{s \to 0+} f(s) = \int_0^\infty a(t)\, dt,$$

provided the integral on the right converges. Another form of the result would be that if*

$$f(s) = \int_0^\infty e^{-st}\, d\alpha(t) \qquad (s > 0),$$

then

$$\lim_{s \to 0+} f(s) = \lim_{t \to \infty} \alpha(t)$$

whenever the limit on the right exists. It is this and similar results which we wish to prove in the present section. We shall assume, unless otherwise stated, that $s$ is a real variable.

---

\* We remind the reader that in Laplace-Stieltjes integrals the determining function $\alpha(t)$ is assumed to be normalized (see Section 8 of Chapter I).

THEOREM 1. *If*

(1) $$f(s) = \int_0^\infty e^{-st} \, d\alpha(t) \qquad (s > 0),$$

*then for any $\gamma \geq 0$ and any constant $A$*

(2) $$\varlimsup_{s \to 0+} |s^\gamma f(s) - A| \leq \varlimsup_{t \to \infty} |\alpha(t)t^{-\gamma}\Gamma(\gamma+1) - A|$$

(3) $$\varlimsup_{s \to \infty} |s^\gamma f(s) - A| \leq \varlimsup_{t \to 0+} |\alpha(t)t^{-\gamma}\Gamma(\gamma+1) - A|.$$

Since the integral (1) converges for $s > 0$, we have by Theorem 2.3a of Chapter II

$$f(s) = s \int_0^\infty e^{-st} \alpha(t) \, dt$$

$$s^\gamma f(s) - A = s^{\gamma+1} \int_0^\infty e^{-st} \left[ \alpha(t) - \frac{At^\gamma}{\Gamma(\gamma+1)} \right] dt.$$

If $T$ is any positive number, then

$$|s^\gamma f(s) - A| \leq s^{\gamma+1} \int_0^T e^{-st} \left| \alpha(t) - \frac{At^\gamma}{\Gamma(\gamma+1)} \right| dt$$
$$+ s^{\gamma+1} \int_T^\infty e^{-st} \left| \alpha(t) - \frac{At^\gamma}{\Gamma(\gamma+1)} \right| dt$$

(4) $$|s^\gamma f(s) - A| \leq \operatorname*{u.b.}_{0 \leq t \leq T} |\alpha(t)\Gamma(\gamma+1)t^{-\gamma} - A|$$
$$+ s^{\gamma+1} \int_T^\infty e^{-st} \left| \alpha(t) - \frac{At^\gamma}{\Gamma(\gamma+1)} \right| dt.$$

For any positive $\epsilon$ we can find* a constant $M$ such that

$$\left| \alpha(t) - \frac{At^\gamma}{\Gamma(\gamma+1)} \right| < Me^{\epsilon t} \qquad (0 \leq t < \infty),$$

whence

$$s^{\gamma+1} \int_T^\infty e^{-st} \left| \alpha(t) - \frac{At^\gamma}{\Gamma(\gamma+1)} \right| dt < \frac{Ms^{\gamma+1}}{s - \epsilon} e^{-(s-\epsilon)T} \qquad (s > \epsilon).$$

The right-hand side of this inequality approaches zero as $s$ becomes infinite, so that we have from (4)

$$\varlimsup_{s \to \infty} |s^\gamma f(s) - A| \leq \operatorname*{u.b.}_{0 \leq t \leq T} |\alpha(t)\Gamma(\gamma+1)t^{-\gamma} - A|.$$

---
* See Theorem 2.2a of Chapter II.

The left-hand side being independent of $T$ we allow $T$ to approach zero and obtain

$$\varlimsup_{s\to\infty} |s^\gamma f(s) - A| \leq \varlimsup_{t\to 0+} |\alpha(t)t^{-\gamma}\Gamma(\gamma+1) - A|.$$

We have thus established (3). In a similar way we have

$$|s^\gamma f(s) - A| \leq s^{\gamma+1}\int_0^T e^{-st}\left|\alpha(t) - \frac{At^\gamma}{\Gamma(\gamma+1)}\right|dt$$
$$+ \operatorname*{u.b.}_{T\leq t<\infty} |\alpha(t)\Gamma(\gamma+1)t^{-\gamma} - A|$$

$$\lim_{s\to 0+} |s^\gamma f(s) - A| \leq \operatorname*{u.b.}_{T\leq t<\infty} |\alpha(t)\Gamma(\gamma+1)t^{-\gamma} - A|,$$

from which (2) follows by allowing $T$ to become infinite.

COROLLARY 1a. *If for some non-negative number $\gamma$*

$$\alpha(t) \sim \frac{At^\gamma}{\Gamma(\gamma+1)} \qquad \begin{array}{l}(t\to\infty)\\(t\to 0+),\end{array}$$

*then*

$$f(s) \sim \frac{A}{s^\gamma} \qquad \begin{array}{l}(s\to 0+)\\(s\to\infty).\end{array}$$

Note that if $A = 0$ we must write

$$\alpha(t) = o(t^\gamma)$$
$$f(s) = o(s^{-\gamma}).$$

In particular we note that the existence of $\alpha(\infty)$ implies $f(0+) = \alpha(\infty)$. Also $f(\infty) = \alpha(0+)$.

COROLLARY 1b. *If (1) converges for $s > \sigma_c$ and if*

$$\alpha(t) \sim \frac{At^\gamma e^{\sigma_c t}}{\Gamma(\gamma+1)} \qquad (t\to\infty),$$

*then*

$$f(s) \sim \frac{A\sigma_c}{(s-\sigma_c)^{\gamma+1}} \qquad (s\to\sigma_c+).$$

For,

(5) $$f(s + \sigma_c) = \int_0^\infty e^{-st}d\beta(t) \qquad (s > 0)$$

$$\beta(t) = \int_0^t e^{-\sigma_c u}d\alpha(u) = \alpha(t)e^{-\sigma_c t} + \sigma_c\int_0^t e^{-\sigma_c u}\alpha(u)\,du.$$

§2]  ABELIAN THEOREMS  183

Under our present hypotheses it is easy to see that

$$\beta(t) \sim \frac{A\sigma_c t^{\gamma+1}}{\Gamma(\gamma+2)} \qquad (t \to \infty),$$

so that Corollary 1a is applicable to the integral (5).

**COROLLARY 1c.** *If* (1) *converges for* $s > 0$, *then*

$$\varliminf_{\substack{t\to\infty \\ t\to 0+}} \alpha(t) \leqq \varlimsup_{\substack{s\to 0+ \\ s\to\infty}} f(s) \leqq \varlimsup_{\substack{t\to\infty \\ t\to 0+}} \alpha(t).$$

The proof of this requires only slight modifications of the argument above and is left to the reader.

## 2. Abelian Theorems for the Stieltjes Transform

We consider here the equation

$$(1) \qquad f(s) = \int_0^\infty \frac{d\alpha(t)}{(s+t)^\rho} \qquad (\rho > 0).$$

It will be shown* in a later chapter that if (1) converges for a single positive value of $s$ it converges uniformly in any closed interval of the positive $s$ axis not including the origin and that then

$$(2) \qquad \alpha(t) = o(t^\rho) \qquad (t \to \infty).$$

In order to study the asymptotic behavior of $f(s)$ at $s = 0$ and at $s = \infty$ as determined by the properties of $\alpha(t)$ we need the following result.

**LEMMA 2.** *If the integral* (1) *converges for* $s > 0$, *and if* $T > 0$, *then*

$$(3) \qquad \lim_{s\to 0+} \int_T^\infty \frac{\alpha(t)}{(s+t)^{\rho+1}} \, dt = \int_T^\infty \frac{\alpha(t)}{t^{\rho+1}} \, dt.$$

It is clear from (2) and the convergence of (1) that the integral on the left-hand side of (3) converges. Since

$$(4) \qquad \int_T^\infty \frac{\alpha(t)}{(s+t)^{\rho+1}} \, dt = \int_0^\infty \frac{\alpha(t+T)}{(s+T+t)^{\rho+1}} \, dt,$$

it follows from the result stated above that (4) converges for $s + T > 0$ and hence uniformly in a neighborhood of $s = 0$, from which (3) follows.

**THEOREM 2a.** *If* (1) *converges for* $s > 0$ *and if* $0 < \gamma \leqq \rho$, *then for any constant* $A$

$$\varlimsup_{\substack{s\to 0+ \\ s\to\infty}} |s^\gamma f(s) - A| \leqq \varlimsup_{\substack{t\to 0+ \\ t\to\infty}} \left| \frac{\alpha(t)}{ct^{\rho-\gamma}} - A \right|,$$

---
* Theorem 2c and Corollary 3a3 of Chapter VIII.

where

$$c = \frac{\Gamma(\rho)}{\Gamma(\gamma)\Gamma(\rho - \gamma + 1)}.$$

For,

$$f(s) = \rho \int_0^\infty \frac{\alpha(t)}{(s+t)^{\rho+1}} dt \qquad (s > 0)$$

$$s^\gamma f(s) - A = s^\gamma \rho \int_0^\infty \frac{\alpha(t) - Act^{\rho-\gamma}}{(s+t)^{\rho+1}} dt.$$

This follows from the familiar formula

$$\int_0^\infty \frac{t^\alpha}{(s+t)^\beta} dt = \frac{\Gamma(\beta - \alpha - 1)\Gamma(\alpha + 1)}{\Gamma(\beta) s^{\beta-\alpha-1}} \qquad \begin{matrix}(\alpha > -1)\\(\beta > \alpha + 1).\end{matrix}$$

Hence if $T > 0$,

$$|s^\gamma f(s) - A| \leq \underset{0 \leq t \leq T}{\text{u.b.}} \left|\frac{\alpha(t)}{ct^{\rho-\gamma}} - A\right| + s^\gamma \rho \left|\int_T^\infty \frac{\alpha(t) - Act^{\rho-\gamma}}{(s+t)^{\rho+1}} dt\right|.$$

Making use of Lemma 2 we have

$$\overline{\lim_{s \to 0+}} |s^\gamma f(s) - A| \leq \underset{0 \leq t \leq T}{\text{u.b.}} \left|\frac{\alpha(t)}{ct^{\rho-\gamma}} - A\right|.$$

Since $T$ was arbitrary it follows that

$$\overline{\lim_{s \to 0+}} |s^\gamma f(s) - A| \leq \overline{\lim_{t \to 0+}} \left|\frac{\alpha(t)}{ct^{\rho-\gamma}} - A\right|.$$

Also,

$$|s^\gamma f(s) - A| \leq s^\gamma \rho \int_0^T \frac{|\alpha(t) - Act^{\rho-\gamma}|}{(s+t)^{\rho+1}} dt + \underset{T \leq t < \infty}{\text{u.b.}} \left|\frac{\alpha(t)}{ct^{\rho-\gamma}} - A\right|.$$

But

$$s^\gamma \rho \int_0^T \frac{|\alpha(t) - Act^{\rho-\gamma}|}{(s+t)^{\rho+1}} dt \leq \frac{\rho}{s^{\rho-\gamma+1}} \int_0^T |\alpha(t) - Act^{\rho-\gamma}| dt$$

$$= o(1) \qquad (s \to \infty).$$

Hence

$$\overline{\lim_{s \to \infty}} |s^\gamma f(s) - A| \leq \underset{T \leq t < \infty}{\text{u.b.}} \left|\frac{\alpha(t)}{ct^{\rho-\gamma}} - A\right|$$

$$\overline{\lim_{s \to \infty}} |s^\gamma f(s) - A| \leq \overline{\lim_{t \to \infty}} \left|\frac{\alpha(t)}{ct^{\rho-\gamma}} - A\right|.$$

This concludes the proof of the theorem.

COROLLARY 2a. *If for* $0 < \gamma \leqq \rho$

$$\alpha(t) \sim Act^{\rho-\gamma} \qquad \begin{matrix}(t \to \infty)\\(t \to 0+),\end{matrix}$$

*then*

$$f(s) \sim \frac{A}{s^\gamma} \qquad \begin{matrix}(s \to \infty)\\(s \to 0+).\end{matrix}$$

The case in which $\gamma = 0$ requires special attention. The mere convergence of (1) implies that $\alpha(t) = o(t^\rho)$ $(t \to \infty)$.

THEOREM 2b. *If* (1) *converges for* $s > 0$ *and if* $\rho > 0$, *then*

$$\lim_{s \to \infty} f(s) = 0.$$

If we set

$$\beta(t) = -\int_t^\infty \frac{d\alpha(u)}{(1+u)^\rho} \qquad (0 \leqq t < \infty),$$

then

$$f(s) = -\frac{\beta(0)}{s^\rho} + \rho(1-s)\int_0^\infty \frac{(t+1)^{\rho-1}}{(s+t)^{\rho+1}} \beta(t)\, dt.$$

For any $T > 0$ and any $s > 1$

$$\left| \rho(1-s) \int_0^\infty \frac{(t+1)^{\rho-1}}{(s+t)^{\rho+1}} \beta(t)\, dt \right|$$

$$\leqq \frac{\rho(s-1)}{s^{\rho+1}} \int_0^T (t+1)^{\rho-1} |\beta(t)|\, dt + \underset{T \leqq t < \infty}{\text{u.b.}} |\beta(t)|.$$

Hence

$$\varlimsup_{s \to \infty} \left| \rho(1-s) \int_0^\infty \frac{(t+1)^{\rho-1}}{(s+1)^{\rho+1}} \beta(t)\, dt \right| \leqq \varlimsup_{t \to \infty} |\beta(t)|.$$

Since $\beta(\infty) = 0$ the theorem is established.

## 3. Tauberian Theorems

The converse of Abel's theorem, stated in Section 1, is not true. That is, the series

(1) $$f(x) = \sum_{n=0}^\infty a_n x^n$$

may converge for $|x| < 1$ and $f(x)$ may tend to a limit as $x$ tends to 1 without having the series

$$\sum_{n=0}^{\infty} a_n$$

convergent. The example $f(x) = (x+1)^{-1}$ shows this.

Another example, showing the failure of the converse of Abel's theorem in its integral form, is

$$f(s) = \frac{s}{s^2+1} = \int_0^{\infty} e^{-st} d(\sin t) \qquad (s > 0).$$

Here

$$\lim_{s \to 0+} f(s) = 0.$$

Yet

$$\int_0^{\infty} d(\sin t)$$

diverges. It was A. Tauber [1897] who first gave a conditional converse of Abel's theorem. He proved that if (1) converges for $|x| < 1$, if $a_n = o(n^{-1})$ $(n \to \infty)$, and if $f(1-) = A$, then

$$A = \sum_{n=0}^{\infty} a_n.$$

We prove the integral analogue of this result.

THEOREM 3a. *If $a(t)$ belongs to $L$ in $(0, R)$ for every $R$ and if the integral*

$$f(s) = \int_0^{\infty} a(t) e^{-st} dt$$

*converges for $s > 0$, then the conditions*

$$\lim_{s \to 0+} f(s) = A$$

(1)
$$a(t) = o\left(\frac{1}{t}\right) \qquad (t \to \infty),$$

*imply*

$$f(0+) = \int_0^{\infty} a(t)\, dt = A.$$

To prove this we note first that under the hypotheses of the theorem we have

$$\frac{1}{t}\int_0^t |a(u)|\, u\, du = o(1) \qquad (t \to \infty).$$

Now consider the difference

$$\int_0^t a(u)\,du - \int_0^\infty a(u)e^{-u/t}\,du = \int_0^t a(u)[1 - e^{-u/t}]\,du - \int_t^\infty a(u)e^{-u/t}\,du$$
$$= I_1 + I_2.$$

Then

$$|I_1| \leq \frac{1}{t}\int_0^t |a(u)|\,u\,du = o(1) \qquad (t \to \infty),$$

$$|I_2| \leq \underset{t \leq u < \infty}{\text{u.b.}} |a(u)u| \int_t^\infty \frac{1}{u}e^{-u/t}\,du$$
$$\leq \underset{t \leq u < \infty}{\text{u.b.}} |a(u)u|.$$

Hence

$$\varlimsup_{t\to\infty} |I_2| \leq \varlimsup_{t\to\infty} |a(t)t| = 0.$$

It follows that

$$\int_0^\infty a(u)\,du = \lim_{s\to 0+} \int_0^\infty a(u)e^{-su}\,du = A.$$

This proves the theorem.

Note that in the example given above $a(t)$ was $\cos t$ which fails to satisfy the condition $a(t) = o(t^{-1})$.

A more general result is contained in:

THEOREM 3b. *Let*

$$f(s) = \int_0^\infty e^{-st}\,d\alpha(t)$$

*converge for $s > 0$, and let*

(2) $$\lim_{s\to 0+} f(s) = A.$$

*Then*

(3) $$\lim_{t\to\infty} \alpha(t) = A$$

*if and only if*

(4) $$\beta(t) = \int_0^t u\,d\alpha(u) = o(t) \qquad (t \to \infty).$$

Note that if

$$\alpha(t) = \int_0^t a(u)\,du,$$

then $a(t) = o(t^{-1})$, $(t \to \infty)$, implies that

$$\beta(t) = \int_0^t ua(u)\, du = o(t) \qquad (t \to \infty),$$

so that Theorem 3a is included in Theorem 3b.

To prove the theorem we assume without loss of generality that $A = 0$. First suppose that (3) holds and prove (4). This is obvious from the equation

$$\beta(t) = t\alpha(t) - \int_0^t \alpha(u)\, du.$$

Conversely, assume (4) and prove (3). Now

(5)
$$\int_1^\infty e^{-su}\, d\alpha(u) = \int_1^\infty \frac{e^{-su}}{u}\, d\beta(u)$$
$$= -\beta(1)e^{-s} + s\int_1^\infty \frac{\beta(u)}{u} e^{-su}\, du + \int_1^\infty \frac{\beta(u)}{u^2} e^{-su}\, du.$$

By Corollary 1a

$$\lim_{s \to 0+} s \int_1^\infty \frac{\beta(u)}{u} e^{-su}\, du = 0.$$

By (2)

$$\lim_{s \to 0+} \int_1^\infty e^{-su}\, d\alpha(u) = -\lim_{s \to 0+} \int_0^1 e^{-su}\, d\alpha(u) = -\alpha(1).$$

Consequently (5) shows that

(6)
$$\lim_{s \to 0+} \int_1^\infty e^{-su} \frac{\beta(u)}{u^2}\, du = \beta(1) - \alpha(1).$$

But by (4) we may apply Theorem 3a to the integral (6) and obtain

$$\int_1^\infty \frac{\beta(u)}{u^2}\, du = \beta(1) - \alpha(1).$$

Integrating by parts, we have

$$\alpha(\infty) = 0,$$

which is what we wished to prove.

If we choose $\alpha(t)$ a step-function with jumps at the positive integers this becomes a familiar result of A. Tauber [1897]. It states that a series $\sum_{k=0}^\infty a_k$ which is summable in the sense of Abel to the number $A$ converges to $A$ if and only if

$$\sum_{k=1}^n k a_k = o(n) \qquad (n \to \infty).$$

## 4. Karamata's Theorem

It was Littlewood [1910] who first showed that the condition $a_n = o(n^{-1})$ in Tauber's theorem could be replaced by the condition $a_n = O(n^{-1})$. We consider the integral analogue of the theorem and give an ingenious proof of J. Karamata [1931].

4.1. We shall prove first the following preliminary result.

LEMMA 4.1. *If $g(x)$ is continuous almost everywhere in $(0, 1)$ and is bounded there, and if $\gamma > 0$, then for every $\epsilon > 0$ there exist polynomials $p(x)$ and $P(x)$ such that*

$$p(x) < g(x) < P(x) \qquad (0 \leq x \leq 1)$$

$$\int_0^\infty e^{-t} t^{\gamma-1} [P(e^{-t}) - p(e^{-t})] \, dt < \epsilon.$$

First suppose that

$$g(x) = 1 \qquad (0 \leq \alpha < x < \beta \leq 1)$$
$$\phantom{g(x)} = 0 \qquad (0 \leq x \leq \alpha,\ \beta \leq x \leq 1).$$

Let $\eta$ be an arbitrary positive number. Clearly there exists a continuous function $h(x)$, which may coincide with $g(x)$ except in the neighborhood of the points $\alpha$ and $\beta$, such that

$$g(x) \leq h(x) \qquad (0 \leq x \leq 1)$$

$$\int_0^\infty e^{-t} [h(e^{-t}) - g(e^{-t})] t^{\gamma-1} \, dt < \eta.$$

By the Weierstrass approximation theorem there exists a polynomial $Q(x)$ such that

$$|Q(x) - h(x)| < \eta \qquad (0 \leq x \leq 1).$$

Set

$$P(x) = Q(x) + \eta;$$

then

$$g(x) \leq h(x) < P(x)$$

$$\int_0^\infty e^{-t} t^{\gamma-1} [P(e^{-t}) - g(e^{-t})] \, dt \leq \int_0^\infty e^{-t} t^{\gamma-1} |Q(e^{-t}) - h(e^{-t})| \, dt$$

$$+ \int_0^\infty e^{-t} t^{\gamma-1} [h(e^{-t}) - g(e^{-t})] \, dt$$

$$+ \eta \int_0^\infty e^{-t} t^{\gamma-1} \, dt$$

$$< \eta + 2\eta \int_0^\infty e^{-t} t^{\gamma-1} \, dt.$$

In a similar way we can determine the polynomial $p(x) < g(x)$ such that

$$\int_0^\infty e^{-t} t^{\gamma-1} [g(e^{-t}) - p(e^{-t})] dt < \eta + 2\eta \int_0^\infty e^{-t} t^{\gamma-1} dt.$$

Since the right-hand sides of these last two inequalities can be made arbitrarily small, our lemma is true for the step-function $g(x)$ just considered. It is consequently true for any step-function with a finite number of jumps, for any such function is a linear combination of functions of the type just treated.

We turn now to the general case. With a positive $\epsilon$ we can determine positive numbers $\delta$ and $R$ so that

(1) $$2M \int_0^\delta e^{-t} t^{\gamma-1} dt + 2M \int_R^\infty e^{-t} t^{\gamma-1} dt < \epsilon/6$$

$$M = \underset{0 \leq x \leq 1}{\text{u.b.}} |g(x)|.$$

Since $g(e^{-t}) e^{-t} t^{\gamma-1}$ is Riemann-integrable in $(\delta, R)$ we see from the definition of an integral that there exist two step-functions $g_1(x)$ and $g_2(x)$, each with a finite number of jumps such that

(2) $$g_1(e^{-t}) \leq g(e^{-t}) \leq g_2(e^{-t}) \qquad (\delta \leq t \leq R)$$

(3) $$\int_\delta^R [g_2(e^{-t}) - g_1(e^{-t})] e^{-t} t^{\gamma-1} dt < \epsilon/6.$$

If we complete the definition of $g_1(x)$ and $g_2(x)$ by the equations

$$g_2(e^{-t}) = M \qquad (0 \leq t < \delta, R < t < \infty)$$
$$g_1(e^{-t}) = -M \qquad (0 \leq t < \delta, R < t < \infty),$$

inequalities (2) hold for all positive $t$, and by the first part of the proof there exist polynomials $p(x)$ and $P(x)$ such that

$$p(x) < g_1(x) \leq g(x) \leq g_2(x) < P(x) \qquad (0 \leq x \leq 1)$$

(4) $$\int_0^\infty e^{-t} [P(e^{-t}) - g_2(e^{-t})] t^{\gamma-1} dt < \epsilon/3$$

(5) $$\int_0^\infty e^{-t} [g_1(e^{-t}) - p(e^{-t})] t^{\gamma-1} < \epsilon/3.$$

From (1) and (3) we have

$$\int_0^\infty e^{-t} [g_2(e^{-t}) - g_1(e^{-t})] t^{\gamma-1} dt < \epsilon/3.$$

Combining this with (4) and (5) there results the required inequality

$$\int_0^\infty e^{-t}[P(e^{-t}) - p(e^{-t})]t^{\gamma-1} dt < \epsilon.$$

4.2. We turn now to the proof of the following theorem due to Karamata.

THEOREM 4.2. *Let $\alpha(t)$ be non-decreasing and such that the integral*

$$f(s) = \int_0^\infty e^{-st} d\alpha(t)$$

*converges for $s > 0$, and for some positive number $\gamma$ let*

(1) $\qquad\qquad f(s) \sim \dfrac{1}{s^\gamma} \qquad\qquad \begin{matrix}(s \to 0+) \\ (s \to \infty).\end{matrix}$

*Let $g(x)$ be of bounded variation in $(0, 1)$. Then*

(2) $\qquad \displaystyle\int_0^\infty e^{-st} g(e^{-st}) \, d\alpha(t) \sim \dfrac{1}{s^\gamma} \dfrac{1}{\Gamma(\gamma)} \int_0^\infty e^{-t} g(e^{-t}) t^{\gamma-1} dt \qquad \begin{matrix}(s \to 0+) \\ (s \to \infty),\end{matrix}$

*$s$ varying through the set of points for which the integral on the left exists.*

Let $\alpha(t)$ be continuous except perhaps in the set of points $x_0$, $x_1$, $x_2$, $\cdots$ and let $g(e^{-t})$ be continuous except perhaps in the points $y_0$, $y_1$, $y_2$, $\cdots$. Denote by $E$ the set of points $y_i/x_j$ ($i, j = 0, 1, 2, \cdots$). If $s$ is not in $E$ the integral on the left of (1) exists since $\alpha(t)$ and $g(e^{-st})$ will have no common discontinuities. Since $E$ is a countable set its complement is dense in $(0, \infty)$ so that $s$ may approach zero or become infinite while remaining in the set of points for which the integral on the left of (2) exists.

Let $\epsilon$ be an arbitrary positive number. Determine polynomials (Lemma 4.1) $p(x)$ and $P(x)$ so that

$$p(x) < g(x) < P(x) \qquad (0 \le x \le 1)$$

(3) $\qquad\qquad \dfrac{1}{\Gamma(\gamma)} \displaystyle\int_0^\infty e^{-t} t^{\gamma-1}[P(e^{-t}) - p(e^{-t})] dt < \epsilon.$

Since $\alpha(t)$ is non-decreasing we have

(4) $\qquad \displaystyle\int_0^\infty e^{-st} p(e^{-st}) \, d\alpha(t) \le \int_0^\infty e^{-st} g(e^{-st}) \, d\alpha(t) \le \int_0^\infty e^{-st} P(e^{-st}) \, d\alpha(t).$

(5) $\qquad \dfrac{1}{\Gamma(\gamma)} \displaystyle\int_0^\infty e^{-t} p(e^{-t}) t^{\gamma-1} dt \le \dfrac{1}{\Gamma(\gamma)} \int_0^\infty e^{-t} g(e^{-t}) t^{\gamma-1} dt$

$$\le \dfrac{1}{\Gamma(\gamma)} \int_0^\infty e^{-t} P(e^{-t}) t^{\gamma-1} dt.$$

If we replace $s$ by $(n+1)s$ in (1) we obtain

$$\int_0^\infty e^{-st}e^{-nst}\,d\alpha(t) \sim \frac{1}{(n+1)^\gamma s^\gamma} = \frac{1}{\Gamma(\gamma)s^\gamma}\int_0^\infty e^{-t}e^{-nt}t^{\gamma-1}\,dt$$

for any positive integer $n$, whence

$$\int_0^\infty e^{-st}P(e^{-st})\,d\alpha(t) \sim \frac{1}{\Gamma(\gamma)s^\gamma}\int_0^\infty e^{-t}P(e^{-t})t^{\gamma-1}\,dt$$

for any polynomial $P(x)$. From (4) we have

(6)
$$\frac{1}{\Gamma(\gamma)}\int_0^\infty e^{-t}p(e^{-t})t^{\gamma-1}\,dt \leqq \varlimsup s^\gamma \int_0^\infty e^{-st}g(e^{-st})\,d\alpha(t)$$
$$\leqq \frac{1}{\Gamma(\gamma)}\int_0^\infty e^{-t}P(e^{-t})t^{\gamma-1}\,dt,$$

$s$ approaching zero or becoming infinite in the set complementary to $E$. From (3), (5) and (6) we see that

$$\lim s^\gamma \int_0^\infty e^{-st}g(e^{-st})\,d\alpha(t) = \frac{1}{\Gamma(\gamma)}\int_0^\infty e^{-t}g(e^{-t})t^{\gamma-1}\,dt,$$

the desired result.

4.3. From this result by specializing $g(x)$ we may obtain the following Tauberian theorem.

THEOREM 4.3. *If $\alpha(t)$ is non-decreasing and such that the integral*

$$f(s) = \int_0^\infty e^{-st}\,d\alpha(t)$$

*converges for $s > 0$, and if for some non-negative number $\gamma$*

$$f(s) \sim \frac{A}{s^\gamma} \qquad \begin{matrix}(s \to 0+)\\(s \to \infty),\end{matrix}$$

*then*

(1) $$\alpha(t) \sim \frac{At^\gamma}{\Gamma(\gamma+1)} \qquad \begin{matrix}(t \to \infty)\\(t \to 0+).\end{matrix}$$

It is no restriction to take $A = 1$. First suppose $\gamma = 0$. Since $\alpha(t)$ is non-decreasing it either approaches a limit or becomes infinite as $t \to \infty$. From Corollary 1c the latter case would imply

$$\lim_{s \to 0+} f(s) = \infty$$

contradicting our hypothesis. The finite limit which $\alpha(t)$ must approach is then certainly unity. A similar argument applies to the case $t \to 0+$, $s \to \infty$.

If $\gamma > 0$ we choose the function $g(e^{-t})$ of Theorem 4.2 as follows:

$$g(e^{-t}) = e^t \qquad (0 \leq t \leq 1)$$
$$= 0 \qquad (1 < t < \infty).$$

Then the conclusion of Theorem 4.2 is precisely (1) provided $t$ approaches its limit through the set of points where $\alpha(t)$ is continuous. Obviously this restriction may be removed since $\alpha(t)$ is monotonic.

4.4. We prove now a very useful theorem due to Hardy and Littlewood [1912] in its original form. We employ the slightly more general form of Landau.*

THEOREM 4.4. *If $f(x)$ has a second derivative in the interval $0 < x < \infty$ and if for some real value of $\alpha$*

(1) $$f(x) = o(x^\alpha) \qquad \begin{array}{l}(x \to 0+) \\ (x \to \infty),\end{array}$$

(2) $$f''(x) < O(x^{\alpha-2}) \qquad \begin{array}{l}(x \to 0+) \\ (x \to \infty),\end{array}$$

*then*

(3) $$f'(x) = o(x^{\alpha-1}) \qquad \begin{array}{l}(x \to 0+) \\ (x \to \infty).\end{array}$$

Let $\delta$ be an arbitrary positive number less than unity and let $x$ be positive. By Taylor's remainder theorem

(4) $$f(x \pm \delta x) - f(x) \mp \delta x f'(x) = \frac{\delta^2 x^2}{2} f''(x \pm \theta \delta x) \qquad (0 < \theta < 1).$$

By (2) there exists a constant $M$ such that

(5) $$f''(x \pm \theta \delta x) < M(1 \pm \theta \delta)^{\alpha-2} x^{\alpha-2} \leq M x^{\alpha-2}(1 + \delta)^{\alpha-2} \qquad (\alpha \geq 2)$$
$$\leq M x^{\alpha-2}(1 - \delta)^{\alpha-2} \qquad (\alpha \leq 2).$$

By (1), (4), and (5) we obtain

$$o(x^\alpha) \mp \delta x f'(x) \leq \frac{M\delta^2}{2} x^\alpha (1 + \delta)^{\alpha-2} \qquad (\alpha \geq 2)$$
$$\leq \frac{M\delta^2}{2} x^\alpha (1 - \delta)^{\alpha-2} \qquad (\alpha \leq 2).$$

* E. Landau [1929] p. 58.

Hence

$$-\frac{M\delta}{2}(1+\delta)^{\alpha-2} \leq \underline{\lim} \, x^{1-\alpha}f'(x) \leq \overline{\lim} \, x^{1-\alpha}f'(x)$$

$$\leq \frac{M\delta}{2}(1+\delta)^{\alpha-2} \qquad (\alpha \geq 2)$$

$$-\frac{M\delta}{2}(1-\delta)^{\alpha-2} \leq \underline{\lim} \, x^{1-\alpha}f'(x) \leq \overline{\lim} \, x^{1-\alpha}f'(x)$$

$$\leq \frac{M\delta}{2}(1-\delta)^{\alpha-2} \qquad (\alpha \leq 2).$$

Since $\delta$ is arbitrary we have in either case

$$\lim x^{1-\alpha}f'(x) = 0,$$

which is equation (3).

It is obvious that in condition (2) the direction of the inequality may be reversed.

COROLLARY 4.4a. *If condition* (1) *is replaced by*

$$f(x) \sim Ax^\alpha$$

*then* (3) *becomes*

$$f'(x) \sim \alpha A x^{\alpha-1}.$$

This follows at once from the theorem by consideration of the function

$$f(x) - Ax^\alpha.$$

Of course if $\alpha = 0$ our conclusion becomes

$$f'(x) = o\,(x^{\alpha-1}).$$

COROLLARY 4.4b. *If $f(x)$ is of class $C'$ in $0 \leq x < \infty$ and if for some non-negative number $\alpha$*

$$F(x) = \int_0^x f(t)\,dt \sim Ax^\alpha \qquad (x \to \infty),$$

$$(xf(x))' \geq 0 \qquad (0 \leq x < \infty),$$

*then*

$$f(x) \sim \alpha A x^{\alpha-1} \qquad (x \to \infty).$$

This is a result of Landau.* It is no restriction to take $A = 0$. Then

$$G(x) = -\int_0^x tf(t)\,dt = -xF(x) + \int_0^x F(t)\,dt = o(x^{\alpha+1}) \qquad (x \to \infty)$$

$$G''(x) = -(xf(x))' \leq 0 \qquad (0 \leq x < \infty).$$

Hence $G(x)$ satisfies conditions (1) and (2). Then (3) becomes

$$xf(x) = o(x^\alpha) \qquad (x \to \infty),$$

which is what we wished to prove.

4.5. In Tauber's theorem, i.e., Theorem 3a, the condition

$$a(t) = o(t^{-1}) \qquad (t \to \infty)$$

is stronger than necessary, as Littlewood first showed for series. We show the analogous result for Laplace integrals in this section.

THEOREM 4.5. *If*

$$f(s) = \int_0^\infty e^{-st}\,d\alpha(t)$$

*converges for* $s > 0$, *if*

$$f(s) \to A \qquad (s \to 0+),$$

*and if there exists a constant $K$ such that the function*

(1) $$\beta(t) = Kt + \int_0^t u\,d\alpha(u)$$

*is a non-decreasing function of $t$ in $(0, \infty)$, then*

(2) $$\alpha(t) \to A \qquad (t \to \infty).$$

We note first that Theorem 3a is included in the present result. For, if

$$\alpha(t) = \int_0^t a(u)\,du$$

$$a(t) = o(t^{-1}) \qquad (t \to \infty),$$

then

(3) $$1 + a(t)t \geq 0$$

at least for large values of $t$. Evidently the behavior of $a(t)$ for small values of $t$ is unimportant in Theorem 3a provided only that $f(s)$ is

---

* See E. Landau [1906] p. 218.

defined for $s > 0$. Hence we may suppose that (3) holds for all $t$, so that $\beta(t)$ is non-decreasing for $K = 1$.

To prove the theorem we have

$$f(s) = \int_0^\infty e^{-st} \, d\alpha(t) \qquad (s > 0)$$

$$f''(s) = \int_0^\infty e^{-st} t^2 \, d\alpha(t) \qquad (s > 0)$$

$$= \int_0^\infty e^{-st} t \, d\beta(t) - \frac{K}{s^2}.$$

Since $\beta(t)$ is non-decreasing we have

$$f''(s) \geq -\frac{K}{s^2} \qquad (s > 0).$$

By Theorem 4.4

$$f'(s) = o\left(\frac{1}{s}\right) \qquad (s \to 0+).$$

Hence

$$\int_0^\infty e^{-st} \, d\beta(t) = \frac{K}{s} - f'(s) \sim \frac{K}{s} \qquad (s \to 0+).$$

By Theorem 4.3

$$\beta(t) \sim Kt$$

$$\int_0^t u \, d\alpha(u) = o(t) \qquad (t \to \infty).$$

By Theorem 3b this implies that

$$\alpha(t) \to A \qquad (t \to \infty),$$

and this is what we were to prove.

COROLLARY 4.5a. *The condition* (1) *of Theorem 3a may be replaced by*

$$a(t) \geq O\left(\frac{1}{t}\right) \qquad (t \to \infty).$$

COROLLARY 4.5b. *If*

$$f(s) = \sum_{n=1}^\infty a_n e^{-ns}$$

*converges for $s > 0$, if*

$$\lim_{s \to 0+} f(s) = A,$$

*and if a constant K exists such that*

(4) $$a_n > -\frac{K}{n} \qquad (n = 1, 2, \cdots),$$

*then*

$$\sum_{n=1}^{\infty} a_n = A.$$

Otherwise expressed this result states that a series whose terms satisfy (4) cannot be summable in the sense of Abel unless it converges. It follows from the theorem since (4) implies (1) if $\alpha(t)$ is a suitable step-function.

4.6. By way of generalizing Theorem 4.3 we add the following result.

THEOREM 4.6. *Let the integral*

$$f(s) = \int_0^\infty e^{-st} \, d\alpha(t)$$

*converge for $s > 0$ and let constants $K$, $A$, and $\gamma > 0$ exist such that the function*

$$\beta(t) = \alpha(t) + Kt^\gamma$$

*is non-decreasing in $0 \leq t < \infty$ and such that*

$$f(s) \sim \frac{A}{s^\gamma} \qquad \begin{array}{l}(s \to 0+)\\(s \to \infty).\end{array}$$

*Then*

$$\alpha(t) \sim \frac{At^\gamma}{\Gamma(\gamma + 1)} \qquad \begin{array}{l}(t \to \infty)\\(t \to 0+).\end{array}$$

To prove this we have only to apply Theorem 4.3 to the function

$$\int_0^\infty e^{-st} \, d\beta(t) = f(s) + \frac{K\Gamma(\gamma + 1)}{s^\gamma}$$

$$\sim \frac{A + K\Gamma(\gamma + 1)}{s^\gamma} \qquad \begin{array}{l}(s \to 0+)\\(s \to \infty).\end{array}$$

The conclusion is

$$\alpha(t) + Kt^\gamma = \beta(t) \sim \frac{A + K\Gamma(\gamma + 1)}{\Gamma(\gamma + 1)} t^\gamma$$

$$\alpha(t) \sim \frac{At^\gamma}{\Gamma(\gamma + 1)} \qquad \begin{array}{l}(t \to \infty)\\(t \to 0+).\end{array}$$

This proves the result.

## 5. Tauberian Theorems for the Stieltjes Transform

In this section we shall prove a Tauberian result concerning the Stieltjes transform which we shall need in a later chapter. It is obtained by application of the foregoing results about the Laplace transform. However, we first prove a lemma about the Stieltjes transform which is due to Hardy and Littlewood.*

**LEMMA 5.** *If $\phi(u)$ belongs to $L$ in $(0, R)$ for every $R > 0$, if the integral*

$$f(s) = \int_0^\infty \frac{\varphi(t)}{s+t} dt$$

*converges, and if*

$$G(t) = \int_0^t u\varphi(u)\, du = o(t) \qquad (t \to \infty),$$

*then the relation*

$$f(s) \sim \frac{A}{s} \qquad (s \to \infty)$$

*implies*

$$\int_0^\infty \varphi(t)\, dt = A.$$

Write

$$\int_0^\infty \frac{s\varphi(t)}{s+t} dt = \int_0^s \varphi(t)\, dt - \int_0^s \frac{t\varphi(t)}{s+t} dt + s \int_s^\infty \frac{\varphi(t)}{s+t} dt$$

$$= I_1(s) + I_2(s) + I_3(s).$$

Since we wish to show that $I_1$ tends to $A$ as $s$ becomes infinite we need only prove that $I_2$ and $I_3$ tend to zero. This results from the following relations

$$-I_2(s) = \frac{G(s)}{2s} + \int_0^X \frac{G(t)}{(s+t)^2} dt + \int_X^s \frac{G(t)}{(s+t)^2} dt \qquad (0 < X < s)$$

$$|I_2| \leq \left|\frac{G(s)}{2s}\right| + \frac{X}{s^2} \underset{0 \leq t \leq X}{\text{u.b.}} |G(t)| + \log \frac{2s}{X+s} \underset{X \leq t \leq s}{\text{u.b.}} \left|\frac{G(t)}{t}\right|.$$

Allowing first $s$ and then $X$ to become infinite, we have

$$\varlimsup_{x \to \infty} |I_2| \leq \log 2 \varlimsup_{t \to \infty} \left|\frac{G(t)}{t}\right| = 0.$$

---
* Hardy and Littlewood [1930] p. 34.

Treating $I_3(s)$ in a similar way we have

$$I_3(s) = -\frac{G(s)}{2s} + s\int_s^\infty \frac{G(t)}{t^2(s+t)}\,dt + s\int_s^\infty \frac{G(t)}{t(s+t)^2}\,dt$$

$$|I_3(s)| \leq \left|\frac{G(s)}{2s}\right| + \log 2 \underset{s\leq t<\infty}{\text{u.b.}}\left|\frac{G(t)}{t}\right| + \frac{1}{2}\underset{s\leq t<\infty}{\text{u.b.}}\left|\frac{G(t)}{t}\right|$$

$$\varlimsup_{s\to\infty}|I_3(s)| \leq \left(\log 2 + \frac{1}{2}\right)\varlimsup_{t\to\infty}\left|\frac{G(t)}{t}\right| = 0.$$

This completes the proof of the lemma. By its use we establish the following result also due to Hardy and Littlewood.*

THEOREM 5a. *Let $\phi(t)$ belong to $L$ in $(0, R)$ for every $R > 0$, and let the integral*

$$f(s) = \int_0^\infty \frac{\varphi(t)}{s+t}\,dt$$

*converge. If constants $K$ and $A$ exist such that*

(1) $\qquad\qquad\qquad \phi(t) > -Kt^{-1} \qquad\qquad (0 \leq t < \infty)$

(2) $\qquad\qquad\qquad f(s) \sim As^{-1} \qquad\qquad\qquad (s \to \infty),$

*then*

$$\int_0^\infty \varphi(t)\,dt = A.$$

Consider the function

$$g(s) = s^2 f(s).$$

It is clear from (2) and (1) that

$$g(s) \sim As \qquad\qquad (s \to \infty)$$

and that

$$g''(s) = 2\int_0^\infty \frac{t^2}{(s+t)^3}\varphi(t)\,dt > -\frac{K}{s}.$$

By use of Theorem 4.4 we obtain at once that

$$g'(s) \to A \qquad\qquad (s \to \infty).$$

That is

$$2sf(s) + s^2 f'(s) \to A \qquad\qquad (s \to \infty),$$

$$s^2 f'(s) \to -A \qquad\qquad (s \to \infty).$$

* Hardy and Littlewood [1930] p. 33.

From this we see that

$$\int_0^\infty \frac{t\varphi(t)}{(s+t)^2} dt = f(s) + sf'(s) = o\left(\frac{1}{s}\right) \qquad (s \to \infty).$$

But the integral on the left can be expressed as the product of two Laplace transforms,

(3) $$\int_0^\infty \frac{t\varphi(t)}{(s+t)^2} dt = \int_0^\infty e^{-su} u\psi(u) \, du$$

$$\psi(u) = \int_0^\infty e^{-ut} t\varphi(t) \, dt.$$

One sees this by interchanging the order of integration, a step which is valid by virtue of (1) and Fubini's theorem.

By (1) we have

$$\psi(u) > -\frac{K}{u} \qquad (0 \leqq u < \infty).$$

Hence we may apply Theorem 4.6 to the integral on the right-hand side of (3) and obtain

(4) $$\frac{1}{u}\int_0^u t\psi(t) \, dt = o(1) \qquad (u \to 0+).$$

The second derivative of this function is

$$H(u) = \frac{2}{u^3}\int_0^u t\psi(t) \, dt - \frac{1}{u}\psi(u) + \psi'(u).$$

By (1) we have

$$-\psi(u) < \frac{K}{u}$$

$$\psi'(u) < \frac{K}{u^2}.$$

This with (4) shows that

$$H(u) < o\left(\frac{1}{u^2}\right) \qquad (u \to 0+),$$

so that we may again apply Theorem 4.4 to the function (4) and obtain

$$\psi(u) - \frac{1}{u^2}\int_0^u t\psi(t) \, dt = o\left(\frac{1}{u}\right) \qquad (u \to 0+)$$

$$\psi(u) = \int_0^\infty e^{-ut} t\varphi(t) \, dt = o\left(\frac{1}{u}\right) \qquad (u \to 0+).$$

Again, using (1) and Theorem 4.6 we have
$$\int_0^u t\varphi(t)\,dt = o(u) \qquad (u \to \infty).$$
Now applying Lemma 5 we obtain the desired result
$$\int_0^\infty \varphi(t)\,dt = A.$$

Another result of this type which we shall need in a later chapter is

THEOREM 5b. *Let $\varphi(t)$ belong to $L$ in $(0, R)$ for every positive $R$, and let the integral*
$$f(s) = \int_0^\infty \frac{\varphi(t)}{(s+t)^2}\,dt$$
*converge. If $\varphi(t)$ is bounded in $(0 \leqq t < \infty)$, then the relation*
$$f(s) \sim \frac{A}{s} \qquad (s \to 0+)$$
*implies*

(5) $$\int_0^t \varphi(u)\,du \sim At \qquad (t \to 0+).$$

Clearly we have by Fubini's theorem that
$$f(s) = \int_0^\infty e^{-sy} y\,dy \int_0^\infty e^{-yt}\varphi(t)\,dt \qquad (0 < s < \infty)$$

(6) $$f(s) = \int_0^\infty e^{-sy} yg(y)\,dy,$$

where

(7) $$g(y) = \int_0^\infty e^{-yt}\varphi(t)\,dt.$$

Let $M$ be an upper bound of $|\varphi(t)|$. Then
$$|yg(y)| < M \qquad (0 < y < \infty).$$
Set
$$\alpha(y) = \int_0^y ug(u)\,du.$$

Then $\alpha(y) + My$ is non-decreasing, and we may apply Theorem 4.6 to the integral (6) to obtain
$$\alpha(y) \sim Ay \qquad (y \to \infty).$$

But

$$\alpha''(y) = \int_0^\infty e^{-yt} \varphi(t)[1 - yt]\, dt$$

$$|\alpha''(y)| < \frac{2M}{y} \qquad (0 < y < \infty).$$

Hence we may apply Theorem 4.4 to obtain

$$\alpha'(y) \sim A$$

$$g(y) \sim \frac{A}{y} \qquad (y \to \infty).$$

By virtue of Theorem 4.6, applied to the integral (7), we now prove (5). This completes the proof of the theorem.

## 6. Fourier Transforms

In the proof of Wiener's general Tauberian theorem we shall need certain facts about Fourier transforms which we collect in this section. The first is a combination of Plancherel's theorem and Parseval's theorem.

**THEOREM 6a.** *If $F(x)$ belongs to $L^2$ in $(-\infty, \infty)$, then there exists a function $f(x)$ belonging to $L^2$ in $(-\infty, \infty)$ such that*

1) $$f(x) = \underset{a \to \infty}{\text{l.i.m.}}^{(2)} \frac{1}{\sqrt{2\pi}} \int_{-a}^a F(t) e^{-ixt}\, dt$$

2) $$f(x) = \frac{1}{\sqrt{2\pi}} \frac{d}{dx} \int_{-\infty}^\infty F(t) \frac{e^{-ixt} - 1}{-it}\, dt$$

3) $$F(x) = \underset{a \to \infty}{\text{l.i.m.}}^{(2)} \frac{1}{\sqrt{2\pi}} \int_{-a}^a f(t) e^{ixt}\, dt$$

4) $$F(x) = \frac{1}{\sqrt{2\pi}} \frac{d}{dx} \int_{-\infty}^\infty f(t) \frac{e^{ixt} - 1}{it}\, dt$$

5) $$\int_{-\infty}^\infty |f(x)|^2\, dx = \int_{-\infty}^\infty |F(x)|^2\, dx.$$

The proof of this is to be found in any text on Fourier transforms. The function $f(x)$ is uniquely determined up to an additive function which is zero almost everywhere.

**THEOREM 6b.** *If $f(x)$ belongs to $L$ in $(-\infty, \infty)$ then $F(x)$, defined by the absolutely convergent integral*

(6) $$F(x) = \frac{1}{\sqrt{2\pi}} \int_{-\infty}^\infty e^{ixt} f(t)\, dt,$$

*is continuous for all finite $x$.*

This follows since

$$\lim_{x \to a} \int_{-\infty}^{\infty} e^{ixt} f(t) \, dt = \int_{-\infty}^{\infty} e^{iat} f(t) \, dt$$

for every real number $a$. It is permissible to take the limit under the integral sign since $f(x)$ belongs to $L$.

**THEOREM 6c.** *If $f(x)$ and $g(x)$ belong to $L$ in $(-\infty, \infty)$ and if $h(x)$ is the resultant,*

$$h(x) = \frac{1}{\sqrt{2\pi}} \int_{-\infty}^{\infty} f(x-t) g(t) \, dt,$$

*then $h(x)$ belongs to $L$ and*

$$\frac{1}{\sqrt{2\pi}} \int_{-\infty}^{\infty} |h(x)| \, dx \leq \frac{1}{\sqrt{2\pi}} \int_{-\infty}^{\infty} |f(x)| \, dx \, \frac{1}{\sqrt{2\pi}} \int_{-\infty}^{\infty} |g(x)| \, dx.$$

For, by use of the Fubini theorem

(7)
$$\frac{1}{\sqrt{2\pi}} \int_{-\infty}^{\infty} |h(x)| \, dx \leq \frac{1}{2\pi} \int_{-\infty}^{\infty} dx \int_{-\infty}^{\infty} |f(x-t)| \, |g(t)| \, dt$$
$$= \frac{1}{2\pi} \int_{-\infty}^{\infty} |g(t)| \, dt \int_{-\infty}^{\infty} |f(x-t)| \, dx,$$

from which the result is immediate if we set $x - t$ equal to a new variable.

**THEOREM 6d.** *If $f(x)$, $g(x)$, and $h(x)$ are defined as in Theorem 6c and if $F(x)$, $G(x)$, $H(x)$ are their respective Fourier transforms as defined by (6), then $H(x) = F(x) G(x)$.*

For, simple computation gives

$$\frac{1}{2\pi} \int_{-\infty}^{\infty} e^{ixt} f(t) \, dt \int_{-\infty}^{\infty} e^{ixy} g(y) \, dy = \frac{1}{2\pi} \int_{-\infty}^{\infty} f(t) \, dt \int_{-\infty}^{\infty} e^{ix(t+y)} g(y) \, dy$$
$$= \frac{1}{2\pi} \int_{-\infty}^{\infty} f(t) \, dt \int_{-\infty}^{\infty} e^{ixu} g(u-t) \, du$$
$$= \frac{1}{2\pi} \int_{-\infty}^{\infty} e^{ixu} \, du \int_{-\infty}^{\infty} f(t) g(u-t) \, dt.$$

The change in the order of integration is again justified by Fubini's theorem.

**THEOREM 6e.** *If $f(x)$ belongs to $L$ in $(-\infty, \infty)$ and if $F(x)$ is its Fourier transform, defined by (6), then for almost all $x$*

$$\lim_{R \to \infty} \frac{1}{\sqrt{2\pi}} \int_{-R}^{R} e^{-ixt} \left(1 - \frac{|t|}{R}\right) F(t) \, dt = f(x).$$

For,

$$\frac{1}{2\pi}\int_{-R}^{R}\left(1-\frac{|t|}{R}\right)e^{-ixt}\,dt\int_{-\infty}^{\infty}e^{iyt}f(y)\,dy$$

$$=\frac{1}{2\pi R}\int_{-\infty}^{\infty}f(y)\left[\frac{\sin\,(y-x)R/2}{(y-x)/2}\right]^{2}dy.$$

But this is essentially the same integral considered in the proof of Theorem 9.3 of Chapter II. It approaches $f(x)$ as $R$ becomes infinite for almost all values of $x$.

COROLLARY 6e. *If $F(x)$ is identically zero, then $f(x)$ is zero almost everywhere.*

By use of this theorem or otherwise one may easily show that if $\Delta(x)$ and $\delta(x)$ are the functions defined below, then each is the Fourier transform of the other.

$$\Delta(x)=1-|x| \qquad |x|\leq 1$$
$$=0 \qquad |x|\geq 1$$
$$\delta(2x)=\frac{1}{\sqrt{2\pi}}\left(\frac{\sin x}{x}\right)^{2} \qquad (-\infty<x<\infty).$$

We shall need this pair of functions later. Note that $\delta(x)$ belongs to $L$. In fact

$$\frac{1}{\sqrt{2\pi}}\int_{-\infty}^{\infty}\delta(x)\,dx=\frac{1}{\pi}\int_{-\infty}^{\infty}\left(\frac{\sin x}{x}\right)^{2}dx=1.$$

## 7. Fourier Transforms of Functions of L

We consider in this section functions which are Fourier transforms of functions of $L$. That is, they can be represented by absolutely convergent Fourier integrals. For brevity we give this class of functions a name, and we introduce an operator on functions of the class by the following definitions.

DEFINITION 7a. *A function $F(x)$ belongs to the class $A$ if and only if*

(1) $$F(x)=\frac{1}{\sqrt{2\pi}}\int_{-\infty}^{\infty}e^{ixt}f(t)\,dt$$

*where*

$$\int_{-\infty}^{\infty}|f(t)|\,dt<\infty.$$

DEFINITION 7b. *If $F(x)$ belongs to the class $A$ and has the representation (1), its norm, $\|F(x)\|$, is*

$$\|F(x)\| = \frac{1}{\sqrt{2\pi}} \int_{-\infty}^{\infty} |f(t)|\, dt.$$

Concerning functions of the class $A$ we prove:

THEOREM 7a. *If $F_1(x)$ and $F_2(x)$ belong to $A$, then*

(2) $\qquad \|F_1(x) + F_2(x)\| \leq \|F_1(x)\| + \|F_2(x)\|$

(3) $\qquad \|F_1(x)F_2(x)\| \leq \|F_1(x)\|\, \|F_2(x)\|.$

The inequality (2) is evident, and (3) follows from §6 (7).

THEOREM 7b. *If $F_1(x)$ and $F_2(x)$ belong to $A$ and if $\|F_2(x)\|$ is less than unity, then*

$$\left\| \frac{F_1(x)}{1 + F_2(x)} \right\| \leq \frac{\|F_1(x)\|}{1 - \|F_2(x)\|}.$$

It must not be supposed that $[1 + F_2(x)]^{-1}$ always belongs to $A$. In fact this is false if $F_2(x)$ is identically zero. Since

$$|F_2(x)| \leq \|F_2(x)\| < 1,$$

we have

(4) $\qquad \dfrac{F_1(x)}{1 + F_2(x)} = \sum_{n=0}^{\infty} (-1)^n F_1(x)[F_2(x)]^n.$

By Theorem 7a

$$(-1)^n F_1(x)[F_2(x)]^n = \frac{1}{\sqrt{2\pi}} \int_{-\infty}^{\infty} e^{ixt} f_n(t)\, dt$$

$$\|(-1)^n F_1(x)[F_2(x)]^n\| \leq \|F_1(x)\|\, \|F_2(x)\|^n,$$

where $f_n(x)$ belongs to $L$. But the series $\sum_{n=0}^{\infty} f_n(t)$ converges in mean to a function $f(t)$ since

$$\frac{1}{\sqrt{2\pi}} \int_{-\infty}^{\infty} \left| \sum_{n=N}^{N+p} f_n(t) \right| dt \leq \sum_{n=N}^{N+p} \|F_1(x)\|\, \|F_2(x)\|^n \qquad (p = 0, 1, \cdots).$$

Hence

$$\frac{F_1(x)}{1 + F_2(x)} = \frac{1}{\sqrt{2\pi}} \int_{-\infty}^{\infty} e^{ixt} f(t) \, dt$$

$$\left\| \frac{F_1(x)}{1 + F_2(x)} \right\| \leq \frac{1}{\sqrt{2\pi}} \int_{-\infty}^{\infty} \sum_{n=0}^{\infty} |f_n(t)| \, dt$$

$$\leq \sum_{n=0}^{\infty} \| F_1(x) \| \, \| F_2(x) \|^n = \frac{\| F_1(x) \|}{1 - \| F_2(x) \|}.$$

THEOREM 7c. *If $F(x)$ belongs to $A$ and $a$ is constant, then*

$$\| aF(x) \| = |a| \, \| F(x) \|$$

$$\| F(ax) \| = \| F(x) \| \qquad (a \neq 0).$$

The proofs of these results may be easily supplied. The function $\Delta(x)$ defined in §6 belongs to $A$. In fact

$$\| \Delta(x) \| = 1.$$

As another example which we shall need, consider the function

$$P_\epsilon(x) = 2\Delta\left(\frac{x}{2\epsilon}\right) - \Delta\left(\frac{x}{\epsilon}\right).$$

Here $\epsilon$ is a positive number. Clearly

$$P_\epsilon(x) = 1 \qquad (|x| \leq \epsilon)$$

$$= 2 - \frac{|x|}{\epsilon} \qquad (\epsilon \leq |x| \leq 2\epsilon)$$

$$= 0 \qquad (|x| \geq 2\epsilon).$$

By Theorems 7a and 7c

(5) $$\| P_\epsilon(x) \| \leq 3,$$

so that the norm of $P_\epsilon(x)$ has a bound independent of $\epsilon$. Since $\Delta(x)$ is the Fourier transform of $\delta(x)$ we have

$$P_\epsilon(x) = \frac{1}{\sqrt{2\pi}} \int_{-\infty}^{\infty} e^{ixt} p_\epsilon(t) \, dt,$$

where

(6) $$p_\epsilon(t) = \epsilon p(\epsilon t) = 4\epsilon \delta(2\epsilon t) - \epsilon \delta(\epsilon t).$$

THEOREM 7d. *If $p_\epsilon(t)$ is defined by (6), then for every real $y$*

(7) $$\lim_{\epsilon \to 0} \int_{-\infty}^{\infty} |p_\epsilon(t - y) - p_\epsilon(t)| \, dt = 0.$$

This follows since $p(t)$, as defined by (6) belongs to $L$. For,*

$$\int_{-\infty}^{\infty} |p_\epsilon(t-y) - p_\epsilon(t)| \, dt = \int_{-\infty}^{\infty} |p(t-\epsilon y) - p(t)| \, dt = o(1) \quad (\epsilon \to 0).$$

## 8. The Quotient of Fourier Transforms

In general the quotient of two functions of class $A$ is not a function of class $A$ even though the denominator does not vanish. For, the quotient of a non-vanishing function by itself is unity, a function which can not belong to $A$, as one sees by use of the Riemann-Lebesgue theorem. If in addition the function in the numerator vanishes outside an interval the quotient does belong to $A$.

THEOREM 8. *If $H(x)$ and $K(x)$ belong to $A$, if $K(x)$ does not vanish and if $H(x)$ vanishes outside an interval, then $H(x)/K(x)$ belongs to $A$.*

To prove this we shall need:

LEMMA 8. *If $F(x)$ and $G(x)$ belong to $A$ and are the Fourier transforms of $f(x)$ and $g(x)$ respectively, then for any constant $a$*

(1)
$$\| [F(x) - F(a)]G(x-a) \| \leq \frac{1}{2\pi} \int_{-\infty}^{\infty} |f(t)| \, dt \int_{-\infty}^{\infty} |g(x-t) - g(x)| \, dx.$$

For, the following are pairs of Fourier transforms

$$G(x-a) \qquad\qquad e^{-iax}g(x)$$

$$F(x)G(x-a) \qquad\qquad \frac{1}{\sqrt{2\pi}} e^{-iax} \int_{-\infty}^{\infty} g(x-t)f(t)e^{iat} \, dt$$

$$F(a)G(x-a) \qquad\qquad \frac{1}{\sqrt{2\pi}} e^{-iax} g(x) \int_{-\infty}^{\infty} e^{iat}f(t) \, dt$$

$$[F(x) - F(a)]G(x-a) \qquad\qquad \frac{1}{\sqrt{2\pi}} \int_{-\infty}^{\infty} e^{ia(t-x)}f(t)[g(x-t) - g(x)] \, dt.$$

From this last pair and the definition of the norm the inequality (1) follows by use of Fubini's theorem.

We return now to the proof of Theorem 8. Suppose that $H(x)$ vanishes when $|x| \geq 2\lambda$. Let $N$ be a positive integer. Set

$$\epsilon = \frac{4\lambda}{3N} \qquad x_n = -2\lambda + 3\epsilon n \quad (n = 0, 1, \cdots, N).$$

Then

$$1 = \sum_{n=0}^{N} P_\epsilon(x - x_n) \qquad (-2\lambda \leq x \leq 2\lambda).$$

* See, for example, N. Wiener [1933] p. 14.

Since $H(x)$ vanishes outside the interval $(-2\lambda, 2\lambda)$, we have for all $x$

(2) $$\frac{H(x)}{K(x)} = \sum_{n=0}^{N} \frac{H(x)P_\epsilon(x - x_n)}{K(x)}.$$

We will show that if $N$ is sufficiently large each term of this series belongs to $A$.
Now

(3) $$\frac{H(x)P_\epsilon(x - x_n)}{K(x)}$$
$$= \frac{H(x)P_\epsilon(x - x_n)}{K(x_n) + [K(x) - K(x_n)]P_{2\epsilon}(x - x_n)} \quad (-\infty < x < \infty).$$

For, if $|x - x_n| \geq 2\epsilon$ both sides of this equation are equal to zero. If $|x - x_n| \leq 2\epsilon$, then $P_{2\epsilon}(x - x_n) = 1$ from which equation (3) is evident. Hence we must show that

$$\frac{H(x)P_\epsilon(x - x_n)}{1 + \dfrac{[K(x) - K(x_n)]P_{2\epsilon}(x - x_n)}{K(x_n)}}$$

belongs to $A$. By Theorem 7b this will follow if

$$\frac{1}{|K(x_n)|} \, ||\,[K(x) - K(x_n)]P_{2\epsilon}(x - x_n)\,|| < 1.$$

By Lemma 8

$$||\,[K(x) - K(x_n)]P_{2\epsilon}(x - x_n)\,||$$
$$\leq \frac{1}{2\pi} \int_{-\infty}^{\infty} |k(t)| \, dt \int_{-\infty}^{\infty} |p_{2\epsilon}(x - t) - p_{2\epsilon}(x)| \, dx = M_\epsilon.$$

By §7 (5) and §7 (7) $M_\epsilon$ approaches zero with $\epsilon$. Hence if $\mu$ is the minimum of $K(x)$ in $(-2\lambda, 2\lambda)$, we can choose $N$ so large that

$$\frac{1}{|K(x_n)|} \, ||\,[K(x) - K(x_n)]P_{2\epsilon}(x - x_n)\,|| \leq \frac{M_\epsilon}{\mu} < 1$$

for $n = 0, 1, 2, \cdots, N$. That is, every term of the series (2) belongs to $A$. Thus our theorem is proved.

## 9. A Special Tauberian Theorem

We now prove a Tauberian theorem of a very special nature, but one that will prove useful in the proof of Wiener's general theorem. We first introduce the functions

$$K_\lambda(x) = \Delta\left(\frac{x}{2\lambda}\right) \qquad (\lambda > 0)$$

$$k_\lambda(x) = 2\lambda\delta(2\lambda x) \geq 0 \qquad (-\infty < x < \infty),$$

where $\Delta(x)$ and $\dot{\delta}(x)$ are the functions which were defined in section 6. By Theorem 7c

$$\| K_\lambda(x) \| = 1.$$

We introduce several classes of functions.

DEFINITION 9a. *The function $f(x)$ belongs to the class of slowly oscillating functions in $(-\infty, \infty)$ if it is defined there and if*

$$\lim_{\substack{y-x \to 0 \\ x \to \infty}} [f(y) - f(x)] = 0.$$

For example, any function which has a bounded derivative belongs to the class.

DEFINITION 9b. *The function $f(x)$ belongs to the class $S$ of slowly decreasing functions in $(-\infty, \infty)$ if it is defined there and if*

$$\lim_{y-x \to 0} [f(y) - f(x)] \geq 0 \qquad (x \to \infty, y = y(x) > x).$$

Any increasing function belongs to $S$. Any function having a derivative which remains greater than some negative constant belongs to $S$, as does any slowly oscillating function.

It is easily seen that if $f(x)$ belongs to $S$ and is greater than a positive number $\delta$ (less than a negative number $-\delta$) for arbitrarily large $x$, then it is greater than $\delta/2$ (less than $-\delta/2$) in infinitely many non-overlapping intervals of fixed length lying on the positive $x$-axis.

DEFINITION 9c. *The function $f(x)$ belongs to the class $B$ if it is bounded in $(-\infty, \infty)$.*

THEOREM 9. *If $f(x)$ belongs\* to $SB$ and if for every $\lambda > 0$*

$$\lim_{x \to \infty} \frac{1}{\sqrt{2\pi}} \int_{-\infty}^{\infty} k_\lambda(x - t) f(t)\, dt = A,$$

*then*

(1) $$\lim_{x \to \infty} f(x) = A.$$

\* This notation means that $f(x)$ belongs both to $S$ and to $B$.

Since

$$\frac{1}{\sqrt{2\pi}} \int_{-\infty}^{\infty} k_\lambda(t) \, dt = 1,$$

it is no restriction to assume that $A = 0$. Suppose that (1) were false. Then there would exist a positive number $\delta$ such that either $f(x) > \delta$ or $f(x) < -\delta$ for arbitrarily large $x$. Let us assume the first of these alternatives and deduce a contradiction. The second case will be left to the reader. By the remark following Definition 9b there exist infinitely many non-overlapping intervals $(x - \zeta, x + \zeta)$ on the positive $x$-axis on which $f(x) > \delta/2$. Then

$$\int_{-\infty}^{\infty} k_\lambda(x - t) f(t) \, dt > \frac{\delta}{2} \int_{x-\zeta}^{x+\zeta} k_\lambda(x - t) \, dt$$

$$- \underset{-\infty \leq t \leq \infty}{\text{u.b.}} |f(t)| \left\{ \int_{-\infty}^{x-\zeta} + \int_{x+\zeta}^{\infty} \right\} k_\lambda(x - t) \, dt$$

$$> \delta \int_0^\zeta k_\lambda(t) \, dt - 2 \, \text{u.b.} \, |f(t)| \int_\zeta^\infty k_\lambda(t) \, dt$$

$$> \delta \int_0^{2\lambda\zeta} \delta(t) \, dt - 2 \, \text{u.b.} \, |f(t)| \int_{2\lambda\zeta}^\infty \delta(t) \, dt.$$

As $\lambda$ becomes infinite the right-hand side tends to $\sqrt{2\pi}\delta/2$. Hence we can determine a $\lambda_0$ such that

$$\frac{1}{\sqrt{2\pi}} \int_{-\infty}^{\infty} k_{\lambda_0}(x - t) f(t) \, dt > \frac{\delta}{4}.$$

Allowing $x$ to become infinite through the infinite set of values of $x$ in question we have

$$0 \geq \delta/4 > 0,$$

a contradiction.

## 10. Pitt's Form of Wiener's Theorem

We are now able to prove Wiener's fundamental Tauberian theorem in a form due to H. R. Pitt [1938a].

**DEFINITION 10a.** *The function $f(x)$ belongs to the class $W$ if it belongs to $L$ in $(-\infty, \infty)$ and if its Fourier transform $F(x)$ does not vanish there.*

For example the function

$$f(x) = e^{-e^x} e^x$$

belongs to $W$. For, it is clearly integrable in $(-\infty, \infty)$ and

$$F(x) = \frac{1}{\sqrt{2\pi}} \int_{-\infty}^{\infty} e^{ixt} f(t) \, dt = \frac{\Gamma(1 + ix)}{\sqrt{2\pi}} \neq 0 \qquad (-\infty < x < \infty).$$

THEOREM 10a. *If $g(x)$ belongs to $W$, if $a(x)$ belongs to $SB$ and if*

(1) $$\lim_{x\to\infty}\int_{-\infty}^{\infty} g(x-t)a(t)\,dt = A\int_{-\infty}^{\infty} g(t)\,dt,$$

*then $a(\infty) = A$.*

It is clearly no restriction to assume that $A = 0$. By Theorem 9 we need only show that

(2) $$\lim_{x\to\infty}\int_{-\infty}^{\infty} k_\lambda(x-t)a(t)\,dt = 0$$

for every $\lambda$. By Theorem 8 we see that $K_\lambda(x)/G(x)$ belongs to $A$. Here $G(x)$ is the Fourier transform of $g(x)$. Let $K_\lambda(x)/G(x)$ be the Fourier transform of $r_\lambda(t)$. Then by Theorem 6d the function $K_\lambda(x)$ is the transform of the resultant

$$\frac{1}{\sqrt{2\pi}}\int_{-\infty}^{\infty} g(x-t)r_\lambda(t)\,dt.$$

But it is also the transform of $k_\lambda(x)$. Hence by Corollary 6e

$$k_\lambda(x) = \frac{1}{\sqrt{2\pi}}\int_{-\infty}^{\infty} g(x-t)r_\lambda(t)\,dt.$$

By the Fubini theorem

(3) $$\int_{-\infty}^{\infty} k_\lambda(x-t)a(t)\,dt = \frac{1}{\sqrt{2\pi}}\int_{-\infty}^{\infty} r_\lambda(t)\,dt \int_{-\infty}^{\infty} g(x-u-t)a(u)\,du.$$

But

$$\left|\int_{-\infty}^{\infty} g(x-u-t)a(u)\,du\right| \leqq \underset{-\infty<x<\infty}{\text{u.b.}}|a(x)|\int_{-\infty}^{\infty}|g(t)|\,dt,$$

so that we may take the limit under the first integral sign on the right-hand side of (3). Applying (1) to (3) we obtain (2), and the proof is complete.

It is convenient to state this result after an exponential change of variable. For this purpose we introduce

DEFINITION 10b. *The function $f(x)$ belongs to the class $S^*$ on $(0, \infty)$ if it is defined there and if*

$$\underline{\lim}\,[f(y) - f(x)] \geqq 0 \qquad \left(x \to \infty, y = y(x) > x, \frac{y}{x} \to 1\right).$$

The new form of the result is contained in

THEOREM 10b. *If $a(x)$ belongs to $S^*B$ on $(0, \infty)$, if $g(x)$ belongs to $L$*

and is such that

$$\int_0^\infty t^{ix} g(t)\, dt \neq 0 \qquad (-\infty < x < \infty),$$

then

$$\lim_{x \to \infty} \frac{1}{x} \int_0^\infty g\left(\frac{t}{x}\right) a(t)\, dt = A \int_0^\infty g(t)\, dt$$

implies $a(\infty) = A$.

## 11. Wiener's General Tauberian Theorem

We now obtain Wiener's general Tauberian theorem.

THEOREM 11a. *If $g_1(x)$ belongs to $W$, $g_2(x)$ to $L$, and $p(t)$ to $B$ in $(-\infty, \infty)$, and if*

(1) $$\lim_{x \to \infty} \int_{-\infty}^\infty g_1(x - t) p(t)\, dt = A \int_{-\infty}^\infty g_1(t)\, dt,$$

*then*

(2) $$\lim_{x \to \infty} \int_{-\infty}^\infty g_2(x - t) p(t)\, dt = A \int_{-\infty}^\infty g_2(t)\, dt.$$

For, we have only to apply Theorem 10a taking $g(x) = g_1(x)$ and

$$a(x) = \int_{-\infty}^\infty g_2(x - t) p(t)\, dt.$$

It is no restriction to assume that $A = 0$. Clearly $a(x)$ is bounded. Our result will be established if we can show that $a(x)$ belongs to $S$ and that §10 (1) holds. But

$$|a(y) - a(x)| \leq \underset{-\infty < t < \infty}{\text{u.b.}} |p(t)| \int_{-\infty}^\infty |g_2(y - x - u) - g_2(-u)|\, du.$$

The right-hand side approaches zero with $y - x$ since* $g_2(x)$ belongs to $L$.

Finally,

$$\int_{-\infty}^\infty g_1(x - t) a(t)\, dt = \int_{-\infty}^\infty g_1(x - t)\, dt \int_{-\infty}^\infty g_2(t - u) p(u)\, du$$

$$= \int_{-\infty}^\infty g_1(x - t)\, dt \int_{-\infty}^\infty g_2(u) p(t - u)\, du$$

$$= \int_{-\infty}^\infty g_2(u)\, du \int_{-\infty}^\infty g_1(x - t) p(t - u)\, dt$$

$$= \int_{-\infty}^\infty g_2(u)\, du \int_{-\infty}^\infty g_1(x - t - u) p(t)\, dt.$$

* See, for example, N. Wiener [1933] p. 14.

By (1) the inner integral approaches zero *boundedly* as $x$ becomes infinite. Hence §10 (1) is established. The conclusion of Theorem 10a gives (2).

Another form of this result is obtained by an exponential change of variable.

THEOREM 11b. *If $g_1(x)$ and $g_2(x)$ belong to L, $p(x)$ to B in $(0, \infty)$, if*

$$\int_0^\infty g_1(t) t^{ix} \, dt \neq 0 \qquad (-\infty < x < \infty),$$

*and if*

$$\lim_{x \to \infty} \frac{1}{x} \int_0^\infty g_1\left(\frac{t}{x}\right) p(t) \, dt = A \int_0^\infty g_1(t) \, dt,$$

*then*

$$\lim_{x \to \infty} \frac{1}{x} \int_0^\infty g_2\left(\frac{t}{x}\right) p(t) \, dt = A \int_0^\infty g_2(t) \, dt.$$

It should be observed that the condition $g_1 \, \varepsilon \, W$ in Theorem 11a is a necessary one. Suppose that it were not satisfied, that

$$\int_{-\infty}^\infty e^{-itx_0} g_1(t) \, dt = 0$$

for some $x_0$. Choose $g_2(t)$ so that it belongs to $W$ and choose $p(t) = e^{ix_0 t}$. Then

$$\int_{-\infty}^\infty g_1(x - t) p(t) \, dt = \int_{-\infty}^\infty g_1(t) e^{ix_0(x-t)} \, dt \equiv 0 \qquad (-\infty < x < \infty).$$

Hence $A = 0$. But

$$\int_{-\infty}^\infty g_2(x - t) p(t) \, dt = \int_{-\infty}^\infty g_2(t) e^{ix_0(x-t)} \, dt$$

$$= e^{ix_0 x} \int_{-\infty}^\infty g_2(t) e^{-ix_0 t} \, dt.$$

This function clearly approaches no limit as $x$ becomes infinite. That is, the conclusion of Theorem 11a does not apply if $g_1(x)$ is not assumed to belong to $W$.

### 12. Tauberian Theorem for the Stieltjes Integral

Another result of Wiener is also contained in Theorem 10a. To state it we must first introduce Wiener's class $M$.

DEFINITION 12. *The function $f(x)$ belongs to the class $M$ if it is continuous in the interval $-\infty < x < \infty$ and if*

$$\sum_{n=-\infty}^\infty \underset{n \leq x < n+1}{\mathrm{u.b.}} |f(x)| < \infty.$$

It is obvious that a function of $M$ belongs to $L$.

THEOREM 12. *If $g_1(x)$ belongs to $MW$, if $g_2(x)$ belongs to $M$, if $\alpha(x)$ is of bounded variation in every finite interval and is such that the function*

$$\int_x^{x+1} |d\alpha(t)|$$

*belongs to $B$, and if*

$$\lim_{x\to\infty} \int_{-\infty}^{\infty} g_1(x-t)\,d\alpha(t) = A \int_{-\infty}^{\infty} g_1(t)\,dt,$$

*then*

$$\lim_{x\to\infty} \int_{-\infty}^{\infty} g_2(x-t)\,d\alpha(t) = A \int_{-\infty}^{\infty} g_2(t)\,dt.$$

To prove this set

$$a(x) = \int_{-\infty}^{\infty} g_2(x-t)\,d\alpha(t).$$

This integral exists for all $x$ since

$$\int_{-\infty}^{\infty} |g_2(x-t)||d\alpha(t)| \leq \sum_{n=-\infty}^{\infty} \operatorname*{u.b.}_{n\leq t<n+1} |g_2(t)| \int_{x-n-1}^{x-n} |d\alpha(t)| < \infty.$$

Moreover

$$a(x+y) - a(x) = \int_{-\infty}^{\infty} [g_2(x+y-t) - g_2(x-t)]\,d\alpha(t)$$

$$|a(x+y) - a(x)| \leq \int_{-\infty}^{\infty} |g_2(y+u) - g_2(u)||d\alpha(x-u)|$$

$$\leq \operatorname*{u.b.}_{-\infty<x<\infty} \int_x^{x+1} |d\alpha(u)| \sum_{n=-\infty}^{\infty} \operatorname*{u.b.}_{n\leq u<n+1} |g_2(y+u) - g_2(u)|.$$

The series on the right converges uniformly for $y$ in some neighborhood of $y = 0$. Each term of the series approaches zero with $y$ so that

$$\lim_{\substack{y\to 0 \\ x\to\infty}} [a(x+y) - a(x)] = 0,$$

so that $a(x)$ is a slowly oscillating function. Now

$$\int_{-\infty}^{\infty} g_1(x-t)a(t)\,dt = \int_{-\infty}^{\infty} g_1(x-t)\,dt \int_{-\infty}^{\infty} g_2(t-u)\,d\alpha(u)$$

$$= \int_{-\infty}^{\infty} d\alpha(u) \int_{-\infty}^{\infty} g_1(x-t)g_2(t-u)\,dt$$

$$= \int_{-\infty}^{\infty} d\alpha(u) \int_{-\infty}^{\infty} g_1(x-u-t)g_2(t)\,dt.$$

The change in the order of integration is justified since

$$\int_{-\infty}^{\infty} |g_1(x-t)| \, dt \int_{-\infty}^{\infty} |g_2(t-u)| \, |d\alpha(u)| < \infty.$$

Again changing the order of integration we have

$$\int_{-\infty}^{\infty} g_1(x-t) a(t) \, dt = \int_{-\infty}^{\infty} g_2(t) \, dt \int_{-\infty}^{\infty} g_1(x-u-t) \, d\alpha(u).$$

By hypothesis

$$\lim_{x \to \infty} \int_{-\infty}^{\infty} g_1(x-u-t) \, d\alpha(u) = A \int_{-\infty}^{\infty} g_1(t) \, dt,$$

and the integral on the left has a bound independent of $x$. Hence, by Lebesgue's convergence theorem

$$\lim_{x \to \infty} \int_{-\infty}^{\infty} g_1(x-t) a(t) \, dt = A \int_{-\infty}^{\infty} g_1(t) \, dt \int_{-\infty}^{\infty} g_2(t) \, dt.$$

All conditions of Theorem 10a are satisfied so that

$$\lim_{t \to \infty} a(t) = A \int_{-\infty}^{\infty} g_2(t) \, dt,$$

and our result is established.

## 13. One-sided Tauberian Condition

For the applications of Theorem 11a, it is frequently convenient to suppose that $p(t)$ is bounded only on one side. That the theorem is no longer true without further hypotheses is seen by the following example.

$$p(x) = e^x$$
$$g_1(x) = e^{-x}(1 - 2x) \qquad\qquad (x > 0)$$
$$\phantom{g_1(x)} = 0 \qquad\qquad (x < 0)$$
$$g_2(x) = 1 \qquad\qquad (0 < x < 1)$$
$$\phantom{g_2(x)} = 0 \qquad\qquad (x \leq 0, x \geq 1).$$

Then $p(x)$ is bounded on one side but not on both, $g_1(x)$ belongs to $W$, $g_2(x)$ to $L$, and

$$h_1(x) = \int_{-\infty}^{\infty} g_1(x-t) p(t) \, dt = \int_0^{\infty} e^{-t}(1 - 2t) e^{x-t} \, dt = 0$$

$$h_2(x) = \int_{-\infty}^{\infty} g_2(x-t) p(t) \, dt = e^x \int_0^1 e^{-t} \, dt.$$

The functions $h_1(x)$ and $h_2(x)$ do not approach a common limit as $x$ becomes infinite.

The case when $p(t)$ is bounded on one side can be treated by the following theorem due to H. R. Pitt [1938b].

**THEOREM 13a.** *Let*

(1) $$g_1(x) \geq 0, \qquad g_1(x) \, \varepsilon \, W \qquad (-\infty < x < \infty)$$

(2) $$p(x) \geq -c$$

(3) $$g_2(x) \, \varepsilon \, M$$

*except that $g_2(x)$ may be discontinuous at a set of points of measure zero, and let the function*

(4) $$h_1(x) = \int_{-\infty}^{\infty} g_1(x - t) p(t) \, dt$$

*exist for all $x$ and belong to $B$. Then*

(5) $$\lim_{x \to \infty} \int_{-\infty}^{\infty} g_1(x - t) p(t) \, dt = A \int_{-\infty}^{\infty} g_1(t) \, dt$$

*implies*

$$\lim_{x \to \infty} \int_{-\infty}^{\infty} g_2(x - t) p(t) \, dt = A \int_{-\infty}^{\infty} g_2(t) \, dt.$$

We note first that it is no restriction to take $A = 0$. Our second remark is that we may assume $g_1(x)$ continuous. For, the function

$$g(x) = \frac{1}{\sqrt{\pi}} \int_{-\infty}^{\infty} e^{-y^2} g_1(x - y) \, dy$$

satisfies conditions (1) and is also continuous. It is clearly positive and integrable. To show that it belongs to $W$ we have

$$\int_{-\infty}^{\infty} e^{-ixy} g(y) \, dy = \frac{1}{\sqrt{\pi}} \int_{-\infty}^{\infty} e^{-ixy} \, dy \int_{-\infty}^{\infty} e^{-(y-t)^2} g_1(t) \, dt$$

$$= \frac{1}{\sqrt{\pi}} \int_{-\infty}^{\infty} g_1(t) \, dt \int_{-\infty}^{\infty} e^{-ixy-(y-t)^2} \, dy$$

$$= e^{-x^2/4} \int_{-\infty}^{\infty} e^{-ixt} g_1(t) \, dt \neq 0.$$

## §13] ONE-SIDED TAUBERIAN CONDITION

To show that (5) is satisfied when $g_1(x)$ is replaced by $g(x)$ we have

$$\int_{-\infty}^{\infty} g(x-y)p(y)\,dy = \frac{1}{\sqrt{\pi}} \int_{-\infty}^{\infty} p(y)\,dy \int_{-\infty}^{\infty} e^{-t^2} g_1(x-y-t)\,dt$$

$$= \frac{1}{\sqrt{\pi}} \int_{-\infty}^{\infty} e^{-t^2}\,dt \int_{-\infty}^{\infty} g_1(x-y-t)p(y)\,dy$$

$$= \frac{1}{\sqrt{\pi}} \int_{-\infty}^{\infty} e^{-t^2} h_1(x-t)\,dt.$$

Since $h_1(x)$ is bounded this last integral exists. By use of (2) and Fubini's theorem the interchange of the order of integration is justified. Since $h_1(x)$ is bounded we may take the limit as $x$ becomes infinite under the sign of the last integral and obtain the limit zero as desired. We now replace $g(x)$ by $g_1(x)$ and assume it to be continuous.

Finally we observe that we may assume that

$$\int_{-\infty}^{\infty} g_1(t)\,dt = 1,$$

for this can always be brought about by multiplying $g_1(t)$ and $g_2(t)$ by suitable constants.

Now consider the function

$$S_\delta(x) = \frac{1}{\delta} \int_x^{x+\delta} p(y)\,dy = \frac{1}{\delta} \int_0^\delta p(x+y)\,dy,$$

where $\delta$ is an arbitrary positive constant. Since $g_1(x)$ belongs to $W$ it is not identically zero. Hence there exists an $x_0$ such that

$$g_1(x_0) = \zeta > 0.$$

By the continuity of $g_1(x)$

$$g_1(x) \geq \zeta/2 \qquad (x_0 - \delta \leq x \leq x_0)$$

for all $\delta$ sufficiently small. By (2)

$$c + h_1(x + x_0) = \int_{-\infty}^{\infty} g_1(y)[p(x + x_0 - y) + c]\,dy$$

$$\geq \frac{\zeta}{2} \int_{x_0-\delta}^{x_0} [p(x + x_0 - y) + c]\,dy$$

$$\geq \frac{\zeta}{2} \int_0^\delta [p(x+t) + c]\,dt \geq \frac{\zeta\delta}{2} S_\delta(x)$$

$$-c \leq S_\delta(x) \leq \frac{2}{\zeta\delta}[c + h_1(x + x_0)].$$

Hence for each sufficiently small positive $\delta$ the function $S_\delta(x)$ is bounded. By (2) it is clear that if $D > \delta > 0$ we have

$$\int_0^D [p(x+y) + c] \, dy = \frac{1}{\delta} \int_0^D [p(x+y) + c] \, dy \int_y^{y+\delta} dt$$

$$\leqq \frac{1}{\delta} \int_0^{D+\delta} dt \int_{t-\delta}^t [p(x+y) + c] \, dy = \int_0^{D+\delta} [S_\delta(x+y-\delta) + c] \, dy$$

$$\int_0^D [p(x+y) + c] \, dy$$

$$\geqq \int_\delta^D [S_\delta(x+y-\delta) + c] \, dy = \int_0^{D-\delta} [S_\delta(x+y) + c] \, dy.$$

Hence

(6)
$$\frac{1}{D} \int_0^{D-\delta} [S_\delta(x+y) + c] \, dy$$

$$\leqq S_D(x) + c \leqq \frac{1}{D} \int_0^{D+\delta} [S_\delta(x+y-\delta) + c] \, dy.$$

But we can show that

(7)
$$\lim_{x \to \infty} \int_{-\infty}^{\infty} g_1(x-y) S_\delta(y) \, dy = 0.$$

For,

$$\int_{-\infty}^{\infty} g_1(x-y)[S_\delta(y) + c] \, dy = \int_{-\infty}^{\infty} g_1(x-y) \, dy \, \frac{1}{\delta} \int_0^\delta [p(u+y) + c] \, du$$

$$= \frac{1}{\delta} \int_0^\delta du \int_{-\infty}^{\infty} [p(u+y) + c] g_1(x-y) \, dy$$

$$= \frac{1}{\delta} \int_x^{x+\delta} [h_1(u) + c] \, du \to c \qquad (x \to \infty),$$

so that (7) is evident.

We are now in a position to apply Theorem 11a to the function $S_\delta(x)$. Choosing the function $g_2(x)$ of that theorem as a suitable step-function we have for every $D > 0$

$$\lim_{x \to \infty} \frac{1}{D} \int_0^D [S_\delta(x+y) + c] \, dy = c.$$

Hence inequalities (6) show that

$$\lim_{x \to \infty} \frac{1}{D} \int_0^D [p(x+y) + c] \, dy = c$$

$$\lim_{x \to \infty} S_D(x) = 0.$$

In particular
$$\lim_{x\to\infty} S_1(x) = \lim_{x\to\infty} \int_x^{x+1} p(t)\,dt = 0.$$

Since
$$\int_{-\infty}^{\infty} |g_2(x-t)|\,[p(t)+c]\,dt$$
$$\leq \underset{-\infty<x<\infty}{\text{u.b.}} \int_x^{x+1} [p(t)+c]\,dt \sum_{n=-\infty}^{\infty} \underset{n\leq x<n+1}{\text{u.b.}} |g_2(x)|,$$

it follows from (3) that
$$h_2(x) = \int_{-\infty}^{\infty} g_2(x-y) p(y)\,dy$$

exists for all $x$.

Let $\epsilon$ be an arbitrary positive number. By (3) we can determine $N = N(\epsilon)$ such that
$$\left\{ \sum_{n=-\infty}^{-N-1} + \sum_{n=N}^{\infty} \right\} \underset{n\leq x<n+1}{\text{u.b.}} |g_2(x)| < \epsilon.$$

Then

(8)
$$\left| h_2(x) - \int_{-N}^{N} g_2(y) p(x-y)\,dy \right|$$
$$\leq \underset{-\infty<x<\infty}{\text{u.b.}} |S_1(x)| \left\{ \sum_{n=-\infty}^{-N-1} + \sum_{n=N}^{\infty} \right\} \underset{n\leq x<n+1}{\text{u.b.}} |g_2(x)|$$
$$< \epsilon \underset{-\infty<x<\infty}{\text{u.b.}} |S_1(x)|.$$

Since $g_2(x)$ is Riemann integrable we can determine step-functions $q_1(x)$ and $q_2(x)$ such that
$$q_1(x) \leq g_2(x) \leq q_2(x) \qquad (-N \leq x \leq N)$$
$$\int_{-N}^{N} [q_2(x) - q_1(x)]\,dx \leq \epsilon.$$

Then
$$\int_{-N}^{N} q_1(y)[p(x-y) + c]\,dy \leq \int_{-N}^{N} g_2(y)[p(x-y) + c]\,dy$$
$$\leq \int_{-N}^{N} q_2(y)[p(x-y) + c]\,dy.$$

By decomposing the extreme integrals into others over intervals in which the functions $q_1(y)$ and $q_2(y)$ are constant and by use of the fact that
$$\lim_{x\to\infty} S_D(x) = 0$$
we obtain
$$\lim_{x\to\infty} \int_{-N}^{N} q_i(y)([p(x-y)+c]\,dy = c\int_{-N}^{N} q_i(y)\,dy \quad (i=1,2)$$

$$c\int_{-N}^{N} q_1(y)\,dy \leq \varlimsup_{x\to\infty} \int_{-N}^{N} g_2(y)[p(x-y)+c]\,dy \leq c\int_{-N}^{N} q_2(y)\,dy$$

$$-c\int_{-N}^{N} [g_2(y) - q_1(y)]\,dy \leq \varlimsup_{x\to\infty} \int_{-N}^{N} g_2(y)p(x-y)\,dy$$

$$\leq c\int_{-N}^{N} [q_2(y) - g_2(y)]\,dy.$$

Since the left extreme is negative, the right positive, and since the two extremes differ by at most the arbitrary positive number $c\epsilon$, we have

(9) $$\lim_{x\to\infty} \int_{-N}^{N} g_2(y)p(x-y)\,dy = 0.$$

But by use of (8) and (9) we see that
$$|h_2(x)| \leq \left| h_2(x) - \int_{N}^{N} g_2(y)p(x-y)\,dy \right| + \left| \int_{-N}^{N} g_2(y)p(x-y)\,dy \right|$$

$$\varlimsup_{x\to\infty} |h_2(x)| \leq \epsilon \underset{-\infty<x<\infty}{\text{u.b.}} |S_1(x)|.$$

Hence
$$\lim_{x\to\infty} h_2(x) = 0.$$

This completes the proof.

For the application of the foregoing result it is convenient to state it in a form in which the interval $(-\infty, \infty)$ is replaced by $(0, \infty)$.

**THEOREM 13b.** *Let*

(10) $$g_1(x) \geq 0, \quad g_1(x) \,\varepsilon\, L \qquad (0 \leq x < \infty),$$

(11) $$\int_{0}^{\infty} t^{ix} g_1(t)\,dt \neq 0 \qquad (-\infty \leq x < \infty),$$

(12) $$p(x) \geq -c \qquad (0 \leq x < \infty),$$

(13) $g_2(x)$ be continuous almost everywhere on $(0, \infty)$,

(14) $$\sum_{n=-\infty}^{\infty} \underset{e^n < x \leq e^{n+1}}{\text{u.b.}} x|g_2(x)| < \infty,$$

(15) $h_1(x) = \dfrac{1}{x} \displaystyle\int_0^\infty g_1\left(\dfrac{t}{x}\right) p(t)\, dt$ exist and be bounded for $(0 < x < \infty)$.

Then if $h_1(\infty)$ exists,

$$\lim_{x \to \infty} \frac{1}{x} \int_0^\infty g_2\left(\frac{t}{x}\right) p(t)\, dt = h_1(\infty) \frac{\displaystyle\int_0^\infty g_2(t)\, dt}{\displaystyle\int_0^\infty g_1(t)\, dt}.$$

This result follows from Theorem 13a by an exponential change of variable.

## 14. Application of Wiener's Theorem to the Laplace Transform

As an example of the way in which Theorem 13b may be applied we prove the following result, which is contained in Theorem 4.3.

**THEOREM 14.** *If $p(t) \geq 0$ for $0 \leq t < \infty$ and is such that the integral*

$$\int_0^\infty e^{-xt} p(t)\, dt$$

*converges for all $x > 0$ and if*

$$\int_0^\infty e^{-xt} p(t)\, dt \sim \frac{A}{x} \qquad (x \to 0+),$$

*then*

$$\int_0^x p(t)\, dt \sim Ax \qquad (x \to \infty).$$

We see by Corollary 1a that

$$\lim_{x \to \infty} \frac{1}{x} \int_0^1 e^{-t/x} p(t)\, dt = 0.$$

Obviously

$$\lim_{x \to \infty} \frac{1}{x} \int_0^1 p(t)\, dt = 0.$$

Hence it is no restriction to suppose that $p(t)$ is identically zero in $(0, 1)$. In Theorem 13b choose

$$g_1(x) = e^{-x} \qquad (0 \leqq x < \infty)$$
$$g_2(x) = 1 \qquad (0 \leqq x \leqq 1)$$
$$= 0 \qquad (1 < x < \infty).$$

Conditions (10), (12) and (13) of Section 13 are obviously satisfied. Condition (11) holds since

$$\int_0^\infty t^{ix} e^{-t} dt = \Gamma(1 + ix) \neq 0 \qquad (-\infty < x < \infty).$$

Condition (14) becomes

$$\sum_{n=-\infty}^{\infty} \operatorname*{u.b.}_{e^n < x \leqq e^{n+1}} x|g_2(x)| = \sum_{n=-\infty}^{-1} e^{n+1} = \frac{e}{e-1}.$$

Finally (15) is satisfied since

$$h_1(x) = \frac{1}{x} \int_1^\infty e^{-t/x} p(t) \, dt$$

is clearly continuous in the interval $(0 < x < \infty)$ and approaches finite limits when $x$ approaches zero or becomes infinite. It approaches $A$ as $x$ becomes infinite, and since

$$h_1(x) < \frac{1}{x} e^{(x-1)/x} \int_1^\infty e^{-t} p(t) \, dt \qquad (0 < x < 1),$$

it is clear that $h_1(x)$ approaches zero with $x$. All conditions of the theorem are satisfied and when we apply it we obtain the desired result. One could easily obtain more general results by altering the choice of $g_1(x)$ and $g_2(x)$. Since no new principles are involved and since we have already obtained the results by Karamata's method, we content ourselves here with the above special case. It should be observed that Karamata's method is much simpler than Wiener's for the particular Laplace kernel $e^{-t/x}$. The great importance of Wiener's methods and results lies in their extreme generality.

## 15. Another Application

As another example of the use of Wiener's theorem let us prove the following special case of Theorem 4.4. Here again the complete theorem could be obtained by Wiener's method, but since we have already obtained the result by a simpler method we content ourselves here with the special case by way of illustration.

## §15] APPLICATION OF WIENER'S THEOREM

THEOREM 15. *If $f(x)$ has a continuous second derivative for positive $x$ such that*

(1) $$f''(x) = O\left(\frac{1}{x^3}\right) \qquad (x \to \infty),$$

*and if*

(2) $$f(x) \sim \frac{A}{x} \qquad (x \to \infty),$$

*then*

(3) $$f'(x) \sim -\frac{A}{x^2} \qquad (x \to \infty).$$

We note first that (1) implies the existence of the integral

$$\int_x^\infty f''(t)\,dt \qquad (x > 0),$$

so that $f'(\infty)$ exists. It must be zero for otherwise we should have

$$\frac{f(x) - f(1)}{x} = \frac{1}{x}\int_1^x f'(t)\,dt \sim f'(\infty) \neq 0 \qquad (x \to \infty),$$

$$f(x) \sim f'(\infty)x \qquad (x \to \infty)$$

contrary to (2). Hence

$$-f'(x) = \int_x^\infty f''(t)\,dt,$$

and from (1)

(4) $$f'(x) = O\left(\frac{1}{x^2}\right) \qquad (x \to \infty).$$

Now consider the equation

$$\int_x^\infty (t-x)f''(t)\,dt = -\int_x^\infty f'(t)\,dt = f(x)$$

obtained by integration by parts, making use of (3) and (4). In Theorem 11b choose

$$g_1(t) = 2(t-1)t^{-3} \qquad (t > 1)$$
$$= 0 \qquad (t \leq 1)$$
$$g_2(t) = 2t^{-3} \qquad (t > 1)$$
$$= 0 \qquad (t \leq 1)$$
$$p(t) = t^3 f''(t) \qquad (0 < t < \infty).$$

Clearly $g_1(t)$ and $g_2(t)$ belong to $L$ on $(0, \infty)$. In fact their integrals over that range are both equal to unity. Also $p(t)$ belongs to $B$ by (1) and

$$\int_0^\infty g_1(t) t^{ix} dt = \frac{2}{2 - 3ix - x^2} \neq 0 \qquad (-\infty < x < \infty).$$

Then

$$\frac{1}{x} \int_0^\infty g_1\left(\frac{t}{x}\right) p(t) dt = 2x \int_x^\infty (t - x) f''(t) dt$$
$$= 2xf(x) \sim 2A \qquad (x \to \infty),$$

and

$$\frac{1}{x} \int_0^\infty g_2\left(\frac{t}{x}\right) p(t) dt = 2x^2 \int_x^\infty f''(t) dt = -2x^2 f'(x) \sim 2A.$$

This gives (3) at once.

### 16. The Prime-number Theorem

As a further application of Wiener's theorem we prove the *prime-number theorem*. This states that the number $\pi(x)$ of primes not greater than $x$ is asymptotic to $x/\log x$ as $x$ becomes infinite. It was conjectured by Gauss and first proved independently and simultaneously by Hadamard and de la Vallée Poussin in 1896. The methods employed then depended on a knowledge of the function

$$\zeta(s) = \sum_{n=1}^\infty \frac{1}{n^s}$$

*in a region* of the complex $s$-plane. Wiener was the first to give a proof of purely Tauberian character and one which involves the behavior of $\zeta(s)$ merely *on the line* $\sigma = 1$. It is this proof which we shall give here.

16.1. We begin by proving several lemmas of elementary nature.

**DEFINITION 16.1a.** *The function $\pi(x)$ is defined by the equation*

$$\pi(x) = \sum_{p \leq x} 1 \qquad (0 \leq x < \infty).$$

The notation means that the number of units to be added together is the number of positive primes $p$ not greater than $x$. The integer one is not considered to be a prime. Thus

$$\pi(1) = 0, \qquad \pi(2) = 1, \qquad \pi(5) = 3, \qquad \pi(5.7) = 3.$$

**DEFINITION 16.1b.** *The function $\Lambda(n)$ is defined by the equations*

$$\begin{aligned}\Lambda(n) &= \log p & n &= p^m \\ &= 0 & n &\neq p^m\end{aligned} \qquad (m = 1, 2, 3, \ldots).$$

§16] THE PRIME-NUMBER THEOREM 225

For example,
$$\Lambda(1) = \Lambda(6) = \Lambda(14) = 0$$
$$\Lambda(3) = \Lambda(27) = \log 3.$$

DEFINITION 16.1c. *The function $\psi(x)$ is defined by the equation*
$$\psi(x) = \sum_{n \leq x} \Lambda(n) \qquad (0 \leq x < \infty).$$

For example,
$$\psi(1) = 0 \qquad \psi(2) = \log 2 \qquad \psi(8) = \log(2^3 \cdot 3 \cdot 5 \cdot 7).$$

DEFINITION 16.1d. *The function $\vartheta(x)$ is defined by the equation*
$$\vartheta(x) = \sum_{p \leq x} \log p \qquad (0 \leq x < \infty).$$

For example,
$$\vartheta(1) = 0, \qquad \vartheta(2) = \log 2, \qquad \vartheta(8) = \log 2 \cdot 3 \cdot 5 \cdot 7.$$

It is clear that

(1) $$\psi(x) = \vartheta(x) + \vartheta(\sqrt{x}) + \cdots + \vartheta(\sqrt[n]{x}),$$

where $n$ is an integer such that

(2) $$2^n \leq x < 2^{n+1}.$$

THEOREM 16.1. *The function $\psi(x)/x$ is bounded,*

(3) $$\psi(x) = O(x) \qquad (x \to \infty).$$

If we can prove that

(4) $$\vartheta(x) = O(x) \qquad (x \to \infty),$$

the theorem will follow from (1) and (2). For, there will exist a constant $A$ such that
$$\psi(x) \leq Ax + A\sqrt{x} + \cdots + A\sqrt[n]{x} < Ax + \frac{A\sqrt{x}\log x}{\log 2},$$
from which (3) is evident.

To prove (4) we observe that if $m$ is an integer, the binomial coefficient
$$\frac{(m+1)(m+2)\cdots(2m)}{m!} = \frac{(2m)!}{m!\,m!}$$
is an integer less than $(1+1)^{2m}$. It is divisible by all the primes between $m+1$ and $2m$. Hence
$$\prod_{m < p \leq 2m} p < 2^{2m},$$

or
$$\prod_{2^{k-1}<p\leq 2^k} p < 2^{2^k}$$

for any integer $k$. That is,

$$\vartheta(2^k) - \vartheta(2^{k-1}) < 2^k \log 2 \qquad (k = 1, 2, \cdots).$$

Summing with respect to $k$ we obtain

$$\vartheta(2^k) < 2^{k+1} \log 2.$$

For any $x$ we can determine an integer $k$ such that $2^{k-1} < x \leq 2^k$. Then

$$\vartheta(x) \leq \vartheta(2^k) < 2^{k+1} \log 2 < 4x \log 2.$$

This completes the proof of the theorem.

16.2. We now show that to prove the prime number theorem it will be sufficient to prove that $\psi(x)$ is asymptotic to $x$ as $x$ becomes infinite.

THEOREM 16.2. *The following relations connect $\pi(x)$ and $\psi(x)$:*

$$\varliminf_{x\to\infty} \frac{\psi(x)}{x} \leq \varliminf_{x\to\infty} \frac{\pi(x) \log x}{x} \leq \varlimsup_{x\to\infty} \frac{\pi(x) \log x}{x} \leq \varlimsup_{x\to\infty} \frac{\psi(x)}{x}.$$

Note first that if we define $[x]$ as the largest integer not greater than $x$ we may write

$$\psi(x) = \sum_{p \leq x} \left[\frac{\log x}{\log p}\right] \log p.$$

For, by the definition of $\psi(x)$,

$$\psi(x) = \sum_{p^m \leq x} \log p,$$

it is clear that for a given prime $p$, $\log p$ is to be added as many times as there are terms in the sequence

$$p, p^2, \cdots, p^r \qquad (p^r \leq x < p^{r+1}),$$

that is $[\log x/\log p]$ times. Since $[x]$ is not greater than $x$ we have

$$\psi(x) < \sum_{p \leq x} \log x = \pi(x) \log x$$

$$\varliminf_{x\to\infty} \frac{\pi(x) \log x}{x} \geq \varliminf_{x\to\infty} \frac{\psi(x)}{x}.$$

To complete the proof of the theorem set

$$y = \frac{x}{(\log x)^2} < x \qquad (x > e).$$

§16] THE PRIME-NUMBER THEOREM 227

Then
$$\pi(x) = \pi(y) + \sum_{y < p \leq x} \frac{\log p}{\log p} \leq \pi(y) + \frac{1}{\log y} \sum_{y < p \leq x} \log p$$
$$\leq y + \frac{\psi(x)}{\log y},$$
$$\frac{\pi(x) \log x}{x} \leq \frac{1}{\log x} + \frac{\psi(x)}{x} \frac{\log x}{\log x - 2 \log \log x},$$
$$\varlimsup_{x \to \infty} \frac{\pi(x) \log x}{x} \leq \varlimsup_{x \to \infty} \frac{\psi(x)}{x}.$$

**16.3.** We now introduce a new function by

**DEFINITION 16.3.** *The function $h(x)$ is defined by the equation*
$$h(x) = \int_{1/2}^{x} \frac{d(\psi(t) - [t])}{t} = \sum_{n \leq x} \frac{\Lambda(n) - 1}{n}.$$

We show in this section that it is bounded.

**THEOREM 16.3.** *The function $h(x)$ is bounded,*
$$h(x) = O(1) \qquad (x \to \infty).$$

To prove this we use the familiar fact that the highest power of a prime $p$ contained in $m!$ is
$$\left[\frac{m}{p}\right] + \left[\frac{m}{p^2}\right] + \cdots.$$

This makes it clear that
$$m! = \prod_{p \leq m} p^{\left[\frac{m}{p}\right] + \left[\frac{m}{p^2}\right] + \cdots}$$
$$\log m! = \sum_{p^n \leq m} \left[\frac{m}{p^n}\right] \log p$$
(1) $$\log m! = \sum_{n \leq m} \left[\frac{m}{n}\right] \Lambda(n) = \int_{1/2}^{m} \left[\frac{m}{t}\right] d\psi(t).$$

By use of Stirling's formula and Theorem 16.1 we have from equation (1)
$$m \int_{1/2}^{m} \frac{d\psi(t)}{t} = m \log m + O(m) \qquad (m \to \infty).$$

But
$$\int_{1/2}^{m} \frac{d[t]}{t} = \frac{[m]}{m} + \int_{1/2}^{m} \frac{[t]}{t^2} dt$$
$$= O(1) + \int_{1/2}^{m} \frac{dt}{t} = O(1) + \log m \qquad (m \to \infty).$$

Hence
$$\int_{/2}^{m} \frac{d\{\psi(t) - [t]\}}{t} = O(1) \qquad (m \to \infty).$$

**16.4.** We next consider the number $\tau(m)$ of divisors of an integer $m$.

**DEFINITION 16.4a.** *The function $\tau(m)$ is defined by the equation*
$$\tau(m) = \sum_{d/m} 1.$$

The notation means that a number of units equal to the number of divisors $d$ (unity and $m$ included) of $m$ is to be added. For example
$$\tau(1) = 1, \qquad \tau(2) = \tau(3) = 2, \qquad \tau(12) = 6.$$

**DEFINITION 16.4b.** *The function $T(m)$ is defined by the equation*
$$T(m) = \sum_{n=1}^{m} \tau(n).$$

**THEOREM 16.4a.** *If $\nu = [\sqrt{m}]$, then*
$$T(m) = 2\left\{ \left[\frac{m}{1}\right] + \left[\frac{m}{2}\right] + \cdots + \left[\frac{m}{\nu}\right] \right\} - \nu^2.$$

We give a graphic proof of this result. Consider the number of points, both of whose coördinates are integers, in the region $R$ of the $xy$-plane defined by the relation
$$xy \leq m, \qquad x > 0, \qquad y > 0.$$

We call these points lattice points. The number of them on the curve $xy = k$ ($k$ an integer) is clearly $\tau(k)$, so that the total number is $T(m)$. We now count them in a different way.

The number of lattice points of $R$ on a line $x = k$ ($k$ an integer) is $[m/k]$. Hence the number in the region
$$R_1 : \quad 1 \leq x \leq \nu, \qquad 1 \leq y \leq m/x$$
is
$$\left[\frac{m}{1}\right] + \left[\frac{m}{2}\right] + \cdots + \left[\frac{m}{\nu}\right];$$
by symmetry there is an equal number in $R_2$:
$$1 \leq y \leq \nu, \qquad 1 \leq x \leq m/y.$$

Since there are no lattice points in the region
$$xy \leq m \qquad x > \nu \qquad y > \nu,$$

(because $(\nu + 1)^2 > m$), we see that the number of lattice points in $R$ is the sum of the numbers in $R_1$ and $R_2$ less the number $\nu^2$ in the square

$$1 \leq x \leq \nu \qquad 1 \leq y \leq \nu.$$

This gives the number stated in the theorem.

**THEOREM 16.4b.** *If we denote Euler's constant by $\gamma$, then*

(1) $$T(n) = n \log n + (2\gamma - 1)n + O(\sqrt{n}) \qquad (n \to \infty).$$

By Theorem 16.4a we have

(2) $$T(n) = 2n \sum_{k=1}^{\nu} \frac{1}{k} + O(\nu) - \nu^2.$$

But it is a familiar fact* that

(3) $$\sum_{k=1}^{n} \frac{1}{k} = \log n + \gamma + O\left(\frac{1}{n}\right) \qquad (n \to \infty).$$

Substituting (3) in (2) we obtain (1).

**16.5.** In this section we show that the function $\zeta(s)$ does not vanish on the line $\sigma = 1$. This is really the vital point of the proof. Later in applying Wiener's theorem it will be necessary to show that a certain kernel belongs to $W$, and it is precisely the non-vanishing of $\zeta(1 + i\tau)$ that will prove this.

**THEOREM 16.5.** *The function $\zeta(s)$ is not zero on the line $\sigma = 1$:*

$$\zeta(1 + i\tau) \neq 0 \qquad (-\infty < \tau < \infty).$$

Consider the function

$$\varphi = |\zeta(1 + \epsilon)|^{3/4} |\zeta(1 + \epsilon + i\tau)| |\zeta(1 + \epsilon + 2i\tau)|^{1/4}.$$

By expanding the $n$th factor of the product

$$\prod_{p_n \leq N} \frac{1}{1 - p_n^{-s}}$$

in powers of $p_n^{-s}$ and multiplying the resulting series together, we see easily that

$$\left| \prod_{p_n \leq N} \frac{1}{1 - p_n^{-s}} - \sum_{n=1}^{N} \frac{1}{n^s} \right| \leq \sum_{n=N+1}^{\infty} \frac{1}{n^\sigma} \qquad (\sigma > 1),$$

so that

$$\zeta(s) = \prod_{n=1}^{\infty} \frac{1}{1 - p_n^{-s}} \qquad (\sigma > 1)$$

---

* Compare Pólya and Szegö [1925] vol. 1, p. 197.

230    TAUBERIAN THEOREMS    [CH. V

and

(1) $$\log \zeta(s) = \sum_{n=1}^{\infty} \log \frac{1}{1 - p_n^{-s}} = \sum_{n=1}^{\infty} \sum_{m=1}^{\infty} \frac{1}{m} p_n^{-ms} \qquad (\sigma > 1).$$

It is thus clear that $\zeta(s)$ has no zeros for $\sigma > 1$.

The real part of $\log \zeta(s)$ is seen to be

$$\sum_{n=1}^{\infty} \sum_{m=1}^{\infty} \frac{1}{m} p_n^{-m\sigma} \cos(m\tau \log p_n).$$

Hence

$$\log \varphi = \sum_{n=1}^{\infty} \sum_{m=1}^{\infty} \frac{1}{m} p_n^{-m(1+\epsilon)} \{\tfrac{3}{4} + \cos(m\tau \log p_n) + \tfrac{1}{4} \cos(2m\tau \log p_n)\}.$$

But

$$\tfrac{3}{4} + \cos x + \tfrac{1}{4} \cos 2x = \tfrac{1}{2}[1 + \cos x]^2 \geq 0,$$

so that

(2) $\qquad\qquad\qquad \log \varphi \geq 0, \qquad \varphi \geq 1.$

Since

$$(1 - 2^{1-s})\zeta(s) = \sum_{n=1}^{\infty} (-1)^{n+1} n^{-s} \qquad (\sigma > 1)$$

$$(1 - 3^{1-s})\zeta(s) = \sum_{n=1}^{\infty} \left\{ \frac{1}{(3n-2)^s} + \frac{1}{(3n-1)^s} - \frac{2}{(3n)^s} \right\} \qquad (\sigma > 1),$$

both series converging to analytic functions for $\sigma > 0$, one sees from the first equation that $\zeta(s)$ is analytic for $\sigma > 0$ except perhaps at

$$s = 1 + \frac{2k\pi i}{\log 2} \qquad (k = 0, \pm 1, \pm 2, \cdots)$$

and from the second that $\zeta(s)$ is analytic for $\sigma > 0$ except perhaps at

$$s = 1 + \frac{2k\pi i}{\log 3} \qquad (k = 0, \pm 1, \pm 2, \cdots).$$

That is, $\zeta(s)$ must be analytic for $\sigma > 0$ except for a simple pole at $s = 1$. If $\zeta(s)$ had a zero at $s = 1 + i\tau_0$, $\tau_0 \neq 0$, we should have

$$|\zeta(1 + \epsilon + i\tau_0)| = O(\epsilon) \qquad (\epsilon \to 0).$$

$$|\zeta(1 + \epsilon + 2i\tau_0)|^{1/4} = O(1)$$

$$|\zeta(1 + \epsilon)|^{3/4} = O(\epsilon^{-3/4})$$

$$\varphi = O(\epsilon^{1/4}),$$

contradicting (2). Hence $\zeta(1 + i\tau)$ vanishes for no value of $\tau$.

## §16] THE PRIME-NUMBER THEOREM

**16.6.** We next introduce a function $f(x)$ defined for $x > 0$ by the Lambert series

$$(1) \qquad f(x) = \sum_{n=1}^{\infty} \frac{\{\Lambda(n) - 1\}e^{-nx}}{1 - e^{-nx}} = \int_0^{\infty} \frac{te^{-xt}}{1 - e^{-xt}} dh(t)$$

and prove

THEOREM 16.6. *The function $f(x)$ defined by (1) satisfies the relation*

$$f(x) \sim \frac{-2\gamma}{x} \qquad (x \to 0+).$$

Expanding the general term of the series (1) in power series we obtain

$$f(x) = \sum_{n=1}^{\infty} \sum_{m=1}^{\infty} \{\Lambda(n) - 1\} e^{-mnx} \qquad (x > 0).$$

Since

$$\Lambda(n) - 1 = O(\log n) \qquad (n \to \infty),$$

it is clear that the double series converges absolutely for $x > 0$. Hence

$$f(x) = \sum_{n=1}^{\infty} c_n e^{-nx}$$

$$c_n = \sum_{d/n} (\Lambda(d) - 1),$$

where the latter summation is extended over all the divisors of $n$. But if $n$ is factored into prime factors

$$n = p_1^{\alpha_1} \cdots p_k^{\alpha_k},$$

then

$$\sum_{d/n} \Lambda(d) = \alpha_1 \log p_1 + \cdots + \alpha_k \log p_k = \log n$$

$$c_n = \log n - \tau(n)$$

$$\sum_{k=1}^{n} c_k = \log n! - T(n)$$

$$= -2\gamma n + O(\sqrt{n})$$

by Stirling's formula and Theorem 16.4b. Now by Corollary 1a our result is established.

**16.7.** We are now ready to apply Theorem 10b. Take

$$g(t) = \frac{d}{dt}\left\{\frac{te^{-t}}{1 - e^{-t}}\right\}.$$

Then integrating by parts gives

$$f(x) = \int_0^\infty \frac{te^{-xt}}{1 - e^{-xt}} dh(t) = -\int_0^\infty h(t)g(xt)\,dt.$$

By Theorem 16.6

$$\frac{1}{x}\int_0^\infty g\left(\frac{t}{x}\right) h(t)\,dt \to 2\gamma \qquad (x \to \infty).$$

Now $g(x)$ belongs to $L$. In fact $g(t)$ is never positive and

$$\int_0^\infty g(t)\,dt = -1.$$

Also

$$\int_0^\infty t^{-xi} g(t)\,dt = \int_0^\infty t^{-xi} \frac{d}{dt}\left(\frac{te^{-t}}{1 - e^{-t}}\right) dt$$

$$= \lim_{\delta \to 0+} \int_0^\infty t^{-xi+\delta} \frac{d}{dt}\left(\frac{te^{-t}}{1 - e^{-t}}\right) dt$$

$$= \lim_{\delta \to 0+} (xi - \delta) \int_0^\infty \frac{e^{-t} t^{-xi+\delta}}{1 - e^{-t}} dt.$$

The last step is obtained by integration by parts. The integrated term vanishes at $t = 0$ by virtue of the factor $t^\delta$, at $t = \infty$ by virtue of the factor $e^{-t}$. By a familiar formula of Riemann

$$\zeta(s)\Gamma(s) = \int_0^\infty \frac{t^{s-1}}{e^t - 1} dt \qquad (\sigma > 1).$$

Hence

$$\int_0^\infty t^{-xi} g(t)\,dt = \lim_{\delta \to 0+} (xi - \delta)\Gamma(1 - xi + \delta)\zeta(1 - xi + \delta)$$

$$= xi\zeta(1 - xi)\Gamma(1 - xi) \qquad (x \neq 0)$$

$$= -1 \qquad (x = 0).$$

Since we have already seen that $h(t)$ belongs to $B$ we need only show that it belongs to $S^*$ to apply Theorem 10b. For this we have

$$h(y) - h(x) = \sum_{x < n \leq y} \frac{\Lambda(n) - 1}{n} > -\sum_x^y \frac{1}{n} \qquad (y > x).$$

By §16.4 (3) it is clear that the right-hand side of this inequality tends to zero as $x$ becomes infinite, $y/x$ approaching unity. Hence

$$\lim \{h(y) - h(x)\} \geq 0 \qquad \left(x \to \infty, \frac{y}{x} \to 1\right).$$

That is, $h(x)$ belongs to $S^*$.

The conclusion of Theorem 10b is that $h(\infty) = -2\gamma$ or
$$\int_{1/2}^{\infty} \frac{1}{t} d\{\psi(t) - [t]\} = -2\gamma.$$
But in the proof of Theorem 3b we saw that if $\alpha(\infty) = A$, then
$$\int_0^t u \, d\alpha(u) = o(t) \qquad (t \to \infty).$$
This result applied here gives
$$\psi(n) - n = o(n) \qquad (n \to \infty)$$
$$\psi(n) \sim n \qquad (n \to \infty).$$
By Theorem 16.2 the proof of the prime-number theorem is now complete.

## 17. Ikehara's Theorem

Another method of approach to the prime number theorem is through a complex variable Tauberian theorem due to Ikehara* [1].

THEOREM 17. *If $\varphi(t)$ is a non-negative, non-decreasing function in $(0 \leq t < \infty)$ such that the integral*
$$f(s) = \int_0^\infty e^{-st} \varphi(t) \, dt \qquad (s = \sigma + i\tau)$$
*converges for $\sigma > 1$, and if for some constant $A$ and some function $g(\tau)$*

(1) $$\lim_{\sigma \to 1+} f(s) - \frac{A}{s - 1} = g(\tau)$$

*uniformly in every finite interval $(-a \leq \tau \leq a)$, then*
$$\lim_{t \to \infty} \varphi(t) e^{-t} = A.$$

Set
$$a(t) = e^{-t} \varphi(t) \qquad (t > 0)$$
$$= 0 \qquad (t \leq 0)$$
$$A(t) = A \qquad (t > 0)$$
$$= 0 \qquad (t \leq 0)$$

(2) $$I_\lambda(x) = \frac{1}{\sqrt{2\pi}} \int_{-\infty}^{\infty} k_\lambda(x - t)[a(t) - A(t)]e^{-\epsilon t} \, dt,$$

*Compare also N. Wiener [1933] p. 127, Theorem 16.

where $\epsilon$ is a positive number and $k_\lambda(x)$ is the function defined in Section 9 for every positive $\lambda$. Since $K_\lambda(x)$ is the Fourier transform of $k_\lambda(x)$, we have by Fubini's theorem

$$I_\lambda(x) = \frac{1}{2\pi} \int_{-\infty}^{\infty} [a(t) - A(t)]e^{-\epsilon t} dt \int_{-2\lambda}^{2\lambda} K_\lambda(y) e^{-iy(x-t)} dy$$

$$= \frac{1}{2\pi} \int_{-2\lambda}^{2\lambda} K_\lambda(y) e^{-ixy} dy \int_{-\infty}^{\infty} [a(t) - A(t)]e^{-\epsilon t + iyt} dt$$

$$= \frac{1}{2\pi} \int_{-2\lambda}^{2\lambda} K_\lambda(y) e^{-ixy} \left[ f(1 + \epsilon - iy) - \frac{A}{\epsilon - iy} \right] dy.$$

By (1)

$$\lim_{\epsilon \to 0+} f(1 + \epsilon - iy) - \frac{A}{\epsilon - iy} = g(y)$$

uniformly in $(-2\lambda \leq y \leq 2\lambda)$. Hence

(3) $$\lim_{\epsilon \to 0+} I_\lambda(x) = \frac{1}{2\pi} \int_{-2\lambda}^{2\lambda} K_\lambda(y) e^{-ixy} g(y)\, dy.$$

But by Corollary 1c

(4) $$\lim_{\epsilon \to 0+} \frac{A}{\sqrt{2\pi}} \int_0^\infty k_\lambda(x - t) e^{-\epsilon t} dt = \frac{A}{\sqrt{2\pi}} \int_0^\infty k_\lambda(x - t)\, dt.$$

Writing the integral (2) as the sum of two integrals we have

(5) $$I_\lambda(x) = \frac{1}{\sqrt{2\pi}} \int_0^\infty k_\lambda(x - t) a(t) e^{-\epsilon t} dt - \frac{A}{\sqrt{2\pi}} \int_0^\infty k_\lambda(x - t) e^{-\epsilon t} dt.$$

We see by (3) and (4) that when $\epsilon$ approaches zero the first integral on the right-hand side of (5) approaches a finite limit, which we shall call $C$. Since the integrand of this integral is non-negative, the integral

$$\int_0^\infty k_\lambda(x - t) a(t)\, dt.$$

must exist. For, otherwise we should have by Corollary 1c that $C = \infty$, contradicting the above result. Applying Corollary 1c to the integral (2) we see that

$$\lim_{\epsilon \to 0} I_\lambda(x) = \frac{1}{\sqrt{2\pi}} \int_{-\infty}^\infty k_\lambda(x - t)[a(t) - A(t)]\, dt,$$

so that by (3)

$$\frac{1}{2\pi} \int_{-2\lambda}^{2\lambda} K_\lambda(y) e^{-ixy} g(y)\, dy = \frac{1}{\sqrt{2\pi}} \int_{-\infty}^\infty k_\lambda(x - t)[a(t) - A(t)]\, dt.$$

By the Riemann-Lebesgue theorem the left-hand integral approaches zero as $x$ becomes infinite. Hence

(6) $$\lim_{x \to \infty} \frac{1}{\sqrt{2\pi}} \int_{-\infty}^{\infty} k_\lambda(x - t)a(t)\,dt = A$$

for *every* positive $\lambda$. If we can show that $a(t)$ belongs to $S$ and to $B$, then by Theorem 9 we shall have

$$\lim_{t \to \infty} a(t) = A$$

and the theorem will be established.

To show that $a(t)$ is slowly decreasing we have for $\delta > 0$

$$a(x + \delta) - a(x) = e^{-x}[e^{-\delta}\varphi(x + \delta) - \varphi(x)]$$
$$\geq e^{-x}\varphi(x)(e^{-\delta} - 1) = a(x)(e^{-\delta} - 1)$$

since $\varphi(x)$ is non-decreasing. If $a(x)$ is bounded then clearly

$$\lim_{\substack{x \to \infty \\ \delta \to 0+}} [a(x + \delta) - a(x)] \geq 0,$$

and $a(x)$ belongs to $S$.

Finally to show $a(x)$ bounded we have by (6) for every positive $\lambda$

$$\frac{1}{\sqrt{2\pi}} \int_{-\infty}^{\infty} a(x - t) 2\lambda \delta(2\lambda t)\,dt \to A \qquad (x \to \infty)$$

$$\frac{1}{\sqrt{2\pi}} \int_{-\infty}^{\infty} a\left(x - \frac{t}{2\lambda}\right)\delta(t)\,dt \to A \qquad (x \to \infty)$$

$$\frac{1}{\sqrt{2\pi}} \int_{-\infty}^{\infty} a\left(x + \frac{1}{\sqrt{\lambda}} - \frac{t}{2\lambda}\right)\delta(t)\,dt \to A \qquad (x \to \infty).$$

Hence there exists a number $x_0$ such that

$$\frac{1}{\sqrt{2\pi}} \int_{-\infty}^{\infty} a\left(x + \frac{1}{\sqrt{\lambda}} - \frac{t}{2\lambda}\right)\delta(t)\,dt < A + 1 \qquad (x > x_0).$$

Since the integrand is positive

$$\frac{1}{\sqrt{2\pi}} \int_{-\sqrt{\lambda}}^{\sqrt{\lambda}} a\left(x + \frac{1}{\sqrt{\lambda}} - \frac{t}{2\lambda}\right)\delta(t)\,dt < A + 1 \qquad (x > x_0).$$

On account of the increasing character of $a(t)e^t$ we have

$$\frac{1}{\sqrt{2\pi}} a(x) \int_{-\sqrt{\lambda}}^{\sqrt{\lambda}} e^{-\frac{1}{\sqrt{\lambda}} + \frac{t}{2\lambda}} \delta(t)\,dt < A + 1 \qquad (x > x_0),$$

or

$$\frac{1}{\sqrt{2\pi}} a(x) e^{-\frac{3}{2\sqrt{\lambda}}} \int_{-\sqrt{\lambda}}^{\sqrt{\lambda}} \delta(t)\,dt < A + 1.$$

Hence

$$a(x) < \frac{\sqrt{2\pi}\,(A+1)e^{\frac{3}{2\sqrt{\lambda}}}}{\int_{-\sqrt{\lambda}}^{\sqrt{\lambda}} \delta(t)\,dt} \qquad (x > x_0).$$

Since the left-hand side is independent of $\lambda$ we may let $\lambda$ become infinite to obtain

$$0 \leqq a(x) \leqq A + 1 \qquad (x > x_0).$$

This completes the proof of our theorem.

This theorem enables us to prove the prime-number theorem very simply.* By differentiating both sides of equation §16.5 (1) we have

$$-\frac{\zeta'(s)}{\zeta(s)} = \sum_{n=1}^{\infty} \sum_{m=1}^{\infty} p_n^{-ms} \log p_n \qquad (\sigma > 1)$$

$$= \sum_{n=1}^{\infty} \Lambda(n) n^{-s}$$

$$= \int_1^{\infty} x^{-s}\, d\psi(x)$$

$$= \int_0^{\infty} e^{-st}\, d\psi(e^t)$$

(7)
$$= s \int_0^{\infty} e^{-st} \psi(e^t)\, dt \qquad (\sigma > 1).$$

But we saw in Section 16.5 that $\zeta(s)$ is analytic for $\sigma \geqq 1$ except for a simple pole at $s = 1$ and has no zeros for $\sigma \geqq 1$. Hence the function

$$-\frac{\zeta'(s)}{s\zeta(s)} - \frac{1}{s-1}$$

approaches a limit uniformly in $(-a \leqq \tau \leqq a)$ for every positive constant $a$ as $\sigma$ approaches unity through values greater than unity. Since $\psi(e^t)$ is surely non-negative and non-decreasing we are in a position to apply Theorem 17 to the integral (7). The conclusion is that

$$\psi(e^t) \sim e^t \qquad (t \to \infty),$$

or

$$\psi(x) \sim x \qquad (x \to \infty).$$

As we have seen, this result is equivalent to the prime-number theorem.

* Compare G. Doetsch [1937b].

# CHAPTER VI

# THE BILATERAL LAPLACE TRANSFORM

## 1. Introduction

By the bilateral transform we mean a Laplace integral whose limits of integration are $-\infty$ and $+\infty$,

$$(1) \qquad f(s) = \int_{-\infty}^{\infty} e^{-st} \, d\alpha(t).$$

Here we assume that $\alpha(t)$ is of bounded variation in every finite interval. In particular, if $\alpha(t)$ is an integral of a function $\phi(t)$, the integral (1) becomes

$$(2) \qquad f(s) = \int_{-\infty}^{\infty} e^{-st} \phi(t) \, dt.$$

We say that the integral (1) converges if and only if the limits

$$\lim_{R \to \infty} \int_{0}^{R} e^{-st} \, d\alpha(t)$$

$$\lim_{R \to \infty} \int_{-R}^{0} e^{-st} \, d\alpha(t)$$

both exist. The sum of these limits is then defined as the value $f(s)$ of the integral (1).

We say that $\alpha(t)$ is normalized in $(-\infty, \infty)$ if and only if $\alpha(0) = 0$ and

$$(3) \qquad \alpha(t) = \frac{\alpha(t+) + \alpha(t-)}{2} \qquad (-\infty < t < \infty).$$

It should be observed that a function normalized in $(0, \infty)$ and zero in $(-\infty, 0)$ is not necessarily normalized* in $(-\infty, \infty)$.

When the integral (1) converges for a given value of $s$ we have

$$(4) \qquad f(s) = \int_{0}^{\infty} e^{-st} \, d\alpha(t) + \int_{0}^{\infty} e^{st} \, d[-\alpha(-t)].$$

---

* This difficulty might be avoided by adopting the convention $\alpha(-\infty) = 0$ instead of $\alpha(0) = 0$. But then we should be excluding the important class of functions for which $\alpha(-\infty)$ does not exist. Moreover, the convention adopted enables us to reduce more easily the study of the bilateral integral to that of the unilateral integral treated in Chapter II.

Thus the study of the bilateral transform is reduced to that of the sum of two unilateral transforms in one of which the variable $s$ has been replaced by $-s$. We can consequently derive results concerning (1) from those of chapter II. However, in some cases the proofs are much more complicated than might be expected from the relation (4). In other cases the proofs are trivial, and the theorems are recorded merely for convenience of reference.

## 2. Region of Convergence

Since the unilateral integral converges in a right half-plane, it is clear from §1 (4) that the region of convergence of the bilateral transform is a vertical strip of the complex $s$-plane or modifications thereof. The bilateral transform is clearly the continuous analogue of the Laurent series

$$(1) \qquad F(z) = \sum_{n=-\infty}^{\infty} a_n z^n,$$

as one sees more clearly after the transformation $z = e^{-s}$. As the region of convergence of (1) is the region between two concentric circles, it is natural to expect that the region of convergence of §1 (1) is a vertical strip. We have in fact:

THEOREM 2. *If the integral*

$$(2) \qquad f(s) = \int_{-\infty}^{\infty} e^{-st} \, d\alpha(t)$$

*converges for two points* $s_1 = \sigma_1 + i\tau_1$ *and* $s_2 = \sigma_2 + i\tau_2 (\sigma_1 < \sigma_2)$, *then it converges in the vertical strip* $\sigma_1 < \sigma < \sigma_2$.

This result follows in a trivial manner from Corollary 1a of Chapter II. It is clear from examples given there that the strip may become a right-half-plane, a left-half-plane, or the entire plane. In fact if both integrals §1 (4) have the same axis of convergence the strip may reduce to this single vertical line. For example, the integral

$$\int_{-\infty}^{\infty} \frac{e^{-st}}{1+t^2} \, dt$$

converges absolutely on the whole line $\sigma = 0$ and nowhere else. Finally the integral may have as its region of convergence certain parts of a vertical line. Thus if $\phi(t) = |t|^{-1/2}$, then the integral §1 (2) converges on the line $\sigma = 0$ except at the origin, and diverges at all points off this line.

If the integral (2) converges in the strip $\sigma'_c < \sigma < \sigma''_c$ and diverges for $\sigma > \sigma''_c$ and for $\sigma < \sigma'_c$ then each of the lines $\sigma = \sigma'_c$ and $\sigma = \sigma''_c$ is called an *axis of convergence* and each of the members $\sigma'_c$ and $\sigma''_c$ is an *abscissa of convergence*.

## 3. Integration by Parts

We here obtain sufficient conditions for the integration by parts of a bilateral Laplace integral. As related results we state first two theorems regarding the behavior of $\alpha(t)$ at $+\infty$ and at $-\infty$.

**THEOREM 3a.** *If the integral*

$$(1) \qquad f(s) = \int_{-\infty}^{\infty} e^{-st}\, d\alpha(t)$$

*converges for* $s = s_0 = \gamma + i\delta$ *with* $\gamma > 0$ *then* $\alpha(-\infty)$ *exists and*

$$\alpha(t) = o(e^{\gamma t}) \qquad (t \to \infty)$$
$$\alpha(t) - \alpha(-\infty) = o(e^{\gamma t}) \qquad (t \to -\infty).$$

This result follows from the decomposition §1 (4) by use of Theorems 2.2a and 2.2b of Chapter II.

**THEOREM 3b.** *If the integral (1) converges for* $s = s_0 = \gamma + i\delta$ *with* $\gamma < 0$ *then* $\alpha(\infty)$ *exists and*

$$\alpha(t) - \alpha(\infty) = o(e^{\gamma t}) \qquad (t \to \infty)$$
$$\alpha(t) = o(e^{\gamma t}) \qquad (t \to -\infty).$$

The proof is similar to that of Theorem 3a. Both theorems fail if $\gamma = 0$.

We give the conditions for integration by parts mentioned above.

**THEOREM 3c.** *If the integral (1) converges for* $s = s_0 = \gamma + i\delta$ *with* $\gamma > 0$ *and if* $\alpha(-\infty) = 0$, *then*

$$(2) \qquad f(s_0) = s_0 \int_{-\infty}^{\infty} e^{-s_0 t} \alpha(t)\, dt.$$

For, by Theorem 2.3a of Chapter II

$$(3) \qquad \int_0^\infty e^{-s_0 t}\, d\alpha(t) = s_0 \int_0^\infty e^{-s_0 t} \alpha(t)\, dt - \alpha(0),$$

and by Theorem 2.3b

$$(4) \qquad \int_0^\infty e^{s_0 t}\, d[-\alpha(-t)] = s_0 \int_0^\infty e^{s_0 t} \alpha(-t)\, dt + \alpha(0).$$

Adding equations (3) and (4) gives equation (2).

A companion result is:

**THEOREM 3d.** *If the integral (1) converges for* $s = s_0 = \gamma + i\delta$ *with* $\gamma < 0$ *and if* $\alpha(\infty) = 0$, *then*

$$f(s_0) = s_0 \int_{-\infty}^{\infty} e^{-s_0 t} \alpha(t)\, dt.$$

This follows from the previous result by a change of variable $t = -u$. Observe that the restrictions regarding $\alpha(\infty)$ and $\alpha(-\infty)$ in these two theorems are not serious ones, since these numbers are known to exist by virtue of Theorems 3a and 3b. Hence the vanishing of the one desired can always be brought about by the addition of a suitable constant to the determining function. Of course such addition has no effect on the generating function. It must not be supposed that (2) holds if $\gamma = 0$. Thus if $\alpha(t)$ is the constant unity for non-negative $t$ and the constant zero for negative $t$, the integral (1) clearly exists for all $s$ and has the value unity. But if $s_0 = 0 + i$

$$s_0 \int_{-\infty}^{\infty} e^{-s_0 t} \alpha(t)\, dt = i \int_0^{\infty} e^{-it}\, dt,$$

a divergent integral.

## 4. Abscissae of Convergence

We might obtain a variety of formulas analogous to those of Section 2, Chapter II, for determining the region of convergence of a bilateral Laplace integral from its determining function. We content ourselves with the most useful of these. The others are easily obtained when needed by use of the decomposition §1 (4).

THEOREM 4. *If*

$$\overline{\lim_{t \to \infty}}\, t^{-1} \log |\alpha(t)| = k \neq 0$$

$$\overline{\lim_{t \to -\infty}}\, t^{-1} \log |\alpha(t)| = l \neq 0$$

*with $k < l$, then the integral*

(1) $$\int_{-\infty}^{\infty} e^{-st}\, d\alpha(t)$$

*converges for $k < \sigma < l$ and diverges for $\sigma < k$ and $\sigma > l$.*

The proof of this result follows in an obvious way from Theorem 2.4a of Chapter II and is omitted.

It is clear that when the integral (1) converges in a proper strip, then it converges uniformly in any closed bounded region inside the strip and not touching the boundary of the strip. Moreover, there is also uniform convergence in a Stolz region corresponding to a point of the boundary of the strip at which (1) converges. Hence it is clear that the integral (1) represents an analytic function $f(s)$ at interior points of the strip, that

$$f^{(k)}(s) = (-1)^k \int_{-\infty}^{\infty} e^{-st} t^k\, d\alpha(t) \qquad (k = 0, 1, 2, \cdots)$$

at such points, and that $f(s)$ is continuous at those boundary points of the strip at which (1) converges.

We say that the integral (1) converges absolutely if the integral

$$\int_{-\infty}^{\infty} e^{-\sigma t} \mid d\alpha(t) \mid \, = \, \int_{-\infty}^{\infty} e^{-\sigma t} \, du(t)$$

converges. Here the function $u(t)$ is the variation of $\alpha(x)$ in the interval $0 \leq x \leq t$ if $t$ is positive, is zero if $t$ is zero and is the negative of the variation of $\alpha(x)$ in the interval $t \leq x \leq 0$ if $t$ is negative. Clearly $u(t)$ is a non-decreasing function in $(-\infty, \infty)$ which vanishes at the origin.

We now define *abscissas of absolute convergence* $\sigma_a'$ and $\sigma_a''$ in the obvious way. With the above definition of $u(t)$ we have:

THEOREM 4b. *If*

$$\varlimsup_{t \to \infty} t^{-1} \log u(t) = k \neq 0$$

$$\lim_{t \to -\infty} t^{-1} \log [-u(t)] = l \neq 0,$$

*and if* $k < l$, *then* $\sigma_a' = k$ *and* $\sigma_a'' = l$.

## 5. Inversion Formulas

We obtain first an inversion formula for the bilateral Laplace-Lebesgue integral §1 (2).

THEOREM 5a. *If $\phi(t)$ belongs to $L$ in every finite interval, if the integral*

(1) $$f(s) = \int_{-\infty}^{\infty} e^{-st} \phi(t) \, dt$$

*converges absolutely on the line $\sigma = c$, and if $\phi(t)$ is of bounded variation in some neighborhood of $t = t_0$, then*

(2) $$\lim_{T \to \infty} \frac{1}{2\pi i} \int_{c-iT}^{c+iT} f(s) e^{st_0} \, ds = \frac{\phi(t_0+) + \phi(t_0-)}{2}.$$

For, as was done in equation §1 (4), write (1) as the sum of two integrals $f_1(s)$ and $f_2(s)$ corresponding to the intervals $(0, \infty)$ and $(-\infty, 0)$, respectively. If $t_0$ is positive we have by Theorem 7.3 of Chapter II

$$\lim_{T \to \infty} \frac{1}{2\pi i} \int_{c-iT}^{c+iT} f_1(s) e^{st_0} \, ds = \frac{\phi(t_0+) + \phi(t_0-)}{2}$$

$$\lim_{T \to \infty} \frac{1}{2\pi i} \int_{c-iT}^{c+iT} f_2(s) e^{st_0} \, ds = 0,$$

from which (2) follows at once.

If $t_0$ is zero, the same theorem gives

$$\lim_{T \to \infty} \frac{1}{2\pi i} \int_{c-iT}^{c+iT} f_1(s)\, ds = \frac{\phi(0+)}{2}$$

$$\lim_{T \to \infty} \frac{1}{2\pi i} \int_{c-iT}^{c+iT} f_2(s)\, ds = \frac{\phi(0-)}{2},$$

and our result is obtained by adding. The case $t_0$ negative follows in a similar way, or by a change of variable.

We turn now to the general Laplace-Stieltjes integral.

THEOREM 5b. *If $\alpha(t)$ is a normalized function of bounded variation in every finite interval, and if the integral*

$$f(s) = \int_{-\infty}^{\infty} e^{-st}\, d\alpha(t)$$

*converges in the strip $k < \sigma < l$, then for all $t$*

(3) $\quad \displaystyle\lim_{T \to \infty} \frac{1}{2\pi i} \int_{c-iT}^{c+iT} \frac{f(s)}{s} e^{st}\, ds = \begin{cases} \alpha(t) - \alpha(-\infty) & (c > 0, k < c < l) \\ \alpha(t) - \alpha(\infty) & (c < 0, k < c < l). \end{cases}$

To prove this result we appeal to Theorems 7.6a and 7.6b, Chapter II. Write $f(s)$ as the sum of the two integrals $f_1(s)$ and $f_2(s)$ as in the previous proof. If $t$ is positive we have

(4) $\quad \displaystyle\lim_{T \to \infty} \frac{1}{2\pi i} \int_{c-iT}^{c+iT} \frac{f_1(s)}{s} e^{st}\, ds = \alpha(t) \qquad (c > 0)$

$\hspace{7.5cm} = \alpha(t) - \alpha(\infty) \qquad (c < 0)$

(5) $\quad \displaystyle\lim_{T \to \infty} \frac{1}{2\pi i} \int_{c-iT}^{c+iT} \frac{f_2(s)}{s} e^{st}\, ds = -\alpha(-\infty) \qquad (c > 0)$

$\hspace{7.5cm} = 0 \qquad (c < 0).$

Adding (4) and (5) we have (3). The case in which $t$ is negative is treated in a similar way or is obtained by a change of variable. Finally, for the case $t = 0$ we have

(6) $\quad \displaystyle\lim_{T \to \infty} \frac{1}{2\pi i} \int_{c-iT}^{c+iT} \frac{f_1(s)}{s}\, ds = \frac{\alpha(0+)}{2} \qquad (c > 0)$

$\hspace{7.5cm} = \dfrac{\alpha(0+)}{2} - \alpha(\infty) \qquad (c < 0)$

(7) $\quad \displaystyle\lim_{T \to \infty} \frac{1}{2\pi i} \int_{c-iT}^{c+iT} \frac{f_2(s)}{s}\, ds = \frac{\alpha(0-)}{2} - \alpha(-\infty) \qquad (c > 0)$

$\hspace{7.5cm} = \dfrac{\alpha(0-)}{2} \qquad (c < 0).$

Adding (6) and (7) gives us (3) with $t = 0$ since $[\alpha(0+) + \alpha(0-)]/2$ is equal to $\alpha(0)$ by our definition of normalization.

As an illustration of this theorem take $f(s) = 1$ so that $\alpha(t)$ may be taken $-1/2$ for $-\infty < t < 0$ and $+1/2$ for $0 < t < \infty$. It is easily seen by Cauchy's theorem, or otherwise, that

$$\lim_{T \to \infty} \frac{1}{2\pi i} \int_{1-iT}^{1+iT} \frac{e^{st}}{s} ds = 1 = \alpha(t) - \alpha(-\infty) \qquad (t > 0)$$

$$= \frac{1}{2} = \frac{\alpha(0+) + \alpha(0-)}{2} - \alpha(-\infty) \qquad (t = 0)$$

$$= 0 = \alpha(t) - \alpha(-\infty) \qquad (t < 0).$$

We have chosen a positive $c$, so that the theorem is verified in this special case. In a similar way we have

$$\lim_{T \to \infty} \frac{1}{2\pi i} \int_{-1-iT}^{-1+iT} \frac{e^{st}}{s} ds = -1 = \alpha(t) - \alpha(\infty) \qquad (t < 0)$$

$$= -\frac{1}{2} = \frac{\alpha(0+) + \alpha(0-)}{2} - \alpha(\infty) \qquad (t = 0)$$

$$= 0 = \alpha(t) - \alpha(\infty) \qquad (t > 0),$$

so that the theorem is again verified for this negative value of $c$.

## 6. Uniqueness

The inversion formulas established in the previous section enable us to discuss the uniqueness of the representation of a function by a bilateral Laplace integral. It is to be noted that Theorem 6.3 of Chapter II does not yield a uniqueness theorem for the bilateral transform, so that a different approach is required.

**Theorem 6a.** *If $\alpha_1(t)$ and $\alpha_2(t)$ are two normalized functions of bounded variation in every finite interval such that*

(1) $$\int_{-\infty}^{\infty} e^{-st} d\alpha_1(t) = \int_{-\infty}^{\infty} e^{-st} d\alpha_2(t)$$

*in a common strip of convergence $k < \sigma < l$, then $\alpha_1(t) = \alpha_2(t)$ for all $t$.*

For, we have from Theorem 5b

(2) $\qquad \alpha_1(t) - \alpha_1(-\infty) = \alpha_2(t) - \alpha_2(-\infty) \qquad (-\infty < t < \infty)$

if the interval $(k, l)$ includes points of the positive axis and

(3) $\qquad \alpha_1(t) - \alpha_1(\infty) = \alpha_2(t) - \alpha_2(\infty) \qquad (-\infty < t < \infty)$

if the interval $(k, l)$ includes points of the negative axis. Since $\alpha_1(0) = \alpha_2(0) = 0$ it follows that $\alpha_1(-\infty) = \alpha_2(-\infty)$ in (2) or $\alpha_1(\infty) = \alpha_2(\infty)$ in (3). This proves the result.

We can obtain a similar result for the Laplace-Lebesgue integral. It is not proved, however, from Theorem 5a, but directly from Theorem 6a.

**THEOREM 6b.** *If $\phi_1(t)$ and $\phi_2(t)$ are of class $L$ in every finite interval and such that*

$$(4) \qquad \int_{-\infty}^{\infty} e^{-st}\phi_1(t)\,dt = \int_{-\infty}^{\infty} e^{-st}\phi_2(t)\,dt$$

*in a common strip of convergence $k < \sigma < l$, then $\phi_1(t)$ is equal to $\phi_2(t)$ for almost all $t$.*

For equation (1) becomes equation (4) if

$$\alpha_1(t) = \int_0^t \phi_1(u)\,du \qquad (-\infty < t < \infty)$$

$$\alpha_2(t) = \int_0^t \phi_2(u)\,du \qquad (-\infty < t < \infty).$$

By equations (2) or (3) we have $\alpha_1'(t) = \alpha_2'(t)$ at all points where these derivatives exist. But for almost all $t$

$$\alpha_1'(t) = \phi_1(t)$$
$$\alpha_2'(t) = \phi_2(t),$$

so that the result is established.

## 7. Summability

Regarding the summability of the inversion integral for a bilateral Laplace transform we have:

**THEOREM 7a.** *If $\phi(t)$ belongs to $L$ in every finite interval and if the integral*

$$f(s) = \int_{-\infty}^{\infty} e^{-st}\phi(t)\,dt$$

*converges in the strip $k < \sigma < l$, then the integral*

$$(1) \qquad \frac{1}{2\pi i}\int_{c-i\infty}^{c+i\infty} f(s)e^{st}\,ds \qquad (k < c < l)$$

*is summable $(C, 1)$ to $[\phi(t+) + \phi(t-)]/2$ for any $t$ where $\phi(t+)$ and $\phi(t-)$ exist.*

This result follows in an obvious way from Theorem 9.2, Chapter II. In particular, (1) is summable to $\phi(t)$ at all points $t$ where $\phi(t)$ is continuous.

**THEOREM 7b.** *The integral (1) is summable $(C, 1)$ to $\phi(t)$ at all points of the Lebesgue set for $\phi(t)$.*

This follows from Theorem 9.3, Chapter II. From this it is clear that (1) is summable to $\phi(t)$ for almost all $t$. In particular it gives a new proof of uniqueness, Theorem 6b.

## 8. Determining Function Belonging to $L^2$

We here extend the results of Section 10, Chapter II to the bilateral transform.

THEOREM 8a. *If for two numbers $k < l$*

$$\int_0^\infty e^{-2kt} |\phi(t)|^2 dt < \infty$$

$$\int_{-\infty}^0 e^{-2lt} |\phi(t)|^2 dt < \infty,$$

*then*

$$\underset{R\to\infty}{\overset{(2)}{\text{l.i.m.}}} \int_{-R}^R e^{-st} \phi(t)\, dt$$

*exists for $k \leqq \sigma \leqq l$ and defines an analytic function $f(s)$ for $k < \sigma < l$. Moreover,*

$$f(s) = \int_{-\infty}^\infty e^{-st} \phi(t)\, dt,$$

*the integral converging absolutely for $k < \sigma < l$, and*

$$\underset{\sigma\to k+}{\overset{(2)}{\text{l.i.m.}}} f(\sigma + i\tau) = f(k + i\tau)$$

$$\underset{\sigma\to l-}{\overset{(2)}{\text{l.i.m.}}} f(\sigma + i\tau) = f(l + i\tau).$$

Here the notation l.i.m. indicates mean square convergence. Thus

$$\underset{R\to\infty}{\overset{(2)}{\text{l.i.m.}}} F_R(s) = F(s)$$

for $k \leqq \sigma \leqq l$ means that for any such $\sigma$

$$\lim_{R\to\infty} \int_{-\infty}^\infty |F_R(\sigma + i\tau) - F(\sigma + i\tau)|^2 d\tau = 0.$$

The theorem follows in an obvious way from Theorem 10, Chapter II.

THEOREM 8b. *If $f(s)$ is defined as in Theorem 8a, then*

$$\underset{T\to\infty}{\overset{(2)}{\text{l.i.m.}}} \frac{e^{-ct}}{2\pi i} \int_{c-iT}^{c+iT} f(s) e^{st}\, ds = \phi(t) e^{-ct} \qquad (k \leqq c \leqq l)$$

and

$$\int_{-\infty}^{\infty} e^{-2\sigma t} |\phi(t)|^2 dt = \frac{1}{2\pi} \int_{-\infty}^{\infty} |f(\sigma + i\tau)|^2 d\tau \quad (k \leq \sigma \leq l).$$

This result is a direct consequence of the Plancherel theorem, or may be obtained in an obvious way from Theorem 10.1, Chapter II.

## 9. The Mellin Transform

This is the transform

$$f(s) = \int_0^\infty t^{s-1} \psi(t) \, dt$$

It was used by Riemann and Cahen but was first put on a rigorous basis by H. Mellin [1902] and now bears his name. It may be obtained by an exponential transformation from the bilateral Laplace transform and hence requires no special treatment here. It is, however, useful for reference to have the results recorded in the Mellin form.

In the integral

(1) $$f(s) = \int_{-\infty}^{\infty} e^{-st} d\alpha(t)$$

we make the change of variable $e^{-t} = u$ and obtain

(2) $$f(s) = \int_{0+}^{\infty} u^s \, d\beta(u),$$

where

$$\beta(u) = -\alpha (\log u^{-1}).$$

We shall refer to (2) as the Mellin-Stieltjes transform. If $\alpha(t)$ is an integral of $\phi(t)$ the integral (2) becomes

(3) $$f(s) = \int_{0+}^{\infty} u^{s-1} \psi(u) \, du,$$

where

$$\psi(u) = \phi (\log u^{-1}).$$

This is the classical form of the Mellin transform except that we are here considering the integral as a Cauchy limit at its lower limit of integration.

THEOREM 9a. *If the integral*

(4) $$f(s) = \int_0^\infty t^{s-1} \psi(t) \, dt$$

converges absolutely on the line $\sigma = c$, and if $\psi(t)$ is of bounded variation in a neighborhood of $t = x$ ($x > 0$), then

$$\lim_{T \to \infty} \frac{1}{2\pi i} \int_{c-iT}^{c+iT} f(s) x^{-s} ds = \frac{\psi(x+) + \psi(x-)}{2}.$$

This follows from Theorem 5a. Since $t^{-1}\psi(t)$ belongs to $L$ on $(0, \infty)$ we may clearly replace (3) by (4).

THEOREM 9b. *If the integral (2) converges for $\sigma_c' < \sigma < \sigma_c''$, then*

$$\lim_{T \to \infty} \frac{1}{2\pi i} \int_{c-iT}^{c+iT} \frac{f(s)t^{-s}}{s} ds = \beta(\infty) - \frac{\beta(t+) + \beta(t-)}{2} \quad (c > 0)$$

$$= \beta(0+) - \frac{\beta(t+) + \beta(t-)}{2} \quad (c < 0)$$

*for any value of $c \neq 0$ between $\sigma_c'$ and $\sigma_c''$.*

This is a corollary of Theorem 5b.

THEOREM 9c. *If $u^k \psi(u)$ belongs to $L^2$ on $(0, 1)$ and $u^l \psi(u)$ to $L^2$ on $(1, \infty)$, with $k < l$, then*

$$\underset{R \to \infty}{\overset{(2)}{\text{l.i.m.}}} \int_{R^{-1}}^{R} u^{s-1} \psi(u) \, du$$

*exists for $k \leq \sigma \leq l$ and defines a function $f(s)$ which is analytic in the strip $k < \sigma < l$. Moreover,*

$$f(s) = \int_0^\infty u^{s-1} \psi(u) \, du \qquad (k < \sigma < l),$$

*the integral converging absolutely, and*

$$\underset{\sigma \to k+}{\overset{(2)}{\text{l.i.m.}}} f(\sigma + i\tau) = f(k + i\tau)$$

$$\underset{\sigma \to l-}{\overset{(2)}{\text{l.i.m.}}} f(\sigma + i\tau) = f(l + i\tau).$$

THEOREM 9d. *If $f(s)$ is defined as in Theorem 9c, then*

$$\lim_{T \to \infty} \int_0^\infty \left| \psi(u) u^c - \frac{u^c}{2\pi i} \int_{c-iT}^{c+iT} f(s) u^{-s} ds \right|^2 \frac{du}{u} = 0 \qquad (k \leq c \leq l)$$

*and*

$$\int_0^\infty u^{2c-1} |\psi(u)|^2 du = \frac{1}{2\pi} \int_{-\infty}^{\infty} |f(c + i\tau)|^2 d\tau \qquad (k \leq c \leq l).$$

These two theorems are corollaries of Theorems 8a and 8b. In particular if $c = 1/2$ we see that the function

$$\frac{1}{2\pi i} \int_{1/2-iT}^{1/2+iT} f(s) u^{-s} ds$$

converges in the mean to $\psi(u)$ in the interval $(0, \infty)$.

### 10. Stieltjes Resultant

Let $\alpha(t)$ and $\beta(t)$ be two functions defined for all real values of $t$. We define their bilateral Stieltjes resultant.

**DEFINITION 10.** *The bilateral Stieltjes resultant of $\alpha(t)$ and $\beta(t)$ is the function*

$$(1) \qquad \gamma(t) = \int_{-\infty}^{\infty} \alpha(t-u)\,d\beta(u) = \int_{-\infty}^{\infty} \beta(t-u)\,d\alpha(u)$$

*when these two integrals exist and are equal.*

If $\alpha(t)$ is continuous for $-\infty < t < \infty$, if

$$\int_{-\infty}^{\infty} |d\beta(u)| < \infty, \qquad \beta(-\infty) = 0,$$

and if $\alpha(\infty)$, $\alpha(-\infty)$ exist with $\alpha(-\infty) = 0$, then the integrals (1) exist and are equal, so that the resultant of $\alpha(t)$ and $\beta(t)$ exists. Any one of the following pairs of end conditions would suffice:

$$\begin{cases} \alpha(-\infty) = 0 \\ \alpha(\infty) = 0, \end{cases} \quad \begin{cases} \alpha(-\infty) = 0 \\ \beta(-\infty) = 0, \end{cases} \quad \begin{cases} \beta(\infty) = 0 \\ \alpha(\infty) = 0, \end{cases} \quad \begin{cases} \beta(\infty) = 0 \\ \beta(-\infty) = 0. \end{cases}$$

Note that it is the Cauchy value of the integral on the right-hand side of (1) which is known to exist under our assumptions. It need not converge absolutely. By symmetry the conditions imposed on $\alpha(t)$ and $\beta(t)$ may be interchanged.

Let us next consider the case in which both $\alpha(t)$ and $\beta(t)$ are of bounded variation in $(-\infty, \infty)$,

$$(2) \qquad \int_{-\infty}^{\infty} |d\alpha(t)| < \infty, \qquad \int_{-\infty}^{\infty} |d\beta(t)| < \infty.$$

Then $\alpha(\pm\infty)$ and $\beta(\pm\infty)$ exist. It is no essential restriction to assume that $\alpha(-\infty) = \beta(-\infty) = 0$.

Let $P_\alpha$ be the countable set of points where $\alpha(t)$ is discontinuous (possibly a null set), and $P_\beta$ the set where $\beta(t)$ is discontinuous. Define $P_{\alpha+\beta}$ as in Section 11 of Chapter II. We now prove:

**THEOREM 10.** *If $\alpha(t)$ and $\beta(t)$ are of bounded variation in $(-\infty, \infty)$ with discontinuities in the sets $P_\alpha$ and $P_\beta$ respectively, and if $\alpha(-\infty) = \beta(-\infty) = 0$, then the Stieltjes resultant $\gamma(t)$ of $\alpha(t)$ and $\beta(t)$ exists for all $t$ not in $P_{\alpha+\beta}$.*

For, if $t$ is not in $P_{\alpha+\beta}$, then $\alpha(t-u)$ and $\beta(u)$ cannot have a common point of discontinuity for any $u$ in $(-\infty, \infty)$. Hence for $t$ not in $P_{\alpha+\beta}$ the integrals

$$\int_S^R \alpha(t-u)\,d\beta(u), \qquad \int_S^R \beta(t-u)\,d\alpha(u)$$

exist for every $R$ and $S$ by Theorem 14, Chapter I. The limits of these integrals as $R$ becomes positively infinite and $S$ becomes negatively infinite exist since

$$\int_{-\infty}^{\infty} |\alpha(t-u)| \, |d\beta(u)| < \infty, \qquad \int_{-\infty}^{\infty} |\beta(t-u)| \, |d\alpha(u)| < \infty.$$

That $\alpha(t)$ and $\beta(t)$ are bounded follows from the inequalities

$$|\alpha(t)| \leq \int_{-\infty}^{t} |d\alpha(u)|, \qquad |\beta(t)| \leq \int_{-\infty}^{t} |d\beta(u)|$$

and (2). Finally one sees that the two integrals (1) are equal by integration by parts. Notice that these two integrals might exist and be unequal but for the assumption that $\alpha(-\infty) = \beta(-\infty) = 0$. This happens, for example, if

$$\alpha(t) = 1 \qquad (-\infty < t < \infty)$$
$$\beta(t) = e^{-t} \qquad (t \geq 0)$$
$$= 1 \qquad (t < 0).$$

Both integrals exist, but they differ by unity.

## 11. Stieltjes Resultant at Infinity

We next investigate the behavior of the resultant of two functions of bounded variation for large values of the variable.

THEOREM 11. *If $\alpha(t)$ and $\beta(t)$ satisfy the conditions of Theorem 10, then*

$$\int_{-\infty}^{\infty} \alpha(t-u) \, d\beta(u) \to \alpha(\infty)\beta(\infty) \qquad (t \to +\infty)$$
$$\to 0 \qquad (t \to -\infty),$$

*the variable $t$ becoming infinite through the set complementary to $P_{\alpha+\beta}$.*

Consider first the case in which $t$ is becoming positively infinite. It is clear from the assumptions §10 (2) that $\beta(\infty)$ and $\alpha(\infty)$ exist. Let $V_\beta(t)$ be the variation of $\beta(u)$ in the interval $-\infty < u \leq t$. Then it will clearly be sufficient to prove that

$$\lim_{t \to \infty} \int_{-\infty}^{\infty} [\alpha(t-u) - \alpha(\infty)] \, d\beta(u) = 0,$$

or that

(1) $$\lim_{t \to \infty} \int_{-\infty}^{\infty} |\alpha(t-u) - \alpha(\infty)| \, dV_\beta(u) = 0.$$

Let $\epsilon$ be an arbitrary positive number. Choose $R$ so large that for all $t$ not in $P_{\alpha+\beta}$

$$\int_R^\infty |\alpha(t-u) - \alpha(\infty)| \, dV_\beta(u) < \epsilon.$$

This is possible since $\beta(t)$ is of bounded variation and $\alpha(t)$ is bounded. Next determine $t_0$ such that for all $t$ not in $P_{\alpha+\beta}$ but greater than $t_0$

$$|\alpha(t-u) - \alpha(\infty)| < \epsilon \qquad (-\infty < u \leqq R).$$

For such $t$

(2) $$\int_{-\infty}^R |\alpha(t-u) - \alpha(\infty)| \, dV_\beta(u) < \epsilon V_\beta(\infty).$$

Combining (1) and (2) we have the desired result.

The case $t \to -\infty$ may be treated by applying what we have just proved to the integral

$$\int_{-\infty}^\infty [\alpha(u+t) - \alpha(\infty)] \, d[\beta(\infty) - \beta(-u)]$$
$$= \alpha(\infty)\beta(\infty) + \int_{-\infty}^\infty \alpha(t-u) \, d\beta(u).$$

COROLLARY 11. *If $\alpha(t)$ and $\beta(t)$ are of bounded variation in $(-\infty, \infty)$, then*

$$\lim \int_{-\infty}^\infty \alpha(t-u) \, d\beta(u) = \begin{cases} \alpha(\infty)[\beta(\infty) - \beta(-\infty)] & (t \to \infty) \\ \alpha(-\infty)[\beta(\infty) - \beta(-\infty)] & (t \to -\infty) \end{cases}$$

This follows by applying the theorem to the functions $\alpha(t) - \alpha(-\infty)$ and $\beta(t) - \beta(-\infty)$.

## 12. Stieltjes Resultant Completely Defined

We show next that the resultant of two functions $\alpha(t)$ and $\beta(t)$ of bounded variation in $(-\infty, \infty)$ can be defined in the set $P_{\alpha+\beta}$ in such a way as to be of bounded variation in $(-\infty, \infty)$.

THEOREM 12. *If $\alpha(t)$, $\beta(t)$ and $\gamma(t)$ are defined as in Theorem 10, then $\gamma(t)$ may be defined in the set $P_{\alpha+\beta}$ so as to be of bounded variation in $(-\infty, \infty)$ and so that the total variation of $\gamma(t)$ is not greater than the product of the total variations of $\alpha(t)$ and $\beta(t)$. Moreover $\gamma(-\infty) = 0$, $\gamma(\infty) = \alpha(\infty)\beta(\infty)$.*

By breaking $\alpha(t)$ and $\beta(t)$ into their real and imaginary parts and then decomposing each of these into the difference of two non-decreasing functions we see that $\gamma(t)$ is a linear combination of integrals of the form

$$c(t) = \int_{-\infty}^\infty a(t-u) \, db(u),$$

where $a(t)$ and $b(t)$ are bounded non-decreasing functions vanishing at $-\infty$ and continuous except in $P_\alpha$ and $P_\beta$, respectively. Since $c(t)$ is non-decreasing, $\gamma(t)$ has right-hand and left-hand limits at all points and may be defined in $P_{\alpha+\beta}$ so that

$$\gamma(t) = \frac{\gamma(t+) + \gamma(t-)}{2} \qquad (-\infty < t < \infty). \tag{1}$$

Then if $t_0 < t_1 < \cdots < t_n$ are any points not in $P_{\alpha+\beta}$ we have, as in the proof of Theorem 11.2b, Chapter II, that

$$\sum_{i=0}^{N-1} |\gamma(t_{i+1}) - \gamma(t_i)| \leq V_\alpha(\infty)V_\beta(\infty).$$

By a limiting process we see by virtue of (1) that the same inequality holds if some or all of the $t_i$ lie in $P_{\alpha+\beta}$. It thus becomes clear that $\gamma(t)$ as now defined is of bounded variation in $(-\infty, \infty)$ and that its total variation is not greater than $V_\alpha(\infty)V_\beta(\infty)$. By Theorem 11

$$\lim_{t \to -\infty} \gamma(t) = 0, \qquad \lim_{t \to \infty} \gamma(t) = \alpha(\infty)\beta(\infty)$$

if the variable is restricted to lie in the complement of $P_{\alpha+\beta}$. But by (1) the same is true if $t$ varies through all real values.

If the restriction $\alpha(-\infty) = \beta(-\infty) = 0$ is omitted, then by Corollary 11

$$\gamma(\infty) = \alpha(\infty)[\beta(\infty) - \beta(-\infty)], \quad \gamma(-\infty) = \alpha(-\infty)[\beta(\infty) - \beta(-\infty)].$$

## 13. Preliminary Results

Before discussing the product of two bilateral Laplace integrals we need several preliminary results.

**THEOREM 13a.** *If for some non-negative constant $c$*

$$\int_{-\infty}^{\infty} e^{-ct} |d\alpha(t)| < \infty, \tag{1}$$

*and if $\alpha(-\infty) = 0$, then $e^{-ct}\alpha(t)$ is of bounded variation in $(-\infty, \infty)$ and vanishes at $-\infty$.*

The result is trivial if $c$ is zero. If $c > 0$, let $V_\alpha(t)$ be the variation of $\alpha(x)$ in $(-\infty < x \leq t)$, vanishing at $-\infty$. By Theorem 3c

$$\int_{-\infty}^{\infty} e^{-ct} dV_\alpha(t) = c \int_{-\infty}^{\infty} e^{-ct} V_\alpha(t) \, dt. \tag{2}$$

By the same theorem

$$e^{-ct}\alpha(t) = \int_{-\infty}^{t} e^{-cu} d\alpha(u) - c \int_{-\infty}^{t} e^{-cu}\alpha(u) \, du, \tag{3}$$

so that $e^{-ct}\alpha(t)$ clearly vanishes at $-\infty$. Since $\alpha(-\infty) = 0$, we have

(4) $$|\alpha(t)| \leq V_\alpha(t) \qquad (-\infty < t < \infty).$$

From (3) and (4) we see that the total variation of $e^{-ct}\alpha(t)$ does not exceed

$$\int_{-\infty}^{\infty} e^{-ct} |d\alpha(t)| + c \int_{-\infty}^{\infty} e^{-ct} V_\alpha(t)\, dt,$$

and by (2) this is equal to twice the integral (1). This completes the proof of the theorem.

**THEOREM 13b.** *If for some non-negative constant $c$*

(5) $$\int_{-\infty}^{\infty} e^{-ct} |d\alpha(t)| < \infty, \qquad \int_{-\infty}^{\infty} e^{-ct} |d\beta(t)| < \infty,$$

*and if $\alpha(-\infty) = \beta(-\infty) = 0$, then the Stieltjes resultant $\gamma(t)$ of $\alpha(t)$ and $\beta(t)$ exists and $e^{-ct}\gamma(t)$ is of bounded variation in $(-\infty, \infty)$, vanishing at $-\infty$.*

If $c = 0$ the result is contained in Theorem 12. If $c > 0$ we see by (5) that the function

$$B(t) = \int_{-\infty}^{t} e^{-cu}\, d\beta(u)$$

is of bounded variation in $(-\infty, \infty)$ and vanishes at $-\infty$. By Theorem 13a the function $\alpha(t)e^{-ct}$ has the same property. Compute the Stieltjes resultant of these two functions. It is

$$\int_{-\infty}^{\infty} e^{-c(t-u)} \alpha(t-u)\, dB(u) = \int_{-\infty}^{\infty} e^{-ct} \alpha(t-u)\, d\beta(u) = e^{-ct}\gamma(t).$$

By Theorem 12 this exists for all $t$ (with the usual definition at the points where the Stieltjes integral is not defined). Also the total variation of $e^{-ct}\gamma(t)$ does not exceed the product of the variation of $\alpha(t)e^{-ct}$ by the variation of $B(t)$, and $e^{-ct}\gamma(t)$ vanishes at $-\infty$ by Theorem 12.

## 14. The Product of Fourier-Stieltjes Transforms

Let $\alpha(t)$ be of bounded variation in $(-\infty, \infty)$. Then its Fourier-Stieltjes transform is

$$F(x) = \int_{-\infty}^{\infty} e^{ixt}\, d\alpha(t),$$

and this integral is clearly absolutely convergent for all real $x$. We wish to discuss the product of two such functions $F(x)$. We shall need*

---
* Compare S. Bochner, [1932], p. 70, Theorem 20.

§14]   FOURIER-STIELTJES TRANSFORMS

LEMMA 14. *If for each positive number $R$ the function $\psi_R(t)$ is non-decreasing and bounded, and if*

(1) $$\lim_{R\to\infty} \psi_R(t) = \gamma(t) \qquad (-\infty < t < \infty)$$

(2) $$\lim_{R\to\infty} \psi_R(\pm\infty) = \gamma(\pm\infty),$$

*then*

(3) $$\lim_{R\to\infty} \int_{-\infty}^{\infty} e^{ixt}\, d\psi_R(t) = \int_{-\infty}^{\infty} e^{ixt}\, d\gamma(t).$$

Clearly $\gamma(t)$ is non-decreasing and bounded so that the integral on the right-hand side of (3) exists. Set

$$I_R = \int_{-\infty}^{\infty} e^{ixt}\, d[\psi_R(t) - \gamma(t)],$$

and express it as the sum of three integrals $I'_R$, $I''_R$, $I'''_R$ corresponding to the three intervals of integration $(-\infty, -A)$, $(-A, A)$ and $(A, \infty)$, respectively. Here $A$ is an arbitrary positive constant. Then

$$|I'_R| \leq \int_{-\infty}^{-A} d[\psi_R(t) + \gamma(t)] = \psi_R(-A) + \gamma(-A) - \psi_R(-\infty) - \gamma(-\infty)$$

$$|I'''_R| \leq \int_{A}^{\infty} d[\psi_R(t) + \gamma(t)] = \psi_R(\infty) + \gamma(\infty) - \psi_R(A) - \gamma(A)$$

Hence

$$|I_R| \leq \left|\int_{-A}^{A} e^{ixt}\, d[\psi_R(t) - \gamma(t)]\right| + \psi_R(-A) + \gamma(-A)$$
$$- \psi_R(-\infty) - \gamma(-\infty) + \psi_R(\infty) + \gamma(\infty) - \psi_R(A) - \gamma(A).$$

Then* by Theorem 16.4 of Chapter I and by (2)

$$\varlimsup_{R\to\infty} |I_R| \leq 2\gamma(-A) - 2\gamma(-\infty) + 2\gamma(\infty) - 2\gamma(A).$$

But the right-hand side of this inequality can be made arbitrarily small by choice of $A$, whereas the left-hand side is independent of $A$. Consequently the latter is the number zero, and our proof is complete.

By use of this lemma we can prove the fundamental theorem regarding the product of two Fourier-Stieltjes transforms.

---

* The Helly-Bray theorem is applicable here, for if $R$ is sufficiently large we have by (2)

$$\gamma(-\infty) - 1 < \psi_R(-\infty) \leq \psi_R(t) \leq \psi_R(\infty) < \gamma(\infty) + 1,$$

so that the set of functions $\psi_R(t)$ is uniformly bounded.

THEOREM 14. *If $\alpha(t)$ and $\beta(t)$ are of bounded variation in $(-\infty, \infty)$ and if $\gamma(t)$ is their Stieltjes resultant, then*

(4) $$\int_{-\infty}^{\infty} e^{ixt}\, d\alpha(t) \int_{-\infty}^{\infty} e^{ixt}\, d\beta(t) = \int_{-\infty}^{\infty} e^{ixt}\, d\gamma(t) \qquad (-\infty < x < \infty).$$

It will be sufficient to consider the case in which $\alpha(-\infty) = \beta(-\infty) = 0$. For, subtracting a constant from $\alpha(t)$ and one from $\beta(t)$ only alters $\gamma(t)$ by the addition of a constant, so that equation (4) is not altered at all. Also by decomposing $\alpha(t)$ and $\beta(t)$ first into real and imaginary parts and then into monotonic functions we see that there is no restriction in assuming that $\alpha(t)$ and $\beta(t)$, and hence $\gamma(t)$, are non-decreasing.

Set

$$F_R(x) = \int_{-R}^{\infty} e^{ixt}\, d\alpha(t) \int_{-R}^{\infty} e^{ixt}\, d\beta(t)$$

$$= e^{-2xRi} \int_{0}^{\infty} e^{ixt}\, d\alpha(t-R) \int_{0}^{\infty} e^{ixt}\, d\beta(t-R).$$

Then by Theorem 11.5 of Chapter II (with $s_0 = -ix$), we have

$$F_R(x) = e^{-2xRi} \int_{0}^{\infty} e^{ixt}\, d\varphi_R(t) = \int_{-2R}^{\infty} e^{ixt}\, d\varphi_R(t + 2R),$$

where

$$\varphi_R(t) = \int_{0}^{t} \alpha(t - R - u)\, d\beta(u - R)$$

(5) $$\varphi_R(t + 2R) = \int_{-R}^{t+R} \alpha(t - y)\, d\beta(y).$$

Set

$$\psi_R(t) = \varphi_R(t + 2R) \qquad (-2R \leqq t)$$
$$= 0 \qquad (t \leqq -2R).$$

It is to be understood of course that we are following the conventions of Chapter II in the definition of $\varphi_R(t)$ at a point $t$ where the integral (5) is undefined. Then

(6) $$F_R(x) = \int_{-\infty}^{\infty} e^{ixt}\, d\psi_R(t).$$

From (5) we have for all $t$

$$\lim_{R \to \infty} \psi_R(t) = \int_{-\infty}^{\infty} \alpha(t - y)\, d\beta(y) = \gamma(t).$$

If we may let $R$ become infinite under the integral sign of (6) our theorem is established. To establish this point we use Lemma 14. From its definition it is obvious that $\psi_R(t)$ is a non-decreasing function of $t$ and that $\psi_R(-\infty) = 0$. Let us compute $\psi_R(\infty)$. By Theorem 11

$$\gamma(-\infty) = 0$$
$$\gamma(\infty) = \alpha(\infty)\beta(\infty).$$

Now

$$\psi_R(\infty) = \varphi_R(\infty) = \lim_{t \to \infty} \int_{-R}^{t+R} \alpha(t-y)\, d\beta(y).$$

Set

$$\beta^*(y) = \beta(y) \qquad y > -R$$
$$= \beta(-R) \qquad y \leqq -R$$
$$\alpha^*(y) = \alpha(y) \qquad y > -R$$
$$= 0 \qquad y \leqq -R.$$

Then

$$\int_{-R}^{t+R} \alpha(t-y)\, d\beta(y) = \int_{-\infty}^{\infty} \alpha^*(t-y)\, d\beta^*(y).$$

By Corollary 11

$$\psi_R(\infty) = \lim_{t \to \infty} \int_{-\infty}^{\infty} \alpha^*(t-y)\, d\beta^*(y) = \alpha^*(\infty)[\beta^*(\infty) - \beta^*(-\infty)]$$
$$= \alpha(\infty)[\beta(\infty) - \beta(-R)].$$

Hence

$$\lim_{R \to \infty} \psi_R(\infty) = \alpha(\infty)\beta(\infty) = \gamma(\infty).$$

That is, hypothesis (2) of the lemma is satisfied, and our proof is complete.

COROLLARY 14. *If $\phi(t)$ and $\psi(t)$ belong to $L$ in $(-\infty, \infty)$, then*

$$\int_{-\infty}^{\infty} e^{ixt}\varphi(t)\, dt \int_{-\infty}^{\infty} e^{ixt}\psi(t)\, dt = \int_{-\infty}^{\infty} e^{ixt}\omega(t)\, dt,$$

where

$$\omega(t) = \int_{-\infty}^{\infty} \varphi(t-u)\psi(u)\, du.$$

This follows if in the theorem we take
$$\alpha(t) = \int_{-\infty}^{t} \varphi(u)\,du, \qquad \beta(t) = \int_{-\infty}^{t} \psi(u)\,du.$$
Then
$$\gamma(t) = \int_{-\infty}^{\infty} \psi(u)\,du \int_{-\infty}^{t-u} \varphi(y)\,dy = \int_{-\infty}^{\infty} \psi(u)\,du \int_{-\infty}^{t} \varphi(y - u)\,dy$$
and by Fubini's theorem
$$\gamma(t) = \int_{-\infty}^{t} \omega(y)\,dy,$$
so that the result is established. It can also be proved very easily directly by Fubini's theorem without appeal to Theorem 14.

### 15. Stieltjes Resultant of Indefinite Integrals

We discuss here the resultant of two functions defined by indefinite integrals of the form
$$A(x) = \int_{-\infty}^{x} e^{-ct}\,d\alpha(t).$$

THEOREM 15. *If $\alpha(t)$, $\beta(t)$ and $\gamma(t)$ are defined as in Theorem 13b, and if*
$$A(x) = \int_{-\infty}^{x} e^{-ct}\,d\alpha(t), \qquad B(x) = \int_{-\infty}^{x} e^{-ct}\,d\beta(t), \qquad C(x) = \int_{-\infty}^{x} e^{-ct}\,d\gamma(t)$$
*for some non-negative constant $c$, then $C(x)$ is the Stieltjes resultant of $A(x)$ and $B(x)$.*

The result is trivial if $c = 0$. If $c > 0$, by §13 (5) the functions $A(x)$ and $B(x)$ are of bounded variation in $(-\infty, \infty)$. Hence their Stieltjes resultant is defined by
$$\int_{-\infty}^{\infty} A(t - u)\,dB(u) = \int_{-\infty}^{\infty} e^{-cu}\,d\beta(u) \int_{-\infty}^{t-u} e^{-cy}\,d\alpha(y)$$
$$= \int_{-\infty}^{\infty} d\beta(u) \int_{-\infty}^{t} e^{-cy}\,d_y \alpha(y - u).$$

Since $\alpha(-\infty) = 0$, we may apply Theorem 3c to the inner integral. Then
$$\int_{-\infty}^{\infty} A(t - u)\,dB(u) = \int_{-\infty}^{\infty} d\beta(u) \left[ \alpha(t - u)e^{-ct} + c \int_{-\infty}^{t} e^{-cy}\alpha(y - u)\,dy \right]$$
(1)
$$= \gamma(t)e^{-ct} + c \int_{-\infty}^{\infty} d\beta(u) \int_{-\infty}^{t} e^{-cy}\alpha(y - u)\,dy$$
(2)
$$= \gamma(t)e^{-ct} + c \int_{-\infty}^{t} e^{-cy}\gamma(y)\,dy$$

if it is permissible to interchange the order of integration in the iterated integral (1). By Théorem 15$d$, Chapter I, this will be so if

(3) $$I = \int_{-\infty}^{\infty} |d\beta(u)| \cdot \int_{-\infty}^{t} e^{-cy} |\alpha(y - u)| dy < \infty.$$

But

$$I = \int_{-\infty}^{\infty} |d\beta(u)| \int_{-\infty}^{t-u} e^{-c(y+u)} |\alpha(y)| dy$$

$$\leqq \int_{-\infty}^{\infty} e^{-cu} |d\beta(u)| \int_{-\infty}^{\infty} e^{-cy} V_\alpha(y) dy = \frac{1}{c} \int_{-\infty}^{\infty} e^{-cu} |d\beta(u)| \int_{-\infty}^{\infty} e^{-ct} |d\alpha(t)|$$

by Theorem 3$c$. Inequality (3) follows by §13 (5).

Since by Theorem 13$b$ the function $\gamma(t)e^{-ct}$ is of bounded variation in $(-\infty, \infty)$, vanishing at $-\infty$, we may integrate by parts in (2) again using Theorem 3$c$. Thus

$$\int_{-\infty}^{\infty} A(t - u) dB(u) = \int_{-\infty}^{t} e^{-cy} d\gamma(y).$$

This shows that the definition of $C(x)$ given in the statement of the theorem has a meaning and that $C(x)$ is the resultant of $A(x)$ and $B(x)$.

## 16. Product of Bilateral Laplace Integrals

We can now express the product of two absolutely convergent bilateral Laplace-Stieltjes integrals as another Laplace-Stieltjes integral.

THEOREM 16$a$. *If the integrals*

$$f(s) = \int_{-\infty}^{\infty} e^{-st} d\alpha(t), \qquad g(s) = \int_{-\infty}^{\infty} e^{-st} d\beta(t)$$

*converge absolutely for a common value of $s$, then for that value*

(1) $$f(s)g(s) = \int_{-\infty}^{\infty} e^{-st} d\gamma(t),$$

*where $\gamma(t)$ is the Stieltjes resultant of $\alpha(t)$ and $\beta(t)$.*

If $s$ is zero or pure imaginary the result is Theorem 14. Let the real part of $s$ be $c$, the imaginary part $-x$, $s = c - ix$. Suppose first that $c$ is positive and define $A(t)$, $B(t)$, $C(t)$ as in Theorem 15. Then

$$f(s)g(s) = \int_{-\infty}^{\infty} e^{ixt} dA(t) \int_{-\infty}^{\infty} e^{ixt} dB(t).$$

By Theorem 15 the function $C(t)$ is the Stieltjes resultant of $A(t)$ and $B(t)$. By Theorem 14

$$f(s)g(s) = \int_{-\infty}^{\infty} e^{ixt}\, dC(t) = \int_{-\infty}^{\infty} e^{-ct+ixt}\, d\gamma(t)$$

$$= \int_{-\infty}^{\infty} e^{-st}\, d\gamma(t),$$

so that (1) is proved for positive $c$.

If $c$ is negative we have by a change of variable

$$f(s) = \int_{-\infty}^{\infty} e^{-(-s)t}\, d[-\alpha(-t)],$$

and since $-c$ is positive we have by the previous case

$$f(s)g(s) = \int_{-\infty}^{\infty} e^{-(-s)t}\, d\omega(t)$$

where $\omega(t)$ is the Stieltjes resultant of $-\alpha(-t)$ and $-\beta(-t)$

$$\omega(t) = \int_{-\infty}^{\infty} \alpha(-t+u)\, d\beta(-u) = -\int_{-\infty}^{\infty} \alpha(-t-u)\, d\beta(u)$$

$$-\omega(-t) = \int_{-\infty}^{\infty} \alpha(t-u)\, d\beta(u) = \gamma(t)$$

Hence

$$f(s)g(s) = -\int_{-\infty}^{\infty} e^{-(-s)t}\, d\gamma(-t) = \int_{-\infty}^{\infty} e^{-st}\, d\gamma(t),$$

so that the theorem is completely established.

The case of the Laplace-Lebesgue integral may be treated as a special case of the Laplace-Stieltjes integral. We prefer to treat it directly by classical methods.

THEOREM 16b. *If the integrals*

$$f(s) = \int_{-\infty}^{\infty} e^{-st}\varphi(t)\, dt, \qquad g(s) = \int_{-\infty}^{\infty} e^{-st}\psi(t)\, dt$$

*converge absolutely for a common value of $s$, then*

$$f(s)g(s) = \int_{-\infty}^{\infty} e^{-st}\omega(t)\, dt,$$

*where*

(2) $$\omega(t) = \int_{-\infty}^{\infty} \varphi(t-u)\psi(u)\, du.$$

For,

$$f(s)g(s) = \int_{-\infty}^{\infty} \varphi(t)\,dt \int_{-\infty}^{\infty} e^{-s(t+y)}\psi(y)\,dy$$

$$= \int_{-\infty}^{\infty} \varphi(t)\,dt \int_{-\infty}^{\infty} e^{-su}\psi(u-t)\,du$$

$$= \int_{-\infty}^{\infty} e^{-su}\,du \int_{-\infty}^{\infty} \psi(u-t)\varphi(t)\,dt$$

$$= \int_{-\infty}^{\infty} e^{-su}\omega(u)\,du.$$

To justify the interchange in the order of integration we use Fubini's theorem, which is applicable since

$$\int_{-\infty}^{\infty} |\varphi(t)|\,dt \int_{-\infty}^{\infty} e^{-\sigma u}|\psi(u-t)|\,du$$

$$= \int_{-\infty}^{\infty} e^{-\sigma t}|\varphi(t)|\,dt \int_{-\infty}^{\infty} e^{-\sigma t}|\psi(t)|\,dt < \infty.$$

It is understood of course that $\omega(t)$ is defined by (2) almost everywhere.

## 17. Resultants in a Special Case

As an example of the usefulness of the methods of the present chapter we compute the successive iterates of the Stieltjes kernel.* As we shall show, the problem is essentially that of computing resultants. Set

(1) $$g(t) = g_1(t) = \frac{1}{1 + e^{-t}}$$

(2) $$g_n(t) = \int_{-\infty}^{\infty} g_{n-1}(t-u)g_1(u)\,du \quad (n = 2, 3, 4, \cdots).$$

It is easy to verify that the integral

$$f(s) = \int_{-\infty}^{\infty} e^{-st}g(t)\,dt$$

converges absolutely for $0 < \sigma < 1$. Hence we are in a position to apply Theorem 16b. It is thus clear that the functions $g_n(t)$ are defined almost everywhere and that

(3) $$[f(s)]^n = \int_{-\infty}^{\infty} e^{-st}g_n(t)\,dt \quad (0 < \sigma < 1).$$

* See D. V. Widder [1937].

We compute the left-hand side of this equation explicitly and obtain a new Laplace integral representation for it with an explicit determination of the determining function. By the uniqueness theorem this latter must be $g_n(t)$.

**THEOREM 17a.** *If $g(t) = (1 + e^{-t})^{-1}$, then*

$$f(s) = \int_{-\infty}^{\infty} e^{-st} g(t)\,dt = \frac{\pi}{\sin \pi s} \qquad (0 < \sigma < 1).$$

For, expanding $g(t)$ in powers of $e^{-t}$ gives us

$$\int_{-\infty}^{\infty} e^{-st} g(t)\,dt = \int_{0}^{\infty} e^{-st} \sum_{k=0}^{\infty} (-1)^k e^{-kt}\,dt - \int_{0}^{\infty} e^{st} \sum_{k=1}^{\infty} (-1)^k e^{-kt}\,dt$$

$$= \frac{1}{s} - \int_{0}^{\infty} [e^{-st} - e^{st}] \lim_{N \to \infty} e^{-t} \frac{1 - (-1)^N e^{-Nt}}{1 + e^{-t}}\,dt.$$

If it is permissible to put the limit sign in front of the integral sign we may integrate term by term to obtain

(4) $$\int_{-\infty}^{\infty} e^{-st} g(t)\,dt = \lim_{N \to \infty} \sum_{k=-N}^{N} \frac{(-1)^k}{s + k} \qquad (0 < \sigma < 1).$$

To show the validity of the above operation we have for $(0 \leq t < \infty)$

$$| e^{-t}(e^{-st} - e^{st})(1 - (-1)^N e^{-Nt})(1 + e^{-t})^{-1} | \leq 2 e^{-t}(e^{-\sigma t} + e^{\sigma t})(1 + e^{-t})^{-1}.$$

But

$$2 \int_{0}^{\infty} \frac{e^{-t}(e^{-\sigma t} + e^{\sigma t})}{1 + e^{-t}}\,dt < \infty \qquad (0 < \sigma < 1),$$

so that we obtain our result by application of Lebesgue's limit theorem. But is is well known* that the right-hand side of (4) is $\pi/\sin \pi s$. This completes the proof of our theorem.

**COROLLARY 17a.** *If $n = 1, 2, \cdots$ and $0 < \sigma < 1$, then*

$$(-1)^n f^{(n)}(s) = (-1)^n \left(\frac{\pi}{\sin \pi s}\right)^{(n)} = \int_{-\infty}^{\infty} e^{-st} t^n g(t)\,dt = n! \sum_{k=-\infty}^{\infty} \frac{(-1)^k}{(s + k)^{n+1}}.$$

This is obvious formally by differentiation. The formal operations are easily justified. Of course, if $n = 0$, the equation still holds if the series is interpreted as in equation (4).

**THEOREM 17b.** *If $F(s)$ is defined as*

(5) $$F(s) = \int_{-\infty}^{\infty} \frac{e^{-st} t}{1 - e^{-t}}\,dt,$$

---

* See, for example, W. F. Osgood [1923] p. 509.

*then*

(6) $$F(s) = [f(s)]^2 = \left(\frac{\pi}{\sin \pi s}\right)^2 = \sum_{k=-\infty}^{\infty} \frac{1}{(s+k)^2} \qquad (0 < \sigma < 1).$$

To show this we compute $g_2(t)$ explicitly. By (2) we have

$$g_2(t) = \int_{-\infty}^{\infty} \frac{1}{1+e^{-(t-u)}} \frac{1}{1+e^{-u}} du.$$

Using the method of partial fractions, this gives

$$g_2(t) = \frac{t}{1-e^{-t}}.$$

But now (3) and (5) give (6).

COROLLARY 17b. *If* $n = 1, 2, \cdots$ *and* $0 < \sigma < 1$, *then*

$$(-1)^{n-1} F^{(n-1)}(s) = \int_{-\infty}^{\infty} \frac{e^{-st} t^n}{1-e^{-t}} dt = n! \sum_{k=-\infty}^{\infty} \frac{1}{(s+k)^{n+1}}.$$

We can now prove our principal result.

THEOREM 17c. *If* $g_n(t)$ *is defined by* (1) *and* (2), *then*

(7) $$g_{2n}(t) = \frac{t}{1-e^{-t}}$$
$$\times \left[ A_{2n,2n} \frac{t^{2n-1}}{(2n-1)!} + A_{2n,2n-2} \frac{t^{2n-3}}{(2n-3)!} + \cdots + A_{2n,2} \right]$$

(8) $$g_{2n+1}(t) = \frac{1}{1+e^{-t}}$$
$$\times \left[ A_{2n+1,2n+1} \frac{t^{2n}}{(2n)!} + A_{2n+1,2n-1} \frac{t^{2n-2}}{(2n-2)!} + \cdots + A_{2n+1,1} \right],$$

*where the constants* $A_{n,n}$ *are defined by the expansion*

(9) $$\left(\frac{\pi s}{\sin \pi s}\right)^n = A_{n,n} + A_{n,n-1} s + A_{n,n-2} s^2 + \cdots.$$

To prove this we make use of the Mittag-Leffler* development of the function $(\pi/\sin \pi s)^n$. The principal part of this function at the origin is obtained at once from the expansion (9). Denote it by $Q_n(s)$:

$$Q_n(s) = \frac{A_{n,n}}{s^n} + \frac{A_{n,n-1}}{s^{n-1}} + \cdots + \frac{A_{n,1}}{s}.$$

---

* See, for example, W. F. Osgood [1923] p. 540.

Then the principal part at the integer $s = k$ is $(-1)^k Q_n(s+k)$ if $n$ is odd and is $Q_n(s+k)$ if $n$ is even. Hence

$$\left(\frac{\pi}{\sin \pi s}\right)^{2n} = \sum_{k=-\infty}^{\infty} Q_{2n}(s+k)$$

$$\left(\frac{\pi}{\sin \pi s}\right)^{2n+1} = \sum_{k=-\infty}^{\infty} (-1)^k Q_{2n+1}(s+k),$$

where

$$\sum_{k=-\infty}^{\infty} (-1)^k \frac{A_{2n+1,1}}{s+k}$$

is understood to mean

$$\lim_{N \to \infty} \sum_{k=-N}^{N} \frac{(-1)^k A_{2n+1,1}}{s+k} = A_{2n+1,1} \frac{\pi}{\sin \pi s}.$$

Hence by use of Corollaries 17a and 17b we have

$$\left(\frac{\pi}{\sin \pi s}\right)^{2n} = \int_{-\infty}^{\infty} \frac{e^{-st}}{1-e^{-t}} t \left[ A_{2n,2n} \frac{t^{2n-1}}{(2n-1)!} + \cdots + A_{2n,2} \right] dt.$$

$$\left(\frac{\pi}{\sin \pi s}\right)^{2n+1} = \int_{-\infty}^{\infty} \frac{e^{-st}}{1+e^{-t}} \left[ A_{2n+1,2n+1} \frac{t^{2n}}{(2n)!} + \cdots + A_{2n+1,1} \right] dt.$$

Finally by virtue of (3) we have two determining functions for the same generating functions. By the uniqueness theorem we may equate these two determining functions to obtain (7) and (8).

## 18. Iterates of the Stieltjes Kernel

The Stieltjes transform

$$f(s) = \int_0^{\infty} \frac{\varphi(t)}{s+k} dt$$

will be discussed in detail in Chapter VIII. If this transform is applied to itself one obtains

$$f_2(s) = \int_0^{\infty} \frac{f(t)}{s+t} dt = \int_0^{\infty} \frac{dt}{s+t} \int_0^{\infty} \frac{\varphi(u)}{t+u} du$$

$$= \int_0^{\infty} \varphi(u) du \int_0^{\infty} \frac{dt}{(s+t)(t+u)}.$$

The inner integral $H_2(u, s)$ is the iterate of the Stieltjes kernel $H(u, s) = (u+s)^{-1}$:

$$H_2(u, s) = \int_0^{\infty} H(u, t) H(t, s) dt.$$

The successive iterates are defined by the equations

$$H_n(u, s) = \int_0^\infty H_{n-1}(u, t) H(t, s)\, dt \quad (n = 3, 4, \cdots).$$

One may easily obtain an explicit expression for the function $H_2(u, s)$ by direct integration of the integral defining it. It is found to be

(1) $$H_2(u, s) = \frac{\log(u/s)}{u - s}.$$

The higher iterates can also be obtained by use of the functions $g_n(t)$ of section 17. Thus

$$H_n(u, s) = \int_0^\infty \frac{H_{n-1}(u, t)}{t + s}\, dt$$

(2) $$H_n(e^x, e^y) = \int_{-\infty}^\infty \frac{e^u H_{n-1}(e^x, e^u)}{e^u + e^y}\, du.$$

In particular we see that

(3) $$H(e^x, e^y) = H_1(e^x, e^y) = e^{-y} g(y - x)$$
$$= e^{-x} g(x - y).$$

Also from (2), with $n = 2$, and (3) we have

$$H_2(e^x, e^y) = \int_{-\infty}^\infty \frac{g(u - x)}{e^u + e^y}\, du$$
$$= e^{-y} \int_{-\infty}^\infty g(y - u) g(u - x)\, du$$
$$= e^{-y} \int_{-\infty}^\infty g(y - x - u) g(u)\, du$$
$$= e^{-y} g_2(y - x) = e^{-x} g_2(x - y).$$

We can now apply induction to prove

(4) $$H_n(e^x, e^y) = e^{-y} g_n(y - x) = e^{-x} g_n(x - y).$$

For assume that (4) holds for $n - 1$ and substitute in (2). We obtain

$$H_n(e^x, e^y) = e^{-y} \int_{-\infty}^\infty g(u - x) g_{n-1}(y - u)\, du$$
$$= e^{-y} \int_{-\infty}^\infty g(u) g_{n-1}(y - x - u)\, du$$
$$= e^{-y} g_n(y - x) = e^{-x} g_n(x - y).$$

Hence (4) is completely established. We have thus proved

**THEOREM 18a.** *If*
$$H_1(x, y) = (x + y)^{-1}$$
$$H_n(x, y) = \int_0^\infty H_{n-1}(x, t) H_1(t, y) \, dt \qquad (n = 2, 3, \cdots)$$
$$g_1(t) = (1 + e^{-t})^{-1}$$
$$g_n(t) = \int_{-\infty}^\infty g_{n-1}(t - u) g_1(u) \, du \qquad (n = 2, 3, \cdots),$$
*then*
$$H_n(e^x, e^y) = e^{-x} g_n(x - y) = e^{-y} g_n(y - x) \qquad (n = 1, 2, \cdots).$$

By use of Theorem 17c we can now prove

**THEOREM 18b.** *If $H_n(x, y)$ is defined as in Theorem 18a, then*

(5) $$H_{2n}(x, y) = \sum_{k=1}^n \frac{A_{2n,2k}}{(2k - 1)!} \frac{[\log (x/y)]^{2k-1}}{x - y} \qquad (n = 1, 2, \cdots),$$

(6) $$H_{2n+1}(x, y) = \sum_{k=0}^n \frac{A_{2n+1\,2k+1}}{(2k)!} \frac{[\log (x/y)]^{2k}}{x + y} \qquad (n = 0, 1, 2, \cdots),$$

where the constants $A_{n,k}$ are defined by the expansion
$$\left(\frac{\pi s}{\sin \pi s}\right)^n = A_{n,n} + A_{n,n-1} s + A_{n,n-2} s^2 + \cdots.$$

For, we have
$$H_{2n}(x, y) = \frac{1}{x} g_{2n}\left(\log \frac{x}{y}\right)$$

and by §17 (7) this gives (5). In a similar way one obtains (6) from §17 (8).

As an example, let us compute $H_2(x, y)$ from formula (5) and compare with (1). It is known* that

$$\frac{\pi s}{\sin \pi s} = \sum_{k=0}^\infty (-1)^{k-1} \frac{(2^{2k} - 2) B_{2k}}{(2k)!} (\pi s)^{2k}$$

(7) $$= 1 + \frac{1}{6} (\pi s)^2 + \frac{7}{360} (\pi s)^4 + \frac{31}{15120} (\pi s)^6 + \cdots,$$

where the constants $B_k$ are the Bernoulli numbers defined by the symbolic equation
$$(B + 1)^n - B^n = 0 \qquad (B^n = B_n).$$

* See, for example, K. Knopp [1928] p. 204.

Squaring (1), we see that $A_{22} = 1$, and from (5)

$$H_2(x, y) = \frac{\log (x/y)}{x - y},$$

and this checks with (1).

Let us also compute $H_3(x, y)$. From (7) we have

$$\left(\frac{\pi s}{\sin \pi s}\right)^3 = 1 + \frac{\pi^2}{2}s^2 + \cdots,$$

so that $A_{3,3}$ is 1 and $A_{3,1}$ is $\pi^2/2$. Hence

$$H_3(x, y) = \frac{A_{3,1}}{x + y} + \frac{A_{3,3}(\log x/y)^2}{2(x + y)}.$$

$$H_3(x, y) = \frac{\pi^2 + (\log x/y)^2}{2(x + y)}.$$

Incidentally we have thus evaluated a certain definite integral:

$$\int_0^\infty \frac{\log (x/t)}{x - t} \cdot \frac{1}{t + y} \, dt = \frac{\pi^2 + (\log (x/y))^2}{2(x + y)}.$$

## 19. Representation of Functions

We derive here certain sufficient conditions that a function $f(s)$ can be represented as a bilateral Laplace integral.*

THEOREM 19a. *Let $f(s)$ be analytic in the strip $\alpha < \sigma < \beta$ and such that*

(1) $$\int_{-\infty}^\infty |f(\sigma + i\tau)| \, d\tau < \infty \qquad (\alpha < \sigma < \beta).$$

*Let*

(2) $$\lim_{|\tau| \to \infty} f(\sigma + i\tau) = 0$$

*uniformly in every closed subinterval of $(\alpha < \sigma < \beta)$, and set*

(3) $$\varphi(x) = \frac{1}{2\pi i} \int_{\sigma - i\infty}^{\sigma + i\infty} f(s) e^{xs} \, ds \qquad (\alpha < \sigma < \beta, \ -\infty < x < \infty).$$

*Then*

$$f(s) = \int_{-\infty}^\infty e^{-sx} \varphi(x) \, dx \qquad (\alpha < \sigma < \beta).$$

We observe first that $\phi(x)$ is really independent of $\sigma$. This follows at once by Cauchy's integral theorem, using the analyticity of $f(s)$ and (2).

* Compare H. Hamburger [1921], p. 416.

Let $s_0 = \sigma_0 + i\tau_0$ be an arbitrary point of the strip $(\alpha < \sigma < \beta)$. Choose $\sigma_1$ and $\sigma_2$ so that $\alpha < \sigma_1 < \sigma_0 < \sigma_2 < \beta$. By (1) and (3)

$$|\varphi(x)| \leq \frac{e^{x\sigma}}{2\pi} \int_{-\infty}^{\infty} |f(\sigma + i\tau)|\, d\tau = e^{x\sigma} A(\sigma)$$

$$(\alpha < \sigma < \beta,\ -\infty < x < \infty),$$

where $A(\sigma)$ is a non-negative function which is finite for each $\sigma$. Set

(4) $$F(s_0) = \int_{-\infty}^{\infty} e^{-s_0 x} \varphi(x)\, dx.$$

The integral converges since

$$\int_0^\infty e^{-s_0 x} \varphi(x)\, dx \ll A(\sigma_1) \int_0^\infty e^{-x(\sigma_0 - \sigma_1)}\, dx$$

$$\int_{-\infty}^0 e^{-s_0 x} \varphi(x)\, dx \ll A(\sigma_2) \int_{-\infty}^0 e^{x(\sigma_2 - \sigma_0)}\, dx.$$

Substituting (3) in (4), choosing $\sigma = \sigma_0$ in (3), we have

$$F(s_0) = \frac{1}{2\pi i} \int_{-\infty}^{\infty} e^{-s_0 x}\, dx \int_{\sigma_0 - i\infty}^{\sigma_0 + i\infty} f(s) e^{xs}\, ds$$

$$= \frac{1}{2\pi} \int_{-\infty}^{\infty} dx \int_{-\infty}^{\infty} f(\sigma_0 + i\tau) e^{ix(\tau - \tau_0)}\, d\tau$$

$$= \lim_{R \to \infty} \frac{1}{2\pi} \int_{-R}^{R} dx \int_{-\infty}^{\infty} f(\sigma_0 + i\tau) e^{ix(\tau - \tau_0)}\, d\tau.$$

By Fubini's theorem this becomes

$$F(s_0) = \lim_{R \to \infty} \frac{1}{\pi} \int_{-\infty}^{\infty} f(\sigma_0 + i\tau) \frac{\sin R(\tau - \tau_0)}{\tau - \tau_0}\, d\tau.$$

By Theorem 7.2 of Chapter II this limit is $f(\sigma_0 + i\tau_0)$, and our result is established.

We prove now two other representation theorems of Hamburger [1920a], one involving the unilateral Laplace integral. These theorems involve the positiveness* of the sequence of derivatives of the functions to be represented. We need a preliminary result.

LEMMA 19b. *If $\alpha(t)$ is non-decreasing in $(0 \leq t < \infty)$, and if for two positive constants $M$ and $\rho$*

$$\int_0^\infty t^n\, d\alpha(t) < Mn!\rho^{-n} \qquad (n = 0, 1, 2, \cdots),$$

---

* For the meaning of this see Definition 9b of Chapter III.

then

$$\int_0^\infty e^{-st}\,d\alpha(t) = \sum_{n=0}^\infty \frac{(-s)^n}{n!}\int_0^\infty t^n\,d\alpha(t) \qquad (|s|<\rho).$$

For, by Theorem 5d of Chapter I we have for every positive $R$ and any $s$

(5) $$\int_0^R e^{-st}\,d\alpha(t) = \sum_{n=0}^\infty \frac{(-s)^n}{n!}\int_0^R t^n\,d\alpha(t).$$

But series (5) is dominated by

$$M\sum_{n=0}^\infty |s|^n \rho^{-n}$$

and is consequently uniformly convergent in $0 \leqq R < \infty$ for any fixed $s$ in modulus less than $\rho$. Hence we may let $R$ become infinite in (5) to obtain the desired result.

THEOREM 19b. *If $f(s)$ is analytic in the strip ($\alpha < \sigma < \beta$) and if for some real $c$ in ($\alpha < \sigma < \beta$) the two sequences*

(6) $$\{(-1)^n f^{(n)}(c)\}_0^\infty \qquad \{(-1)^n f^{(n)}(c)\}_1^\infty$$

*are positive, then $f(s)$ is analytic in the half-plane $\sigma > \alpha$ and*

(7) $$f(s) = \int_0^\infty e^{-st}\,d\alpha(t) \qquad (\sigma > \alpha),$$

*where $\alpha(t)$ is a non-decreasing function.*

By Theorem 13a, Chapter III, there exists by virtue of the positiveness of the sequences (6) a non-decreasing function $\beta(t)$ such that

$$(-1)^n f^{(n)}(c) = \int_0^\infty t^n\,d\beta(t).$$

Since $f(s)$ is analytic at $s = c$ there exist positive constants $M$ and $\rho$ such that

(8) $$|f^{(n)}(c)| < Mn!\rho^{-n} \qquad (n = 0, 1, 2, \cdots),$$

and the Taylor's series

$$f(s) = \sum_{n=0}^\infty f^{(n)}(c)\frac{(s-c)^n}{n!}$$

$$= \sum_{n=0}^\infty \frac{(c-s)^n}{n!}\int_0^\infty t^n\,d\beta(t)$$

converges to $(f)s$ for $|s - c| < \rho$. But by Lemma 19b this sum is equal for $|s - c| < \rho$ to the integral

$$(9) \qquad \int_0^\infty e^{(c-s)t} \, d\beta(t) = \int_0^\infty e^{-st} \, d\alpha(t)$$

$$\alpha(t) = \int_0^t e^{cu} \, d\beta(u) \qquad (0 \leqq t < \infty).$$

The integral (9) clearly converges for $\sigma > c - \rho$, but by Theorem 5b of Chapter II, it must also converge for $\sigma > \alpha$. Then we conclude our proof by analytic extension.

Conversely, it is evident from Theorem 13a, Chapter III, that if (7) holds then $f(s)$ is analytic for $\sigma > \alpha$ and the sequences (6) are positive for every $c$ greater than $\alpha$.

COROLLARY 19b. *If the sequences* $\{f^{(n)}(c)\}_0^\infty$ *and* $\{f^{(n+1)}(c)\}_0^\infty$ *are positive, then*

$$f(s) = \int_0^\infty e^{st} \, d\alpha(t) \qquad (-\infty < \sigma < \beta).$$

For under the present hypotheses $f(-s)$ satisfies the conditions of Theorem 19b in the interval $(-\beta, -\alpha)$.

THEOREM 19c. *If $f(s)$ is analytic in the strip $\alpha < \sigma < \beta$ and if for some real $c$ in $(\alpha < \sigma < \beta)$ the sequence*

$$\{(-1)^n f^{(n)}(c)\}_0^\infty$$

*is positive, then*

$$f(s) = \int_{-\infty}^\infty e^{-st} \, d\alpha(t) \qquad (\alpha < \sigma < \beta),$$

*where $\alpha(t)$ is a non-decreasing function.*

By Theorem 10 of Chapter III there exists a non-decreasing function $\beta(t)$ such that

$$(-1)^n f^n(c) = \int_{-\infty}^\infty t^n \, d\beta(t) \qquad (n = 0, 1, 2, \cdots)$$

As in the previous proof we have (8). Then

$$(10) \qquad \int_0^\infty t^{2n} \, d\beta(t) \leqq \int_{-\infty}^\infty t^{2n} \, d\beta(t) \leqq M(2n)! \rho^{-2n} \qquad (n = 0, 1, 2, \cdots).$$

Also by the inequality of Schwarz and (10)

$$\int_0^\infty t^{2n+1} \, d\beta(t) \leqq \left[ \int_0^\infty t^{2n} \, d\beta(t) \int_0^\infty t^{2n+2} \, d\beta(t) \right]^{1/2}$$

$$\leqq M(2n + 1)! \rho^{-(2n+1)} \left( \frac{2n + 2}{2n + 1} \right)^{1/2}$$

$$\leqq 2M(2n + 1)! \rho^{-(2n+1)} \qquad (n = 0, 1, \cdots).$$

That is,

(11) $$\int_0^\infty t^n \, d\beta(t) \leq 2Mn!\rho^{-n} \qquad (n = 0, 1, \cdots),$$

and in like manner

(12) $$\int_{-\infty}^0 (-t)^n \, d\beta(t) \leq 2Mn!\rho^{-n} \qquad (n = 0, 1, \cdots).$$

By Lemma 19b the inequalities (11) and (12) enable us to conclude

$$\int_0^\infty e^{(c-s)t} \, d\beta(t) = \sum_{n=0}^\infty \frac{(c-s)^n}{n!} \int_0^\infty t^n \, d\beta(t) \qquad (|s-c| < \rho)$$

$$\int_{-\infty}^0 e^{(c-s)t} \, d\beta(t) = \sum_{n=0}^\infty \frac{(c-s)^n}{n!} \int_{-\infty}^0 t^n \, d\beta(t) \qquad (|s-c| < \rho),$$

respectively. Adding and setting

(13) $$\alpha(t) = \int_0^t e^{cu} \, d\beta(u) \qquad (-\infty < t < \infty),$$

we have

$$\int_{-\infty}^\infty e^{-st} \, d\alpha(t) = \sum_{n=0}^\infty \frac{(s-c)^n}{n!} f^{(n)}(c) = f(s) \qquad |s-c| < \rho.$$

As in the previous proof this equation must hold throughout the strip of analyticity. Again it is clear that the converse of the theorem is trivial.

COROLLARY 19c. *If $\{f^{(n)}(c)\}_0^\infty$ is a positive sequence, the result holds.*

For the sequences $\{f^{(n)}(c)\}_0^\infty$ and $\{(-1)^n f^{(n)}(c)\}_0^\infty$ are both positive if one is, as is easily verified by an examination of the defining quadratic forms.

Following the same order of ideas we prove:

THEOREM 19d. *If $f(s)$ is analytic at a real point $s = c$ where the sequence*

(14) $$\{(-1)^n f^{(n)}(c)\}_0^\infty$$

*is completely monotonic, then for all complex $s$*

(15) $$f(s) = \int_0^1 e^{-st} \, d\alpha(t),$$

*where $\alpha(t)$ is non-decreasing and bounded in $(0 \leq t \leq 1)$.*

For, by Theorem 4a of Chapter III there exists a non-decreasing bounded function $\beta(t)$ such that

$$(-1)^n f^{(n)}(c) = \int_0^1 t^n \, d\beta(t) \qquad (n = 0, 1, 2, \cdots).$$

Since $f(s)$ is analytic at $s = c$

$$f(s) = \sum_{n=0}^{\infty} f^{(n)}(c) \frac{(s-c)^n}{n!} = \sum_{n=0}^{\infty} \frac{(c-s)^n}{n!} \int_0^1 t^n \, d\beta(t)$$

for some neighborhood, $|s - c| < \rho$, of $c$. By Theorem 5d of Chapter I this gives

$$f(s) = \int_0^1 e^{-st} \, d\alpha(t),$$

where $\alpha(t)$ is defined by (13) in $(0 \leq t \leq 1)$. But this integral defines an entire function, so that we can complete our proof by analytic extension. It is obvious conversely that if $f(s)$ has the representation (15) the sequence (14) is completely monotonic for every real $c$.

## 20. Kernels of Positive Type

In order to obtain necessary and sufficient conditions for the representation of a function as a bilateral Laplace integral we need a preliminary discussion of kernels of positive type.* These are the continuous analogues of positive or semidefinite quadratic forms.

DEFINITION 20. *A real function $k(x, y)$ which is continuous in the square $(a \leq x \leq b, a \leq y \leq b)$ is of positive type there if for every real function $\phi(x)$ continuous in $(a \leq x \leq b)$*

(1) $$J(\phi) = \int_a^b \int_a^b k(x, y) \phi(x) \phi(y) \, dx \, dy \geq 0.$$

As an example take $k(x, y) = g(x)g(y)$ where $g(x)$ is any function continuous in $(a \leq x \leq b)$. Note that the integral (1) may vanish without having $\phi(x)$ identically zero. Thus in our example we have only to choose $\phi(x)$ orthogonal to $g(x)$ on $(a, b)$.

A kernel is said to be *positive definite* if it is of positive type and if integral (1) can vanish for no real continuous function $\phi(x)$ except $\phi(x) \equiv 0$. As an example take $a = 0$, $b = \pi$ and

$$k(x, y) = \sum_{n=0}^{\infty} e^{-n} \cos nx \cos ny.$$

The integral (1) becomes

(2) $$\sum_{n=0}^{\infty} e^{-n} a_n^2$$

$$a_n = \int_0^\pi \phi(x) \cos nx \, dx \qquad (n = 0, 1, 2, \cdots).$$

---

* See J. Mercer [1909], p. 242.

But (2) cannot be zero unless all the $a_n$ are zero. But by the completeness of the cosine set on $(0, \pi)$ this implies that $\phi(x)$ is identically zero.

We now prove an important result of J. Mercer [1909] which brings out the connection between kernels and quadratic forms.

THEOREM 20. *A continuous kernel $k(x, y)$ is of positive type if and only if for every finite sequence $\{x_i\}_0^n$ of distinct numbers of $(a \leq x \leq b)$ the quadratic form*

$$(3) \qquad Q_n = \sum_{i=0}^{n} \sum_{j=0}^{n} k(x_i, x_j) \xi_i \xi_j$$

*is positive (definite or semidefinite).*

Suppose first that the quadratic forms (3) are positive. Let us prove that $k(x, y)$ is of positive type. Choose an arbitrary continuous function $\phi(x)$. The integral $J(\phi)$ can be expressed as the limit of a sum

$$J(\phi) = \lim_{n \to \infty} \frac{(b-a)^2}{n^2} \sum_{i=0}^{n-1} \sum_{j=0}^{n-1} k(x_i, x_j) \phi(x_i) \phi(x_j)$$

$$x_i = a + \frac{i}{n}(b-a) \qquad (i = 0, 1, \ldots, n).$$

Choosing

$$\xi_i = \phi(x_i) \frac{(b-a)}{n}$$

in (3) we see that $J(\phi)$ is the limit of a non-negative function of $n$, is therefore itself non-negative.

Let us turn next to the converse. Suppose that we could find some form (3) which is not positive, with $a < x_0 < x_1 < \cdots < x_n < b$. Then we can choose the $\xi_i$ so that $Q_n$ has a negative value $-A$. Now define an auxiliary function $\theta_{\epsilon,\eta}(x, c)$. Let $c$ be a point of $(a < x < b)$. Let $\epsilon$ and $\eta$ be so small that $c - \epsilon - \eta$ and $c + \epsilon + \eta$ are points of the same interval. Now $\theta_{\epsilon,\eta}(x, c)$ is defined as zero in $(a \leq x \leq c - \epsilon - \eta)$ and $(c + \epsilon + \eta \leq x \leq b)$ as unity in $(c - \eta \leq x \leq c + \eta)$. In the rest of the interval it is to be linear and such that it is continuous in $(a, b)$. Set

$$\theta(x) = \sum_{i=0}^{n} \xi_i \theta_{\epsilon,\eta}(x, x_i)$$

where $\epsilon$ and $\eta$ have been chosen so small that no two intervals $(-\epsilon - \eta + x_i, x_i + \epsilon + \eta)$ overlap and such that $a < x_0 - \epsilon - \eta$ and $x_n + \epsilon + \eta < b$. Set

$$F_n(x, y) = \sum_{i=0}^{n} \sum_{j=0}^{n} k(x_i + x, y_j + y) \xi_i \xi_j.$$

Then we may easily compute $J(\theta)$:

$$J(\theta) = \int_{-\eta}^{\eta}\int_{-\eta}^{\eta} F_n(x, y)\, dx\, dy + J_1$$

where

$$J_1 = \sum_{i=0}^{n}\sum_{j=0}^{n}\iint_{q_{ij}} k(x, y)\theta(x)\theta(y)\, dx\, dy.$$

Here $q_{ij}$ is the region between the square ($x_i - \epsilon - \eta \leq x \leq x_i + \epsilon + \eta$, $x_j - \epsilon - \eta \leq y \leq x_j + \epsilon + \eta$) and the square ($x_i - \eta \leq x \leq x_i + \eta$, $x_j - \eta \leq y \leq x_j + \eta$). Since

$$|\theta(x)\theta(y)| \leq |\xi_i \xi_j|$$

for $(x, y)$ in $q_{ij}$ we have easily

$$|J_1| \leq M 4\epsilon(2\eta + \epsilon)\left(\sum_{i=0}^{n} |\xi_i|\right)^2,$$

where $M$ is the maximum of $|k(x, y)|$ in the whole square. Since $F_n(0, 0) = -A$ we can find $\eta$ so small that

$$F_n(x, y) < -A/2 \qquad (|x| < \eta, |y| < \eta).$$

Then

$$\int_{-\eta}^{\eta}\int_{-\eta}^{\eta} F_n(x, y)\, dx\, dy < -2A\eta^2,$$

and

$$J(\theta) < -2A\eta^2 + M 4\epsilon(2\eta + \epsilon)\left(\sum_{i=0}^{n}|\xi_i|\right)^2,$$

for all $\epsilon$ sufficiently small. We can choose $\epsilon$ so small that the right-hand side of this inequality is negative. For this continuous function $\theta(x)$ we see that $J(\theta)$ is negative, contradicting (1). Hence there can be no $Q_n$, with the $x_i$ interior points of $(a, b)$, which can have a negative value. Neither can there be a $Q_n$ having a negative value even if $x_0$ or $x_n$ is allowed to be an end point. This is clear by the continuity of $k(x, y)$. The proof of the theorem is complete.

A kernel which is continuous in an open square ($a < x < b, a < y < b$) is said to be of positive type there if it is of positive type in every closed square interior to the open square.

## 21. Necessary and Sufficient Conditions for Representation

By way of making the results of the present section plausible let us make an analogy with the moment problem. We have already compared the Hausdorff moment problem with the representation problem for the unilateral Laplace integral. Here we shall then expect to compare the Stieltjes moment problem with the bilateral representation problem. We recall that the system of equations

(1) $$\mu_n = \int_0^\infty t^n \, d\beta(t) \qquad (n = 0, 1, 2, \cdots)$$

has a non-decreasing solution $\beta(t)$ if and only if the quadratic forms

(2) $$\sum_{i=0}^n \sum_{j=0}^n \mu_{i+j} \xi_i \xi_j, \quad \sum_{i=0}^n \sum_{j=0}^n \mu_{i+j+1} \xi_i \xi_j \qquad (n = 0, 1, 2, \cdots)$$

are all positive. If in (1) the integer $n$ is replaced by a continuous variable $x$ and if $t = e^{-u}$ we obtain

$$\mu(x) = \int_{-\infty}^\infty e^{-xu} \, d\alpha(u),$$

where

$$\alpha(u) = -\beta(e^{-u}).$$

Clearly $\alpha(u)$ is non-decreasing when $\beta(t)$ is.

When one changes from the discrete to the continuous one would expect the quadratic forms (2) to coalesce into a single double integral

$$\int\int \mu(x+y)\xi(x)\xi(y) \, dx \, dy$$

which would be required to be non-negative for all continuous functions $\xi(x)$. That is, the kernel $\mu(x+y)$ would be of positive type. We are thus led to the following* result.

THEOREM 21. *A necessary and sufficient condition that the function $f(x)$ can be represented in the form*

(3) $$f(x) = \int_{-\infty}^\infty e^{-xt} \, d\alpha(t),$$

*where $\alpha(t)$ is non-decreasing and the integral converges for $a < x < b$, is that $f(x)$ should be analytic there and that the kernel $f(x+y)$ should be of positive type in the square $(a < 2x < b, a < 2y < b)$.*

First suppose that (3) holds. Then if $a < \alpha < \beta < b$ we will show that $f(x+y)$ is of positive type in the square $(\alpha \leq 2x \leq \beta, \alpha \leq 2y \leq \beta)$.

\* Compare S. Bochner [1932] p. 76, M. Mathias [1923], and D. V. Widder [1934b].

Clearly $f(x)$ is analytic in the interval $a < x < b$. For any continuous function $\phi(x)$ we have

$$\int_{\alpha/2}^{\beta/2} \int_{\alpha/2}^{\beta/2} f(x+y)\phi(x)\phi(y)\,dx\,dy$$

$$= \int_{\alpha/2}^{\beta/2} \int_{\alpha/2}^{\beta/2} \phi(x)\phi(y)\,dx\,dy \int_{-\infty}^{\infty} e^{-(x+y)t}\,d\alpha(t)$$

$$= \int_{-\infty}^{\infty} \left[\int_{\alpha/2}^{\beta/2} e^{-xt}\phi(x)\,dx\right]^2 d\alpha(t) \geqq 0.$$

To justify the interchange in the order of integration we have only to observe that the integral (3) converges uniformly in $(\alpha \leqq x \leqq \beta)$. This completes the proof of the necessity of the condition.

Conversely, if $f(x+y)$ is of positive type in the square $(\alpha \leqq 2x \leqq \beta, \alpha \leqq 2y \leqq \beta)$, then by Theorem 20, choosing $x_i = (c/2) + i\delta$ for some number $c$ of $(\alpha \leqq x < \beta)$, the quadratic forms

(4) $$\sum_{i=0}^{n} \sum_{j=0}^{n} f(c + i\delta + j\delta)\xi_i \xi_j \qquad (n = 0, 1, 2, \cdots)$$

are positive, provided that $\delta$ is chosen for each $n$ so that

$$\alpha \leqq c < c + 2n\delta \leqq \beta.$$

In particular, if we set

(5) $$\xi_i = \sum_{k=i}^{n} (-1)^{k+i}\binom{k}{i}\eta_k \qquad (i = 0, 1, \cdots, n)$$

and recall that

$$\Delta_\delta^k f(a) = \sum_{i=0}^{k} (-1)^{k+i}\binom{k}{i} f(a + i\delta),$$

we have

$$\sum_{j=0}^{n} \xi_j \sum_{i=0}^{n} f(c + i\delta + j\delta) \sum_{k=i}^{n} (-1)^{k+i}\binom{k}{i}\eta_k$$

$$= \sum_{j=0}^{n} \xi_j \sum_{k=0}^{n} \eta_k \sum_{i=0}^{k} (-1)^{k+i}\binom{k}{i} f(c + i\delta + j\delta)$$

(6) $$= \sum_{j=0}^{n} \xi_j \sum_{k=0}^{n} \eta_k \Delta_\delta^k f(c + j\delta).$$

If the value of $\xi_j$ from (5) is substituted in (6) the latter reduces in a similar way to

(7) $$\sum_{i=0}^{n} \sum_{j=0}^{n} \Delta_\delta^{i+j} f(c) \eta_i \eta_j.$$

Since this form was obtained from (4) it is itself positive. Now replace $\eta_i$ by $\eta_i/\delta^i$ in (7) and let $\delta$ approach zero. We obtain

$$\sum_{i=0}^{n}\sum_{j=0}^{n} f^{(i+j)}(c)\eta_i\eta_j \geqq 0 \qquad (n = 0, 1, 2, \cdots).$$

Now by Corollary 19c

$$f(x) = \int_{-\infty}^{\infty} e^{-xt}\, d\alpha(t) \qquad (a < x < b)$$

for some non-decreasing function $\alpha(t)$. This completes the proof of the theorem. We observe that it would be sufficient to assume that $f(x)$ is continuous in $(a < x < b)$. For, it could be shown that this with the fact that $f(x + y)$ is of positive type would insure the analyticity\*
of $f(x)$.

\* See R. P. Boas, Jr. and D. V. Widder [1940b] and R. P. Boas, Jr. [1941].

# CHAPTER VII

## INVERSION AND REPRESENTATION PROBLEMS FOR THE LAPLACE TRANSFORM

### 1. Introduction

We have seen that the Laplace transform may be regarded as a generalization of Taylor's series. Thus the series

(1) $$F(z) = \sum_{n=0}^{\infty} a_n z^n$$

and the integral

(2) $$F(z) = \int_0^{\infty} a(t) z^t \, dt$$

may be regarded as the discrete and continuous aspects, respectively, of the same Stieltjes integral

(3) $$F(z) = \int_0^{\infty} z^t \, d\alpha(t).$$

Making a change of variable we obtain the Laplace-Stieltjes integral

(4) $$F(e^{-s}) = f(s) = \int_0^{\infty} e^{-st} \, d\alpha(t).$$

One familiar determination of the coefficients of (1)

(5) $$\frac{F^{(k)}(0)}{k!} = a_k \qquad (k = 0, 1, 2, \cdots)$$

involves a knowledge of the derivatives of $F(z)$ at $z = 0$. By analogy we should expect the existence of an inversion formula for (4) which depends on a knowledge of the derivatives of $f(s)$ in a neighborhood of $s = \infty$, say along the positive real axis, since such a neighborhood is the transform in the $z$-plane under the transformation $z = e^{-s}$ of a right-handed real neighborhood of the origin in the $z$-plane. Such an inversion formula was discovered by E. L. Post [1930] for the case when $\alpha(t)$ is the integral of a continuous function $\varphi(t)$. The germ of the formula is contained in a letter from Stieltjes to Hermite dated August 29, 1893 (letter 383, pp. 332–334, Vol. 2 of the collected correspondence, Paris, 1905).

The general case (4) was first treated by the author [1934a]. Post's formula is

(6) $$\phi(t) = \lim_{k\to\infty} \frac{(-1)^k}{k!} f^{(k)}\left(\frac{k}{t}\right)\left(\frac{k}{t}\right)^{k+1}$$

Since a limit process is involved it is clear that one needs to know derivatives of $f(s)$ only for large real values of $s$. In this sense (6) is analogous to (5).

The present chapter will be devoted to an elaboration of formula (6) and to the representation results which grow out of the formula.

## 2. Laplace's Asymptotic Evaluation of an Integral

In our development of an inversion formula it will be convenient to apply an asymptotic method of Laplace [1820]. For the reader's convenience we prove here the results which we shall use.

LEMMA 2. *If $a < b$, $0 < \gamma$, then*

$$I_k = \int_a^b e^{-k\gamma(x-a)^2} dx \sim \frac{1}{2}\left(\frac{\pi}{k\gamma}\right)^{1/2} \qquad (k \to \infty).$$

For, simple changes of variable give

$$I_k = \int_0^{b-a} e^{-k\gamma t^2} dt = \frac{1}{\sqrt{k\gamma}} \int_0^{(b-a)\sqrt{k\gamma}} e^{-u^2} du$$

so that

$$I_k \sim \frac{1}{(k\gamma)^{1/2}} \int_0^\infty e^{-u^2} du = \frac{1}{2}\left(\frac{\pi}{k\gamma}\right)^{1/2} \qquad (k \to \infty).$$

THEOREM 2a. *If*
1. $a < a + \eta < b$;
2. $h(x) \in C^2$ $(a \leq x \leq a + \eta)$, $h'(a) = 0$, $h''(a) < 0$, $h(x)$ *is non-increasing* $(a \leq x \leq b)$; *then*

$$\int_a^b e^{kh(x)} dx \sim e^{kh(a)} \left(\frac{-\pi}{2kh''(a)}\right)^{1/2} \qquad (k \to \infty).$$

Let $\epsilon$ be an arbitrary number such that $0 < \epsilon < -h''(a)$. Then we may choose $\delta$ less than $\eta$ and so small that

(1) $\quad h''(a) - \epsilon < h''(x) < h''(a) + \epsilon < 0 \qquad (a \leq x \leq a + \delta).$

Consider the integral

$$I_k = \sqrt{k} \int_a^b e^{k[h(x) - h(a)]} dx = I_k' + I_k'',$$

where $I'_k$ and $I''_k$ correspond to the intervals $(a, a + \delta)$ and $(a + \delta, b)$, respectively. Then

$$|I''_k| \leq \sqrt{k}\, e^{k[h(a+\delta)-h(a)]}(b - a - \delta).$$

Since $h''(a) < 0$ it is clear that $h(a + \delta) - h(a) < 0$. Hence $I''_k$ tends to zero with $1/k$ for any positive $\delta$. By Taylor's formula with remainder

$$I'_k = \sqrt{k} \int_a^{a+\delta} e^{kh'(\xi)(x-a)^2/2}\, dx \qquad (a < \xi < a + \delta).$$

By (1)

$$\sqrt{k} \int_a^{a+\delta} e^{k[h''(a)-\epsilon](x-a)^2/2} < I'_k < \sqrt{k} \int_a^{a+\delta} e^{k[h''(a)+\epsilon](x-a)^2/2}\, dx.$$

By use of Lemma 2

$$\left(\frac{-\pi}{2[h''(a) - \epsilon]}\right)^{1/2} \leq \varliminf_{k\to\infty} I_k \leq \left(\frac{-\pi}{2[h''(a) + \epsilon]}\right)^{1/2}.$$

Since $\epsilon$ is arbitrary we have

$$\lim_{k\to\infty} I_k = \left(\frac{-\pi}{2h''(a)}\right)^{1/2},$$

and our theorem is proved.

**THEOREM 2b.** *If in addition to the hypotheses 1. and 2. of the previous theorem we have*

3. $\phi(x) \, \varepsilon \, L \, (a \leq x \leq b)$, $\phi(a) \neq 0$,

(2) $$\alpha(x) = \int_a^x [\phi(u) - \phi(a)]\, du = o(x - a) \qquad (x \to a+);$$

*then*

$$\int_a^b \phi(x) e^{kh(x)}\, dx \sim \phi(a) e^{kh(a)} \left(\frac{-\pi}{2kh''(a)}\right)^{1/2} \qquad (k \to \infty).$$

Set

$$I_k = \sqrt{k} \int_a^b [\phi(x) - \phi(a)] e^{k[h(x)-h(a)]}\, dx.$$

By Theorem 2a we need only show that $I_k$ tends to zero when $k$ becomes infinite. Define $\epsilon$ as in the proof of Theorem 2a. Choose $\delta$ so that (1) holds and so that

$$|\alpha(x)| < \epsilon(x - a) \qquad (a \leq x \leq a + \delta).$$

This is possible by hypothesis 3. As in the proof of Theorem 2a, break the integral $I_k$ into two parts $I_k'$ and $I_k''$ corresponding to the intervals $(a, a + \delta)$ and $(a + \delta, b)$, respectively. Then

$$|I_k''| \leq \sqrt{k}\, e^{k[h(a+\delta)-h(a)]} \int_{a+\delta}^{b} |\phi(x) - \phi(a)|\, dx$$

$$= o(1) \qquad (k \to \infty).$$

Integrating the integral $I_k'$ by parts we obtain

$$I_k' = k^{1/2}\alpha(a + \delta)e^{k[h(a+\delta)-h(a)]} - k^{3/2} \int_a^{a+\delta} \alpha(x) e^{kh''(\xi)(x-a)^2/2} h'(x)\, dx$$

$$= o(1) + I_k''' \qquad (k \to \infty).$$

But

$$h'(x) = h''(\xi)(x - a) \qquad (a < \xi < x < a + \delta),$$

so that by use of (1) we have

$$|I_k'''| \leq k^{3/2} \epsilon \int_a^{a+\delta} (x - a)^2 e^{k[h''(a)+\epsilon](x-a)^2/2} [-h''(a) + \epsilon]\, dx.$$

Making the change of variable

$$k^{1/2}[-h''(a) - \epsilon]^{1/2}(x - a) = t$$

we obtain

$$|I_k'''| \leq \epsilon \frac{-h''(a) + \epsilon}{[-h''(a) - \epsilon]^{3/2}} \int_0^\infty t^2 e^{-t^2/2}\, dt.$$

Hence it is clear that

$$\overline{\lim_{k \to \infty}} |I_k| \leq \epsilon \frac{-h''(a) + \epsilon}{[-h''(a) - \epsilon]^{3/2}} \int_0^\infty t^2 e^{-t^2/2}\, dt,$$

or, since $\epsilon$ is arbitrary, that

$$\lim_{k \to \infty} I_k = 0,$$

and this establishes our result.

COROLLARY 2b.1. *If the conditions of the theorem hold except that* $\phi(a) = 0$, *then*

$$\int_a^b \phi(x) e^{kh(x)}\, dx = o(k^{-1/2} e^{kh(a)}) \qquad (k \to \infty).$$

COROLLARY 2b.2. *If*
1. $a < b - \eta < b$;
2. $h(x)\ \varepsilon\ C^2\ (b - \eta \leq x \leq b)$, $h'(b) = 0$, $h''(b) < 0$, $h(x)$ *is non-decreasing in* $(a \leq x \leq b)$;

3. $\phi(x) \, \varepsilon \, L \, (a \leq x \leq b), \phi(b) \neq 0$

$$\alpha(x) = \int_x^b [\phi(u) - \phi(b)] \, du = o(b - x) \qquad (x \to b-);$$

then

$$\int_a^b \phi(x) e^{kh(x)} \, dx \sim \phi(b) e^{kh(b)} \left(\frac{-\pi}{2kh''(b)}\right)^{1/2} \qquad (k \to \infty).$$

COROLLARY 2b.3. *Equation* (2) *of Theorem* 2b *may be replaced by*

$$\phi(a+) = \phi(a).$$

This is clear since the continuity of $\phi(x)$ on the right at $x = a$ implies (2). It is in the form of Corollary 2b. 3 that the theorem is ordinarily stated, but we shall have need for the more general result.

## 3. Applications of the Laplace Method

We proceed at once to apply the results of the previous section to an integral which will lead directly to the inversion formula desired.

THEOREM 3a. *If*

1. $\phi(x) \, \varepsilon \, L \, (0 < t \leq x \leq R)$ *for a fixed $t$ and every larger $R$;*

2. $\int_t^\infty e^{-cx} \phi(x) \, dx$ *converges for a fixed positive $c$;*

3. $\int_t^x [\phi(u) - \phi(t)] \, du = o(x - t) \qquad (x \to t+);$

then

$$\lim_{k \to \infty} \frac{1}{k!} \left(\frac{k}{t}\right)^{k+1} \int_t^\infty e^{-ku/t} u^k \phi(u) \, du = \frac{\phi(t)}{2}.$$

Choose any positive number $\delta$ and set

$$\alpha(x) = \int_{t+\delta}^x e^{-cu} \phi(u) \, du \qquad (x \geq t + \delta).$$

Then by hypothesis 2. there exists a constant $M$ such that

$$|\alpha(x)| \leq M \qquad (t \leq x < \infty).$$

Integration by parts gives

$$I_k' = \frac{1}{k!}\left(\frac{k}{t}\right)^{k+1} \int_{t+\delta}^\infty e^{-ku/t} u^k \phi(u) \, du = \lim_{u \to \infty} \frac{1}{k!}\left(\frac{k}{t}\right)^{k+1} \alpha(u) e^{-(k-ct)u/t} u^k$$

$$- \frac{1}{k!}\left(\frac{k}{t}\right)^{k+1} \int_{t+\delta}^\infty \alpha(u) \, d[e^{-(k-ct)u/t} u^k]$$

if $k > tc$. In this case the above limit is zero. The function $u^k e^{(ct-k)u/t}$

## §3] APPLICATIONS OF LAPLACE METHOD

considered as a function of $u$ has a maximum at $u = kt/(k - ct)$, a number less than $t + \delta$ for $k$ sufficiently large, say for $k > k_0$. Hence

$$|I'_k| \leq \frac{M}{k!} e^{-(t+\delta)(k-ct)/t}(t+\delta)^k \left(\frac{k}{t}\right)^{k+1} = l_k \qquad (k > k_0).$$

But

$$\frac{l_{k+1}}{l_k} = \left(1 + \frac{1}{k}\right)^{k+1} e^{-(t+\delta)/t}\left(1 + \frac{\delta}{t}\right) \to e^{-\delta/t}\left(1 + \frac{\delta}{t}\right) \qquad (k \to \infty).$$

Since

$$e^{-\delta/t}\left(1 + \frac{\delta}{t}\right) < 1,$$

we have

$$\lim_{k \to \infty} I'_k = \lim_{k \to \infty} l_k = 0.$$

We apply Theorem 2b to the integral

$$I''_k = \frac{1}{k!}\left(\frac{k}{t}\right)^{k+1} \int_t^{t+\delta} e^{-ku/t} u^k \phi(u)\, du$$

taking

$$h(x) = \log x - \frac{x}{t}.$$

We have

$$h'(x) = \frac{1}{x} - \frac{1}{t} < 0 \qquad (t < x)$$

$$h'(t) = 0$$

$$h''(t) = -\frac{1}{t^2} < 0,$$

so that the desired hypotheses are satisfied. Hence

$$I'_k + I''_k \sim \phi(t) \frac{k^{k+1}}{k!} \left(\frac{\pi}{2k}\right)^{1/2} e^{-k} \qquad (k \to \infty),$$

and by Stirling's formula

$$I'_k + I''_k \to \frac{\phi(t)}{2} \qquad (k \to \infty),$$

which is what we wished to prove.

A result of a similar nature is contained in:

**THEOREM 3b.** *If*

1. $\phi(x) \in L$ $(0 < \epsilon \leq x \leq t)$ *for a fixed t and every smaller positive* $\epsilon$;
2. $\int_{0+}^{t} x^r \phi(x)\, dx$ *converges for a fixed real constant r;*
3. $\int_{t}^{x} [\phi(u) - \phi(t)]\, du = o(t - x)$ \hfill $(x \to t-)$;

*then*

$$\lim_{k \to \infty} \frac{1}{k!} \left(\frac{k}{t}\right)^{k+1} \int_{0+}^{t} e^{-ku/t} u^k \phi(u)\, du = \frac{\phi(t)}{2}.$$

Choose a positive $\delta$ less than $t$ and set

$$\alpha(x) = \int_{x}^{t-\delta} u^r \phi(u)\, du \qquad (0 < x \leq t - \delta).$$

Then by hypothesis 2. there exists a constant $M$ such that

$$|\alpha(x)| \leq M \qquad (0 < x \leq t - \delta).$$

Integration by parts gives for $k > r$

$$I'_k = \frac{1}{k!} \left(\frac{k}{t}\right)^{k+1} \int_{0+}^{t-\delta} e^{-ku/t} u^k \phi(u)\, du$$

$$= \frac{1}{k!} \left(\frac{k}{t}\right)^{k+1} \int_{0+}^{t-\delta} \alpha(u)\, d[e^{-ku/t} u^{k-r}].$$

The function in brackets considered as a function of $u$ has its only maximum at $u = (k - r)t/k$ and hence is non-decreasing in $0 \leq u \leq t - \delta$ for $k$ sufficiently large, say for $k > k_0$. Hence

$$|I'_k| = Me^{-k(t-\delta)/t}(t - \delta)^{k-r} \frac{k^{k+1}}{k!} \frac{1}{t^{k+1}} \qquad (k > k_0).$$

Using the same device as in the proof of the previous theorem we see that $I'_k$ tends to zero with $1/k$ since

$$e^{\delta/t}\left(1 - \frac{\delta}{t}\right) < 1.$$

By use of Corollary 2b.2 we have

$$I''_k = \frac{1}{k!}\left(\frac{k}{t}\right)^{k+1} \int_{t-\delta}^{t} e^{-ku/t} u^k \phi(u)\, du \to \frac{\phi(t)}{2} \qquad (k \to \infty),$$

so that

$$I'_k + I''_k \to \frac{\varphi(t)}{2} \qquad (k \to \infty).$$

This completes the proof of the theorem.

**THEOREM 3c.** *If*
1. $\phi(x) \, \varepsilon \, L \, (R^{-1} \leq x \leq R)$ *for every* $R > 1$;
2. $\int_1^\infty \phi(x) e^{-cx} \, dx$ *converges for a fixed* $c > 0$;
3. $\int_{0+}^1 \phi(x) x^r \, dx$ *converges for a fixed* $r$;
4. $\int_t^x [\phi(u) - \phi(t)] \, du = o(|x - t|)$ \hfill $(x \to t)$;

*then*

$$\lim_{k \to \infty} \frac{1}{k!} \left(\frac{k}{t}\right)^{k+1} \int_{0+}^\infty e^{-ku/t} u^k \phi(u) \, du = \phi(t).$$

The proof is obtained in an obvious way using Theorem 3a and Theorem 3b.

**COROLLARY 3c.1.** *If hypothesis 4. is replaced by*

4'   $$\phi(t) = \frac{\phi(t+) + \phi(t-)}{2},$$

*the conclusion holds.*

Of course hypothesis 4.' implies the existence of limits on the right and on the left of $\phi(u)$ at $u = t$.

**COROLLARY 3c.2.** *If hypothesis 4. is omitted the conclusion holds for almost all positive* $t$.

For, it is known that condition 1. implies 4. for the points $t$ of the Lebesgue set for the function $\phi(u)$.

## 4. Uniform Convergence

In this section we extend the Laplace method of asymptotic evaluation of an integral to the case in which one function of the integrand involves a parameter.

**THEOREM 4a.** *If*
1. $h(x) \, \varepsilon \, C^2 \, (1 \leq x \leq 1 + \eta, \eta > 0)$, $h'(1) = 0$, $h''(1) < 0$, $h(x)$ *is non-increasing in* $(1 \leq x \leq 1 + \eta)$;
2. $\phi(x) \, \varepsilon \, C \, (0 \leq a \leq x \leq b)$, $\phi(x) \, \varepsilon \, L \, (a \leq x \leq b(1 + \eta))$;

*then*

(1)   $$\lim_{k \to \infty} \left(\frac{-2kh''(1)}{\pi}\right)^{1/2} \int_1^{1+\eta} e^{k[h(x) - h(1)]} \phi(tx) \, dx = \phi(t)$$

*uniformly in* $a \leq t \leq b'$, *where* $b' < b$.

If $a \leq t \leq b$ and $1 \leq x \leq 1 + \eta$, then $a \leq tx \leq b(1 + \eta)$ and the integral (1) exists by virtue of hypothesis 2. Let $\epsilon$ be arbitrary except that $0 < \epsilon < -h''(1)$. Choose $\delta < \eta$ and so small that

(2) $\quad h''(x) < h''(1) + \epsilon, \quad |\phi(tx) - \phi(t)| < \epsilon \quad (a \leq t \leq b')$

when $1 \leq x \leq 1 + \delta$. The second inequality follows by the uniform continuity of $\phi(x)$. For, if

$$\delta \leq \frac{b}{b'} - 1$$

then $a \leq tx \leq b$, so that $tx$ is in the region of continuity of $\phi(x)$. Since $|tx - t| \leq \delta b'$ for all $x$ in $(1 \leq x \leq 1 + \delta)$, it is clear that $\delta$ may be chosen so small that (2) holds.

Set

$$I_k = \sqrt{k} \int_1^{1+\eta} e^{k[h(x)-h(1)]}[\phi(tx) - \phi(t)]\, dx,$$

and write $I_k$ as the sum of two integrals $I_k'$ and $I_k''$ corresponding to the intervals $(1, 1 + \delta)$ and $(1 + \delta, 1 + \eta)$, respectively. Then

$$|I_k''| \leq \sqrt{k}\, e^{k[h(1+\delta)-h(1)]} \int_{1+\delta}^{1+\eta} |\phi(tx) - \phi(t)|\, dx$$

$$\leq \sqrt{k}\, e^{k[h(1+\delta)-h(1)]} A,$$

where $A$ is independent of $t$ in $(a \leq t \leq b')$. To show its existence we have

(3) $\quad \displaystyle\int_{1+\delta}^{1+\eta} |\phi(tx) - \phi(t)|\, dx \leq \frac{1}{t} \int_{(1+\delta)t}^{(1+\eta)t} |\phi(x)|\, dx + |\phi(t)|\,(\eta - \delta).$

The right-hand side is clearly a continuous function of $t$ in $a \leq t \leq b'$ if $a > 0$. If $a = 0$ the same is true, for then $\phi(x)$ is continuous at $x = 0$ and

$$\lim_{t \to 0+} \frac{1}{t} \int_{(1+\delta)t}^{(1+\eta)t} |\phi(x)|\, dx = (\eta - \delta)\,|\phi(0)|.$$

Hence $A$ may be taken as the maximum of the right-hand side of (3). Since $h''(1) < 0$, we have $h(1 + \delta) - h(1) < 0$ and

$$\lim_{k \to \infty} I_k'' = 0$$

uniformly in $(a \leq t \leq b')$ for any value of $\delta$ under consideration.

On the other hand by (2)

$$|I_k'| \leq \epsilon \sqrt{k} \int_1^{1+\delta} e^{k[h''(1)+\epsilon](x-1)^2/2}\, dx \leq \epsilon \left(\frac{-\pi}{2[h''(1)+\epsilon]}\right)^{1/2}$$

for all $t$ in $(a \leq t \leq b')$. Hence

$$\varlimsup_{k\to\infty} |I_k| \leq \epsilon \left(\frac{-\pi}{2[h''(1) + \epsilon]}\right)^{1/2}$$

$$\lim_{k\to\infty} I_k = 0$$

uniformly in $a \leq t \leq b'$. The proof is now completed by use of Theorem 2a.

**THEOREM 4b.** *If*
1. $h(x) \, \varepsilon \, C^2 \, (0 < 1 - \eta \leq x \leq 1)$, $h'(1) = 0$, $h''(1) < 0$, $h(x)$ *is non-decreasing in* $(1 - \eta \leq x \leq 1)$;
2. $\phi(x) \, \varepsilon \, C(0 \leq a \leq x \leq b)$, $\phi(x) \, \varepsilon \, L \, (a(1 - \eta) \leq x \leq b)$;

*then*

$$\lim_{k\to\infty} \left(\frac{-2kh(1)}{\pi}\right)^{1/2} \int_{1-\eta}^{1} e^{k[h(x)-h(1)]} \phi(tx) \, dx = \phi(t)$$

*uniformly in* $a' \leq t \leq b$, *where* $a' > a$ *if* $a > 0$, $a' = a$ *if* $a = 0$.

The proof is similar to that of Theorem 4a with a slight modification in case $a = 0$. In this case it is clear that if $0 \leq t \leq b$ and $1 - \eta \leq x \leq 1$, then $tx$ lies in $(0, b)$, the interval of continuity of $\phi(x)$. Since $|tx - t| \leq \delta b$ for all $x$ in $1 - \delta \leq x \leq 1$, it is clear that $\delta$ may be chosen so small that (2) holds as before. The rest of the proof follows *mutatis mutandis*.

## 5. Uniform Convergence; Continuation

We now apply the result of Section 4 to the Laplace integral inversion formula under consideration.

**THEOREM 5a.** *If*
1. $\phi(u) \, \varepsilon \, L$ *in* $(R^{-1} \leq u \leq R)$ *for every* $R > 1$;
2. $\displaystyle\int_1^\infty e^{-cu} \phi(u) \, du$ *exists for some positive* $c$;
3. $\displaystyle\int_{0+}^1 u^r \phi(u) \, du$ *exists for some positive* $r$;
4. $\phi(u) \, \varepsilon \, C$ *in* $(a \leq u \leq b)$ *for some positive* $a$;

*then*

$$I_k(t) = \left(\frac{k}{t}\right)^{k+1} \frac{1}{k!} \int_{0+}^\infty e^{-ku/t} u^k \phi(u) \, du \to \phi(t) \qquad (k \to \infty)$$

*uniformly in* $a' \leq t \leq b'$, *where* $a < a' < b' < b$.

Set

$$I_k(t) - \phi(t) = \frac{k^{k+1}}{k!} \int_{0+}^\infty e^{-ku} u^k [\phi(tu) - \phi(t)] \, du = I'_k(t) + I''_k(t) + I'''_k(t),$$

where the three integrals $I'_k$, $I''_k$, $I'''_k$ correspond respectively to the intervals $(0, 1 - \eta)$, $(1 - \eta, 1 + \eta)$, $(1 + \eta, \infty)$. Here $\eta$ is any positive number less than unity.

By Theorems 4a and 4b the integral $I''_k(t)$ tends to zero uniformly in the interval $a' \leq t \leq b'$ as $k$ becomes infinite.

Set

$$\alpha(x, t) = \int_x^{1-\eta} [\phi(tu) - \phi(t)] u^r \, du = \frac{1}{t^{r+1}} \int_{xt}^{(1-\eta)t} [\phi(u) - \phi(t)] u^r \, du.$$

Then $\alpha(x, t)$ is continuous in the rectangle $0 \leq x \leq 1$, $a \leq t \leq b$ and consequently has an upper bound $M$ there. But

$$I'_k(t) = -\frac{k^{k+1}}{k!} \int_{0+}^{1-\eta} e^{-ku} u^{k-r} d_u \alpha(u, t)$$

$$= \frac{k^{k+1}}{k!} \int_{0+}^{1-\eta} \alpha(u, t) \, d_u [e^{-ku} u^{k-r}] \qquad (k > r).$$

The function $e^{-ku} u^{k-r}$ is increasing in $(0, 1 - \eta)$ if $k$ is sufficiently large, so that we have for such values of $k$

$$|I'_k(t)| \leq M e^{-k(1-\eta)} (1 - \eta)^{k-r} \frac{k^{k+1}}{k!} = o(1) \qquad (k \to \infty).$$

That is, $I'_k$ approaches zero uniformly in $(a, b)$ as $k$ becomes infinite. Finally, set

$$\beta(x, t) = \int_{1+\eta}^{x} [\phi(tu) - \phi(t)] e^{-cut} \, du = \frac{1}{t} \int_{(1+\eta)t}^{xt} [\phi(u) - \phi(t)] e^{-cu} \, du.$$

By virtue of hypothesis 2. it is clear that $\beta(x, t)$ has an upper bound $M'$ in the region $1 + \eta \leq x < \infty$, $a \leq t \leq b$. Hence

$$I'''_k(t) = -\frac{k^{k+1}}{k!} \int_{1+\eta}^{\infty} \beta(u, t) \, d_u [e^{-(k-ct)u} u^k] \qquad (k > cb).$$

For any $t$ in $(a, b)$ the function in brackets is a decreasing function of $u$ in $(1 + \eta, \infty)$ provided $k$ is so large that

$$\frac{k}{k - cb} < 1 + \eta.$$

Then

$$|I'''_k(t)| \leq \frac{k^{k+1}}{k!} M' e^{-(k-ct)(1+\eta)} (1 + \eta)^k \leq \frac{k^{k+1}}{k!} M' e^{-(k-cb)(1+\eta)} (1 + \eta)^k,$$

so that $I'''_k(t)$ also tends to zero uniformly in $(a, b)$ as $k$ becomes infinite. It follows that the same must be true of $I_k(t)$ in the interval $a' \leq t \leq b'$, and our theorem is proved.

§5]  UNIFORM CONVERGENCE  287

We wish to treat next the case in which $\phi(t)$ is continuous in $0 \leq t < \infty$ and tends to a limit as $t$ becomes infinite. In this case the corresponding Laplace integral is absolutely convergent. This fact simplifies our discussions considerably. We can also avoid completely the use of Theorems 4a and 4b if we employ the following preliminary result.

LEMMA 5. *If $\phi(t)$ is continuous in $0 \leq t < \infty$ and approaches a limit as $t$ becomes infinite, then to an arbitrary positive $\epsilon$ there corresponds a number $\eta$ such that*

$$|\phi(tx) - \phi(t)| < \epsilon$$

*for $0 \leq t < \infty$, $0 < 1 - \eta \leq x \leq 1 + \eta$.*

With $\epsilon$ given we first determine $R$ such that

(1) $\qquad |\phi(t') - \phi(t'')| < \epsilon \qquad (t', t'' \geq R).$

We next determine $\delta$ so that

(2) $\qquad |\phi(t') - \phi(t'')| < \epsilon \qquad (|t' - t''| < \delta)$

for any pair of numbers $t'$ and $t''$ in $0 \leq t \leq 3R$. We consider the range of $t$ in two parts, $(0, 2R)$ and $(2R, \infty)$. For any $\eta < 1/2$ and $t$ in the second interval, $(2R, \infty)$, we have for $-\eta \leq x - 1 \leq \eta$ both $tx \geq R$ and $t \geq R$, so that (1) is applicable. If $t$ is in $(0, 2R)$ then $tx$ and $t$ are both in $(0, 3R)$ where (2) is applicable, and if we now choose $\eta < 1/2$ and $\eta < \delta/(2R)$, we have

$$|tx - t| \leq 2R\eta < \delta \qquad (|x - 1| \leq \eta),$$

and (2) gives

$$|\phi(tx) - \phi(t)| < \epsilon.$$

This completes the proof of the lemma.

THEOREM 5b. *If $\phi(t)$ is continuous for $0 \leq t < \infty$ and approaches a limit as $t$ becomes infinite, then*

$$\lim_{k \to \infty} \left(\frac{k}{t}\right)^{k+1} \frac{1}{k!} \int_0^\infty e^{-ku/t} u^k \phi(u)\, du = \phi(t)$$

*uniformly in $0 \leq t < \infty$.*

For, given $\epsilon$ we determine $\eta$ as indicated in the lemma, and for this choice of $\eta$ we define the integrals $I_k'(t)$, $I_k''(t)$, $I_k'''(t)$ as in the proof of Theorem 5a. Then for $0 \leq t < \infty$ we have·

$$|I_k''(t)| \leq \epsilon \frac{k^{k+1}}{k!} \int_{1-\eta}^{1+\eta} e^{-ku} u^k\, du < \epsilon.$$

288    INVERSION OF LAPLACE TRANSFORM    [Ch. VII

Also

$$|I_k'(t)| \leq 2M \frac{k^{k+1}}{k!} \int_0^{1-\eta} e^{-ku} u^k \, du = o(1) \qquad (k \to \infty)$$

$$|I_k'''(t)| \leq 2M \frac{k^{k+1}}{k!} \int_{1+\eta}^\infty e^{-ku} u^k \, du = o(1) \qquad (k \to \infty),$$

where

$$M = \operatorname*{u.b.}_{0 \leq t \leq \infty} |\phi(t)|.$$

The last three inequalities are clearly sufficient to prove the uniform convergence desired.

### 6. The Inversion Operator for the Laplace-Lebesgue Integral

We now define the following operator, which will serve to invert the Laplace integral.

DEFINITION 6. *An operator $L_{k,t}[f(x)]$ is defined by the equation*

$$L_{k,t}[f(x)] = (-1)^k f^{(k)}\left(\frac{k}{t}\right)\left(\frac{k}{t}\right)^{k+1}$$

*for any real positive number $t$ and any positive integer $k$.*

If the operator is to be applicable to a function $f(x)$ for a given $k$ and $t$, the function must possess a derivative of order $k$ in a neighborhood of $x = k/t$. Actually we shall be applying the operator only to functions which have derivatives of all orders for all $x$ sufficiently large.

As an example, take $f(x) = x^{-1}$. Then

$$L_{k,t}[f(x)] = 1$$

for all positive $t$ and all positive integers $k$.

THEOREM 6a. *If $\phi(u) \, \varepsilon \, L$ in $0 \leq u \leq R$ for every positive $R$, and if the integral*

(1)  $$f(x) = \int_0^\infty e^{-xu} \phi(u) \, du$$

*converges for some $x$, then*

(2)  $$\lim_{k \to \infty} L_{k,t}[f(x)] = \phi(t)$$

*for all positive $t$ in the Lebesgue set for $\phi(u)$.*

For, simple computation gives

$$L_{k,t}[f(x)] = \frac{1}{k!}\left(\frac{k}{t}\right)^{k+1} \int_0^\infty e^{-ku/t} u^k \phi(u) \, du.$$

We have now only to apply Theorem 3c to this integral to obtain the desired result. The number $r$ of that theorem may be taken as zero here and $c$ any number greater than the abscissa of convergence of (1). The Lebesgue set for $\phi(u)$ is the set of numbers $t_0$ for which

$$\int_{t_0}^{t} |\phi(u) - \phi(t_0)|\, du = o(|t - t_0|) \qquad (t \to t_0).$$

For such numbers $t_0$ hypothesis 4. of Theorem 3c is satisfied.

COROLLARY 6a.1. *Equation* (2) *holds for almost all positive* $t$.

COROLLARY 6a.2. *Equation* (2) *holds at points $t$ where $\phi(u)$ is continuous.*

COROLLARY 6a.3. *At a point $t > 0$ in a neighborhood of which $\phi(u)$ is of bounded variation*

$$\lim_{k \to \infty} L_{k,t}[f(x)] = \frac{\phi(t+) + \phi(t-)}{2}.$$

We shall have occasion to use the following result.*

THEOREM 6b. *If $\alpha(u)$ is a normalized function of bounded variation in $0 \leq u \leq R$ for every positive $R$ and if the integral*

$$f(x) = \int_0^\infty e^{-xu} \alpha(u)\, du$$

*converges absolutely for some $x$, then*

$$\lim_{k \to \infty} \frac{(-1)^k}{k!} f^{(k)}(x_k) x_k^{k+1} = \alpha(t) \qquad (t > 0)$$

$$x_k = (k + \theta_k)/t$$
$$0 \leq \theta_k \leq 1 \qquad (k = 1, 2, \cdots).$$

Observe that the result is evident from Corollary 6a.3 if $\theta_k = 0$ for all $k$. We prove first that

(3) $$\lim_{k \to \infty} \left(1 + \frac{\theta_k}{k}\right)^k e^{-\theta_k} = 1.$$

By use of the inequality

$$1 - x^{-1} \leq \log x \leq x - 1 \qquad (0 < x < \infty),$$

we have

$$\frac{-\theta_k^2}{k + \theta_k} \leq k \log\left(1 + \frac{\theta_k}{k}\right) - \theta_k \leq 0.$$

* Compare E. L. Post [1930] and H. Pollard [1940].

The result now follows immediately by allowing $k$ to become infinite. We have seen that

$$L_{k,t}[f(x)] = \frac{k^{k+1}}{k!} \int_0^\infty e^{-ku} u^k \alpha(tu)\, du.$$

Set

$$I_k = e^{\theta_k} \frac{k^{k+1}}{k!} \int_0^\infty e^{-(k+\theta_k)u} u^k \alpha(tu)\, du.$$

By (3)

$$(-1)^k \frac{f^{(k)}(x_k)}{k!} x_k^{k+1} \sim I_k \qquad (k \to \infty),$$

so that it will be sufficient to show that

$$L_{k,t}[f(x)] - I_k = \frac{k^{k+1}}{k!} \int_0^\infty e^{-ku} u^k [1 - e^{(1-u)\theta_k}] \alpha(tu)\, du$$

approaches zero with $1/k$. For, then $I_k$ will have the same limit as that ascribed to $L_{k,t}[f(x)]$ by Corollary 6a.3. But

$$|1 - e^{(1-u)\theta_k}| \leq |1 - e^{1-u}| \qquad (0 \leq u < \infty),$$

so that we need only show

$$\lim_{k\to\infty} \frac{k^{k+1}}{k!} \int_0^\infty e^{-ku} u^k |(1 - e^{1-u})\alpha(tu)|\, du = 0.$$

Since the function $|(1 - e^{1-u})\alpha(tu)|$ vanishes at $u = 1$ we have our result by an application of Theorem 3c.

### 7. The Inversion Operator for the Laplace-Stieltjes Integral

By use of the operator $L_{k,t}[f(x)]$ defined in the previous section we can also invert the Laplace-Stieltjes integral.

THEOREM 7a. *If $\alpha(t)$ is a normalized function of bounded variation in $0 \leq t \leq R$ for every positive $R$, and if the integral*

(1)
$$f(x) = \int_0^\infty e^{-xt}\, d\alpha(t)$$

*converges for some $x$, then*

$$\lim_{k\to\infty} \int_0^t L_{k,u}[f(x)]\, du = \alpha(t) - \alpha(0+).$$

As in the previous section we have for $(k/u) > \sigma_c$

$$L_{k,u}[f(x)] = \frac{1}{k!}\left(\frac{k}{u}\right)^{k+1} \int_0^\infty e^{-ky/u} y^k\, d\alpha(y)$$

§7] LAPLACE-STIELTJES INTEGRAL 291

Since $\alpha(t)$ is normalized, $\alpha(0) = 0$, and integration by parts gives
$$L_{k,u}[f(x)] = -\frac{1}{k!}\left(\frac{k}{u}\right)^{k+1}\int_0^\infty \alpha(y)\frac{\partial}{\partial y}[e^{-ky/u}y^k]\,dy$$
for $(k/u) > \sigma_c$ and $u > 0$. By Euler's theorem on homogeneous functions

$$y\frac{\partial}{\partial y}\left[e^{-ky/u}\frac{y^k}{u^{k+1}}\right] = -\frac{\partial}{\partial u}\left[e^{-ky/u}\frac{y^k}{u^k}\right].$$

Hence

$$L_{k,u}[f(x)] = \frac{k^{k+1}}{k!}\int_0^\infty \frac{\alpha(y)}{y}\frac{\partial}{\partial u}\left[e^{-ky/u}\frac{y^k}{u^k}\right]dy.$$

This integral converges uniformly in any interval $r \leqq u \leqq t$, where $r$ is any positive number and $k > \sigma_c t$. This is seen by use of the relation

$$\alpha(y) = O(e^{cy}) \qquad (c > \sigma_c,\ c > 0,\ y \to \infty).$$

Integrating under the integral sign gives

$$\int_r^t L_{k,u}[f(x)]\,du = \frac{k^{k+1}}{k!}\int_0^\infty \frac{\alpha(y)}{y}e^{-ky/t}\frac{y^k}{t^k}\,dy - \frac{k^{k+1}}{k!}\int_0^\infty \frac{\alpha(y)}{y}e^{-ky/r}\frac{y^k}{r^k}\,dy.$$

Letting $r$ approach zero,

(2) $$\int_{0+}^t L_{k,u}[f(x)]\,du = \frac{k^{k+1}}{k!}\int_0^\infty \frac{\alpha(y)}{y}e^{-ky/t}\frac{y^k}{t^k}\,dy - \alpha(0+).$$

To prove this it is enough to show that

(3) $$\lim_{r \to 0+} \frac{k^{k+1}}{k!}\frac{1}{r^k}\int_0^\infty [\alpha(y) - \alpha(0+)]e^{-ky/r}y^{k-1}\,dy = 0.$$

But if $\epsilon$ is an arbitrary positive number we can determine $\delta$ so small that $|\alpha(y) - \alpha(0+)| < \epsilon$ when $0 \leqq y \leqq \delta$. Then

$$\left|\frac{k^{k+1}}{k!}\frac{1}{r^k}\int_0^\delta [\alpha(y) - \alpha(0+)]e^{-yk/r}y^{k-1}\,dy\right| < \epsilon \qquad (0 < r < \infty).$$

Also

$$\left|\frac{k^{k+1}}{k!}\frac{1}{r^k}\int_\delta^\infty [\alpha(y) - \alpha(0+)]e^{-yk/r}y^{k-1}\,dy\right|$$
$$\leqq \frac{k^{k+1}}{k!}\frac{e^{-\delta(k-cr)/r}}{r^k}\int_\delta^\infty |\alpha(y) - \alpha(0+)|e^{-cy}y^{k-1}\,dy,$$

where $c > \sigma_c$, $c > 0$ and $cr < k$. The right-hand side approaches zero with $r$ so that

$$\varlimsup_{r \to 0+} \left| \frac{k^{k+1}}{k!} \frac{1}{r^k} \int_0^\infty [\alpha(y) - \alpha(0+)] e^{-yk/r} y^{k-1} dy \right| \leq \epsilon,$$

from which (3) follows at once.

Now let $k$ become infinite in (2) and apply Corollary 3c.1 to the right-hand integral, taking $\alpha(x)/x$ equal to the function $\phi(x)$ of the corollary. Since the integrals

$$\int_1^\infty \frac{\alpha(u)}{u} e^{-cu} du, \qquad \int_{0+}^1 \alpha(u) du$$

exist, the hypotheses of the corollary are satisfied.

It remains to show that

$$\int_{0+}^t L_{k,u}[f(x)] du = \int_0^t L_{k,u}[f(x)] du.$$

This will be true if

(4) $$\int_R^\infty |f^{(k)}(u)| u^{k-1} du < \infty \qquad (k = 1, 2, \cdots)$$

for some positive $R$. But if $x > c$, $x > 0$

$$f^{(k)}(x) = x \int_1^\infty e^{-xt}(-t)^k \alpha(t) dt - \alpha(1)(-1)^k e^{-x}$$

$$+ k \int_1^\infty e^{-xt}(-t)^{k-1} \alpha(t) dt + \int_0^1 e^{-xt}(-t)^k d\alpha(t),$$

the integrals converging absolutely. Hence by Fubini's theorem

$$\int_R^\infty |f^{(k)} u| u^{k-1} du \leq \int_1^\infty t^k |\alpha(t)| dt \int_R^\infty e^{-ut} u^k du + |\alpha(1)| \int_R^\infty e^{-u} u^{k-1} du$$

$$+ k \int_1^\infty t^{k-1} |\alpha(t)| dt \int_R^\infty e^{-ut} u^{k-1} du$$

$$+ \int_0^1 t^k |d\alpha(t)| \int_R^\infty e^{-ut} u^{k-1} du \qquad (R > c, R > 0)$$

if the iterated integrals on the right exist. But

$$\int_R^\infty e^{-ut} u^{k-1} du = e^{-Rt} P_k(1/t),$$

where $P_k(x)$ is a polynomial of degree $k$. Hence

$$\int_R^\infty |f^{(k)}(u)| u^{k-1} du \leq \int_1^\infty e^{-Rt} |\alpha(t)| t^k P_{k+1}(1/t) dt$$

$$+ |\alpha(1)| e^{-R} P_k(1) + k \int_1^\infty e^{-Rt} |\alpha(t)| t^{k-1} P_k(1/t) dt$$

$$+ \int_0^1 e^{-Rt} t^k P_k(1/t) |d\alpha(t)|.$$

If $R > c$ one sees by inspection that the integrals on the right exist, so that (4) is established. This completes the proof of the theorem.

In addition to $L_{k,t}[f(x)]$ it is sometimes useful to introduce a further operator.

DEFINITION 7. *An operator $S_{k,t}[f(x)]$ is defined by the equation*

$$S_{k,t}[f(x)] = f(\infty) + (-1)^{k+1} \int_{k/t}^\infty f^{(k+1)}(u) \frac{u^k}{k!} du \qquad (k = 1, 2, \cdots).$$

We shall be applying this operator only to functions $f(x)$ which have derivatives of all orders for large $x$ and for which $f(\infty)$ exists. For example, if $f(x) = e^{-x}$ we have

(5) $$S_{k,t}[e^{-x}] = \frac{1}{k!} \int_{k/t}^\infty e^{-u} u^k du.$$

We shall show that $S_{k,t}[f(x)]$ also tends to $\alpha(t)$ as $k$ becomes infinite under the conditions of Theorem 7a. Thus in particular

$$\lim_{k \to \infty} S_{k,t}[e^{-x}] = 1 \qquad (1 < t)$$

$$= \tfrac{1}{2} \qquad (1 = t)$$

$$= 0 \qquad (0 < t < 1).$$

This could easily be shown directly* by use of the integral (5). We now prove:

THEOREM 7b. *If $f(x)$ has derivatives of all orders in $0 \leq c \leq x < \infty$ and if the integrals*

(6) $$\int_c^\infty u^k f^{(k+1)}(u) du \qquad (k = 0, 1, 2, \cdots)$$

*converge, then*

$$L_{k,t}[f(x)/x] = S_{k,t}[f(x)] \qquad (k > tc;\ k = 1, 2, \cdots).$$

* Compare D. V. Widder [1934a] p. 114.

For, we have by integration by parts for any $R > x$

(7) $\quad \int_x^R \dfrac{u^k}{k!} f^{(k+1)}(u)\, du = \dfrac{R^k f^{(k)}(R)}{k!} - \dfrac{x^k f^{(k)}(x)}{k!} - \int_x^R \dfrac{u^{k-1}}{(k-1)!} f^{(k)}(u)\, du.$

Since these two integrals tend to limits as $R$ becomes infinite the same is true of $R^k f^{(k)}(R)$. In particular if $k = 0$ we see that $f(\infty)$ exists. That is, we have

$$f(x) - f(\infty) = o(1)$$
$$f^{(k)}(x) = O(x^{-k}) \qquad (x \to \infty\,;\, k = 1, 2, \cdots).$$

By Theorem 4.4 of Chapter V,

$$f^{(k)}(x) = o(x^{-k}) \qquad (x \to \infty\,;\, k = 1, 2, \cdots).$$

Hence from (7) we have when $x$ is greater than $c$

$$(-1)^{k+1} \int_x^\infty \dfrac{u^k}{k!} f^{(k+1)}(u)\, du$$
$$= (-1)^k \dfrac{f^{(k)}(x) x^k}{k!} + (-1)^k \int_x^\infty \dfrac{u^{k-1}}{(k-1)!} f^{(k)}(u)\, du$$
$$= \sum_{j=0}^k (-1)^j \dfrac{f^{(j)}(x) x^j}{j!} - f(\infty)$$
$$= \dfrac{(-1)^k}{k!} [f(x)/x]^{(k)} x^{k+1} - f(\infty).$$

Finally, setting $x = k/t$ our result is established.

COROLLARY 7b. *Conditions (6) may be replaced by the condition that the following limits exist*:

(8) $\qquad\qquad\qquad \lim\limits_{x \to \infty} x^k f^{(k)}(x) \qquad\qquad (k = 0, 1, 2, \cdots).$

For, the existence of these limits shows first that

$$\int_c^\infty f'(u)\, du$$

converges. Then by use of the equation

$$\int_c^R u^k f^{(k+1)}(u)\, du = R^k f^{(k)}(R) - c^k f^{(k)}(c) - k \int_c^R u^{k-1} f^{(k)}(u)\, du$$

and by induction we see that all the integrals (6) converge.

THEOREM 7c. *If $f(x)$ is defined as in Theorem 7a, then*

$$\lim_{k \to \infty} S_{k,t}[f(x)] = \alpha(t) \qquad (0 < t < \infty)$$

If $x$ is sufficiently large,
$$f(x) = x \int_0^\infty e^{-xt} \alpha(t)\, dt.$$
Since $f(\infty) = \alpha(0+)$ and
$$\lim_{x \to \infty} f^{(k)}(x) x^k = 0 \qquad (k = 1, 2, \cdots),$$
Corollary 7b is applicable and
$$S_{k,t}[f(x)] = L_{k,t}[f(x)/x].$$
By Corollary $6a_{,}3$
$$\lim_{k \to \infty} L_{k,t}[f(x)/x] = \alpha(t) \qquad (t > 0),$$
so that the theorem is established.

We can now obtain an inversion formula of a somewhat different nature.*

THEOREM 7d. *If $f(x)$ is defined as in Theorem 7a, then*
$$\lim_{x \to \infty} \sum_{n=0}^{[xt]} \frac{(-x)^n}{n!} f^{(n)}(x) = \alpha(t) \qquad (0 < t < \infty). \tag{9}$$

Here $[xt]$ means the largest integer contained in $xt$. Note that the summation is the partial sum of the Taylor development of $f(y)$ about the point $y = x$ evaluated at $y = 0$.

Clearly, for $x$ sufficiently large
$$F(x) = \frac{f(x)}{x} = \int_0^\infty e^{-xt} \alpha(t)\, dt$$
the integral converging absolutely. By Leibniz's rule
$$\frac{(-1)^k}{k!} F^{(k)}(x) x^{k+1} = \sum_{n=0}^{k} \frac{(-x)^n}{n!} f^{(n)}(x). \tag{10}$$
But equation (9) may clearly be written as
$$\lim_{k \to \infty} \sum_{n=0}^{k} \frac{(-x_k)^n}{n!} f^{(n)}(x_k) = \alpha(t),$$
where
$$x_k = (k + \theta_k)/t$$
$$0 \leqq \theta_k \leqq 1 \qquad (k = 1, 2, \cdots).$$

* Compare M. J. Dubourdieu [1939] and W. Feller [1939]. Note that both authors make the unnecessary restriction that the determining function is non-decreasing. See H. Pollard [1940].

By (10) we have only to show that

$$\lim_{k \to \infty} \frac{(-1)^k}{k!} F^{(k)}(x_k) x_k^{k+1} = \alpha(t) \qquad (0 < t < \infty).$$

But this is true by Theorem 6b.

## 8. Laplace Method for a New Integral

In Section 2 we obtained an asymptotic evaluation of the integral

$$\int_a^b \phi(x) e^{kh(x)} \, dx$$

for large $k$. We assumed that $h(x)$ had a flat maximum at $x = a$ and that $\phi(a) \neq 0$. In this section we consider the case in which $\phi(x)$ has a certain type of zero at $x = a$.

**THEOREM 8a.** *If*
1. $a < a + \eta < b$;
2. $h(x) \in C^2$ $(a \leq x \leq a + \eta)$, $h'(a) = 0$, $h''(a) < 0$, $h(x)$ is non-increasing in $(a \leq x \leq b)$;
3. $\phi(x) \in L$ $(a \leq x \leq b)$, $\phi(a) \neq 0$, $\phi(a+)$ exists and equals $\phi(a)$;

*then*

$$\int_a^b e^{kh(x)}(x-a)\phi(x)\,dx \sim -\frac{\phi(a) e^{kh(a)}}{k h''(a)} \qquad (k \to \infty).$$

It is clearly no restriction to assume $\phi(a) > 0$. Given an arbitrary positive number $\epsilon$ less than $\phi(a)$ and less than $-h''(a)$; we determine $\delta$ so small that

(1) $\qquad 0 < \phi(a) - \epsilon < \phi(x) < \phi(a) + \epsilon$

(2) $\qquad h''(a) - \epsilon < h''(x) < h''(a) + \epsilon < 0$

for all $x$ in $(a \leq x \leq a + \delta)$. Set

$$I_k = \int_a^b e^{k[h(x)-h(a)]}(x-a)\phi(x)\,dx$$

and write this integral as the sum of two integrals $I_k'$ and $I_k''$ corresponding to the intervals $(a, a + \delta)$ and $(a + \delta, b)$ respectively. Then

$$|I_k''| \leq e^{k[h(a+\delta)-h(a)]} \int_{a+\delta}^b |\phi(x)| (x-a) \, dx.$$

Since $h(a + \delta) - h(a) < 0$, it is clear that

(3) $\qquad\qquad\qquad I_k'' = O(\alpha^k) \qquad\qquad (k \to \infty),$

where $\alpha$ is a number between zero and unity. By use of (1) and (2) we have

$$[\phi(a) - \epsilon] \int_a^{a+\delta} e^{kh''(\xi)(x-a)^2/2}(x - a)\, dx < I_k'$$
$$< [\phi(a) + \epsilon] \int_a^{a+\delta} e^{kh''(\xi)(x-a)^2/2}(x - a)\, dx$$

where $a < \xi < a + \delta$. Using (2) we only strengthen these inequalities by replacing $h''(\xi)$ by $h''(a) + \epsilon$ on the right, by $h''(a) - \epsilon$ on the left. The resulting integrals may easily be evaluated, whence

$$\frac{[\phi(a) - \epsilon][e^{k[h''(a)-\epsilon]\delta^2/2} - 1]}{k[h''(a) - \epsilon]} < I_k' < \frac{[\phi(a) + \epsilon][e^{k[h''(a)+\epsilon]\delta^2/2} - 1]}{k[h''(a) + \epsilon]}$$

$$\frac{\phi(a) - \epsilon}{-h''(a) + \epsilon} \leq \varliminf_{k \to \infty} kI_k' \leq \varlimsup_{k \to \infty} kI_k' \leq -\frac{\phi(a) + \epsilon}{h''(a) + \epsilon}.$$

By (3)
$$\varlimsup_{k \to \infty} kI_k = \varlimsup_{k \to \infty} kI_k'$$
$$\varliminf_{k \to \infty} kI_k = \varliminf_{k \to \infty} kI_k'.$$

Since $\epsilon$ was arbitrary we see that

$$\lim kI_k = -\frac{\phi(a)}{h''(a)},$$

and this proves our theorem.

**THEOREM 8b.** *If*
1. $a < b - \eta < b$;
2. $h(x)\,\varepsilon\, C^2\,(b - \eta \leq x \leq b)$, $h'(b) = 0$, $h''(b) < 0$, $h(x)$ is non-decreasing $(a \leq x \leq b)$;
3. $\phi(x)\,\varepsilon\, L\,(a \leq x \leq b)$, $\phi(b) \neq 0$, $\phi(b-)$ exists and equals $\phi(b)$;

*then*

(4) $$\int_a^b \phi(x) e^{kh(x)} (b - x)\, dx \sim -\frac{\phi(b) e^{kh(b)}}{kh''(b)} \qquad (k \to \infty).$$

The proof is obtained by applying Theorem 8a to the integral (4) after a change of variable $x = -t$.

As an example of Theorem 8a take the integral

(5) $$\int_1^\infty e^{-ku} u^{k-1} [u - 1]\, du.$$

This has the value $e^{-k}/k$, as one sees by differentiating the equation

$$e^{-x} = x \int_1^\infty e^{-xu} du$$

$k$ times and setting $x = k$. Here $\phi(u) = u^{-1}$, $h(u) = -u + \log u$. Then by Theorem 8a the integral (5) should be asymptotic to $e^{-k}/k$ as $k$ becomes infinite, and this is checked by the computed value of the integral. In a similar way the equation

$$-e^{-k} = k \int_0^1 e^{-ku} u^{k-1}[u - 1] du$$

serves to check Theorem 8b.

## 9. The Jump Operator

We now define an operator similar to $L_{k,t}[f(x)]$ which serves to compute the saltus of the determining function at a point of discontinuity in terms of the generating function.

DEFINITION 9. *An operator $l_{k,t}[f(x)]$ is defined by the equation*

$$l_{k,t}[f(x)] = \left(-\frac{e}{t}\right)^k f^{(k)}\left(\frac{k}{t}\right) \qquad (k = 1, 2, \cdots; t > 0).$$

THEOREM 9. *If*

(1) $$f(x) = \int_0^\infty e^{-xt} d\alpha(t),$$

*the integral converging for some value of $x$, then*

$$\lim_{k \to \infty} l_{k,t}[f(x)] = \alpha(t+) - \alpha(t-) \qquad (t > 0).$$

If $t$ is a fixed positive number, and if $k$ is so large that $k/t$ lies in the region of convergence of (1), then

$$l_{k,t}[f(x)] = \left(\frac{e}{t}\right)^k \int_0^\infty e^{-ku/t} u^k d\alpha(u)$$

$$= -\left(\frac{e}{t}\right)^k \int_0^\infty e^{-ku/t} ku^{k-1}\left[1 - \frac{u}{t}\right] \alpha(u) du.$$

Making the change of variable $u = ty$, this becomes

(2) $$l_{k,t}[f(x)] = -ke^k \int_0^\infty e^{-ky} \alpha(ty) \cdot (1 - y) y^{k-1} dy.$$

Let $\eta$ be a positive number less than unity. Write (2) as the sum of four integrals, $I_1, I_2, I_3, I_4$ corresponding to the intervals $(0, 1 - \eta)$,

$(1-\eta, 1)$, $(1, 1+\eta)$, $(1+\eta, \infty)$, respectively. By Theorems 8a and 8b we have

$$I_2 \sim \alpha(t-)$$
$$I_3 \sim \alpha(t+) \qquad (k \to \infty).$$

For $I_1$ we have

$$|I_1| \leq k e^k [e^{-1+\eta}(1-\eta)]^{k-1} \int_0^{1-\eta} e^{-y} |\alpha(ty)| (1-y) \, dy,$$

and since $e^\eta(1-\eta) < 1$ the right-hand side approaches zero with $1/k$. Since the integral (2) converges absolutely for large $k$, say for $k \geq k_0$, we have

$$|I_4| \leq k e^k [e^{-1-\eta}(1+\eta)]^{k-k_0} \int_{1+\eta}^\infty e^{-k_0 y} |\alpha(ty)| (y-1) y^{k_0-1} \, dy,$$

and since $e^{-\eta}(1+\eta) < 1$, the right-hand side approaches zero with $1/k$. This completes the proof of the theorem.

As a simple example take

$$e^{-x} = \int_0^\infty e^{-xt} \, d\alpha(t)$$

$$\alpha(t) = 0 \qquad (0 \leq t < 1)$$
$$= 1 \qquad (1 < t < \infty)$$

Then

$$l_{k,t}[e^{-x}] = \left(\frac{e}{t}\right)^k e^{-k/t}.$$

This is equal to unity for all $k$ when $t = 1$. For any other positive $t$ it tends to zero with $1/k$. This checks with the fact that $\alpha(t)$ is continuous except at $t = 1$ where the saltus is unity.

## 10. The Variation of the Determining Function

In this section we obtain a formula which gives the variation of the determining function $\alpha(t)$ in any interval $(0, R)$ in terms of the generating function $f(s)$.

**THEOREM 10.** *If $\alpha(t)$ is a normalized function of bounded variation in every finite interval, if*

(1) $$f(x) = \int_0^\infty e^{-xt} \, d\alpha(t),$$

the integral converging for some $x$, and if $V(t)$ is the variation of $\alpha(x)$ in the interval $0 \leq x \leq t$, then

$$V(t) = \lim_{k \to \infty} \int_0^t |L_{k,u}[f(x)]|\,du + |f(\infty)|$$

$$= \lim_{k \to \infty} \int_{k/t}^\infty |f^{(k)}(u)| \frac{u^{k-1}}{(k-1)!}\,du + |f(\infty)|.$$

Let $S$ and $R$ be any two positive numbers, $R > S$. Set

$$f_1(x) = \int_0^S e^{-xt}\,d\alpha(t)$$

$$f_2(x) = \int_S^\infty e^{-xt}\,d\alpha(t).$$

Then

(2) $\quad \int_0^R |L_{k,t}[f(x)]|\,dt \leq \int_0^R |L_{k,t}[f_1(x)]|\,dt + \int_0^R |L_{k,t}[f_2(x)]|\,dt.$

Simple computation gives

$$\int_0^R |L_{k,t}[f_1(x)]|\,dt \leq \int_0^R L_{k,t}\left[\int_0^S e^{-xu}\,dV(u)\right]dt.$$

By Theorem 7a the right-hand side of this equality tends to $V(R) - V(0+)$ as $k$ becomes infinite, so that

(3) $\quad \overline{\lim_{k \to \infty}} \int_0^R |L_{k,t}[f_1(x)]|\,dt \leq V(R) - V(0+).$

On the other hand we can show that

(4) $\quad \lim_{k \to \infty} \int_0^R |L_{k,t}[f_2(x)]|\,dt = 0.$

For, as in Section 7, we have

$$L_{k,u}[f_2(x)] = \frac{k^{k+1}}{k!} \int_S^\infty \frac{\alpha(y) - \alpha(S)}{y} \frac{\partial}{\partial u}\left[e^{-ky/u}\frac{y^k}{u^k}\right]dy.$$

If $0 \leq u \leq R < S \leq y$, then the function $e^{-ky/u}(y/u)^k$ is an increasing function of $u$, so that

$$|L_{k,u}[f_2(x)]| \leq \frac{k^{k+1}}{k!} \int_S^\infty |\alpha(y) - \alpha(S)|\,y^{-1} \frac{\partial}{\partial u}\left[e^{-ky/u}\frac{y^k}{u^k}\right]dy$$

$$\int_0^R |L_{k,u}[f_2(x)]|\,du \leq \frac{k^{k+1}}{k!} \int_S^\infty |\alpha(y) - \alpha(S)|\,y^{-1} e^{-ky/R} y^k R^{-k}\,dy.$$

§10]  VARIATION OF DETERMINING FUNCTION  301

Now by Corollary 3c.1 the right-hand side of the latter inequality tends to zero as $k$ becomes infinite. Hence equation (4) is proved.

Combining (2), (3), and (4) we have

(5) $$\varlimsup_{k\to\infty} \int_0^R |L_{k,t}[f(x)]|\, dt \leq V(R) - V(0+).$$

Now set
$$\alpha_k(t) = \int_0^t L_{k,u}[f(x)]\, du \qquad (0 \leq t < \infty).$$

By Theorem 7a
$$\lim_{k\to\infty} \alpha_k(t) = \alpha(t) - \alpha(0+) \qquad (0 < t < \infty).$$

If
$$0 = t_0 < t_1 < \cdots < t_n = R,$$
then
$$\sum_{i=0}^{n-1} |\alpha_k(t_{i+1}) - \alpha_k(t_i)| \leq \int_0^R |L_{k,u}[f(x)]|\, du.$$

Letting $k$ become infinite this inequality becomes
$$|\alpha(t_1) - \alpha(0+)| + \sum_{i=1}^{n-1} |\alpha(t_{i+1}) - \alpha(t_i)| \leq \varlimsup_{k\to\infty} \int_0^R |L_{k,u}[f(x)]|\, du.$$

The left-hand side can be brought as close to $V(R) - V(0+)$ as desired by suitable choice of position and number of the points $t_i$, whereas the right-hand side is independent of this choice. Hence

$$V(R) - V(0+) \leq \varlimsup_{k\to\infty} \int_0^R |L_{k,u}[f(x)]|\, du.$$

This combined with (5) gives

$$V(R) - V(0+) = \lim_{k\to\infty} \int_0^R |L_{k,u}[f(x)]|\, du.$$

But
$$V(0+) = |\alpha(0+)| = |f(\infty)|,$$
so that the theorem is established.

COROLLARY 10.  *If $V(\infty) < \infty$, then*

$$V(\infty) = \lim_{k\to\infty} \int_0^\infty |L_{k,u}[f(x)]|\, du + |f(\infty)|$$
$$= \lim_{k\to\infty} \int_0^\infty |f^{(k)}(u)| \frac{u^{k-1}}{(k-1)!}\, du + |f(\infty)|.$$

For, suppose first that $f(\infty) = \alpha(0+) = 0$. Since

$$\int_0^R |L_{k,u}[f(x)]|\, du \leq \int_0^\infty |L_{k,u}[f(x)]|\, du,$$

we have by Theorem 10

(6)
$$V(R) \leq \varliminf_{k\to\infty} \int_0^\infty |L_{k,u}[f(x)]|\, du$$

$$V(\infty) \leq \varliminf_{k\to\infty} \int_0^\infty |L_{k,u}[f(x)]|\, du.$$

On the other hand

$$\int_0^\infty |L_{k,u}[f(x)]|\, du \leq \int_0^\infty \frac{u^{k-1}}{(k-1)!}\, du \int_0^\infty e^{-tu} t^k\, dV(t) = V(\infty)$$

(7) $$\varlimsup_{k\to\infty} \int_0^\infty |L_{k,u}[f(x)]|\, du \leq V(\infty),$$

so that the result is proved by combining (6) and (7).

If $\alpha(0+) \neq 0$, we apply the result just established to the function $f(x) - f(\infty)$, whose normalized determining function will have total variation equal to that of $\alpha(t)$ decreased by $|f(\infty)|$. On the other hand

$$L_{k,t}[f(x)] = L_{k,t}[f(x) - f(\infty)] \qquad (k = 1, 2, \cdots)$$

so that

$$V(\infty) - |f(\infty)| = \lim_{k\to\infty} \int_0^\infty |L_{k,t}[f(x)]|\, dt,$$

and the corollary is proved.

## 11. A General Representation Theorem

Since

$$\lim_{k\to\infty} L_{k,t}[f(x)] = \phi(t)$$

when $f(x)$ is the generating function of $\phi(t)$ and when $\phi(t)$ satisfies suitable conditions, we would have formally that

(1) $$\lim_{k\to\infty} \int_0^\infty e^{-xt} L_{k,t}[f(x)]\, dt = \int_0^\infty e^{-xt} \phi(t)\, dt = f(x).$$

This would give us a representation of $f(x)$ in terms of the $L$-operator, operating on $f(x)$. Thus in the end result the determining function $\phi(t)$ does not appear. We shall obtain conditions on $f(x)$ alone for the validity of (1). These conditions except for continuity requirements will depend only on the behavior of $f(x)$ at $x = 0$ and at $x = \infty$, and

§11]   GENERAL REPRESENTATION THEOREM   303

will not require that $f(x)$ should be a generating function. We need the following preliminary result.

LEMMA 11. *If $k$ is a positive integer and $x$ and $u$ are positive variables, then*

$$\frac{\partial^k}{\partial u^k}[e^{-x/u}u^{k-1}] = e^{-x/u}x^k u^{-k-1}.$$

For, since $e^{-x/u}(u/x)^{k-1}$ is homogeneous of order zero, we have by Euler's theorem

$$\frac{\partial}{\partial u}\left[e^{-x/u}\frac{u^{k-1}}{x^k}\right] = -\frac{\partial}{\partial x}\left[e^{-x/u}\frac{u^{k-2}}{x^{k-1}}\right]$$

$$\frac{\partial^2}{\partial u^2}\left[e^{-x/u}\frac{u^{k-1}}{x^k}\right] = \frac{\partial^2}{\partial x^2}\left[e^{-x/u}\frac{u^{k-3}}{x^{k-2}}\right]$$

$$\frac{\partial^k}{\partial u^k}\left[e^{-x/u}\frac{u^{k-1}}{x^k}\right] = (-1)^k\frac{\partial^k}{\partial x^k}[e^{-x/u}u^{-1}] = \frac{e^{-x/u}}{u^{k+1}},$$

from which the result is evident.

THEOREM 11a. *If for each non-negative integer $k$ and for some positive number $c$*

1. $f^{(k)}(u) = o(u^{-k}) \ (u \to \infty)$;
2. $f^{(k)}(u) = O(c^{c/u}) \ (u \to 0+)$;

*then*

$$\lim_{k \to \infty} \int_{0+}^{\infty} e^{-xt} L_{k,t}[f(x)] \, dt = f(x) \qquad (0 < x < \infty).$$

Let $x$ be any fixed number greater than zero. It is understood, of course, that the hypotheses 1. and 2. imply that $f(x)$ is continuous with all its derivatives in $0 < x < \infty$. Choose $k$ so large that $xk > c$. Set

$$I_k = \int_{0+}^{\infty} e^{-xt} L_{k,t}[f(x)] \, dt = \frac{1}{k!}\int_{0+}^{\infty} e^{-xt}(-1)^k (k/t)^{k+1} f^{(k)}(k/t) dt.$$

It will be seen in the course of the argument that this integral exists. By a change of variable it becomes

$$I_k = \frac{(-1)^k}{(k-1)!}\int_{0+}^{\infty} e^{-xk/u}f^{(k)}(u)u^{k-1}\,du.$$

Integrating by parts,

$$I_k = (-1)^k e^{-xk/u} f^{(k-1)}(u) \frac{u^{k-1}}{(k-1)!}\bigg|_{0+}^{\infty}$$

$$+ \frac{(-1)^{k-1}}{(k-1)!}\int_{0+}^{\infty} f^{(k-1)}(u)\frac{\partial}{\partial u}[e^{-xk/u}u^{k-1}]\,du.$$

The integrated part is clearly zero by 1. and 2. Again integrating by parts

$$I_k = \frac{(-1)^{k-1}}{(k-1)!} f^{(k-2)}(u) \frac{\partial}{\partial u}[e^{-xk/u} u^{k-1}]\Big|_{0+}^{\infty}$$
$$+ \frac{(-1)^{k-2}}{(k-1)!} \int_{0+}^{\infty} f^{(k-2)}(u) \frac{\partial^2}{\partial u^2}[e^{-xk/u} u^{k-1}] \, du.$$

Except for constant factors the integrated part is

$$e^{-xk/u} f^{(k-2)}(u)[(k-1)u^{k-2} + kxu^{k-3}],$$

and by 1. and 2. this approaches zero when $u$ becomes infinite and when $u$ approaches zero. Continuing this process we obtain

$$I_k = \frac{1}{(k-1)!} \int_{0+}^{\infty} f(u) \frac{\partial^k}{\partial u^k}[e^{-xk/u} u^{k-1}] \, du.$$

By Lemma 11

$$I_k = \frac{k^k}{(k-1)!} \int_{0+}^{\infty} f(u) e^{-xk/u} \frac{x^k}{u^{k+1}} \, du.$$

It is clear by hypotheses 1. and 2. that this integral exists. In fact it converges absolutely, so that the lower limit $0+$ could be replaced by 0. It follows that all the previous integrals also exist. By another change of variable

$$I_k = \frac{x^k k^{k+1}}{k!} \int_0^{\infty} e^{-xkt} t^{k-1} f(1/t) \, dt.$$

Now apply Theorem 3c to this integral, taking the function $\phi(x)$ of that theorem equal to $f(x^{-1})x^{-1}$. The number $r$ of Theorem 3c may be taken equal to unity; the number $c$ of that theorem, any number greater than the number $c$ of Theorem 11a. Condition 4. of Theorem 3c is satisfied since $f(x^{-1})x^{-1}$ is continuous in $(0 < x < \infty)$. Hence

$$\lim_{k \to \infty} I_k = f(x),$$

and our result is established.

COROLLARY 11a. *The result of the theorem holds if hypothesis 2 is replaced by*

3. $\qquad\qquad f^{(k)}(u) = O(u^{g(k)}) \qquad\qquad (u \to 0+)$

*where $g(k)$ is any real function of $k$.*

For, it is clear that hypothesis 3. implies hypothesis 2. for any positive $c$.

§11]  GENERAL REPRESENTATION THEOREM  305

The sufficient conditions 1. and 2. for the validity of Theorem 11a may be replaced by others which will prove useful.

**THEOREM 11b.** *If for each positive integer $k$*

$$\int_{0+}^{x} L_{k,t}[f(x)]\, dt = O(x) \qquad (x \to \infty),$$

*then $f(\infty)$ exists and*

$$\lim_{k \to \infty} \int_{0+}^{\infty} e^{-xt} L_{k,t}[f(x)]\, dt = f(x) - f(\infty) \quad (0 < x < \infty).$$

For, the hypothesis is equivalent to

(2) $$\int_{1/x}^{\infty} f^{(k)}(t) t^{k-1}\, dt = O(x) \qquad (x \to \infty;\, k = 1, 2, \cdots).$$

We saw in the proof of Theorem 7b that the existence of these integrals implies the existence of $f(\infty)$ and that

$$[f(x) - f(\infty)]^{(k)} = o(x^{-k}) \qquad (x \to \infty;\, k = 0, 1, 2, \cdots).$$

Hence hypothesis 1. of Theorem 11a is applicable to the function $f(x) - f(\infty)$.

Integration by parts gives

(3) $$\int_{1/x}^{\infty} f^{(k+1)}(t) t^k\, dt = -f^{(k)}\left(\frac{1}{x}\right)\frac{1}{x^k} - k \int_{1/x}^{\infty} f^{(k)}(t) t^{k-1}\, dt,$$

and, since these two integrals are $O(x)$ as $x$ becomes infinite by (2), we have

$$f^{(k)}\left(\frac{1}{x}\right) = O(x^{k+1}) \qquad (x \to \infty)$$

(4) $$f^{(k)}(x) = O(x^{-k-1}) \quad (x \to 0+,\, k = 1, 2, \cdots).$$

In particular, if $k = 0$, equation (3) becomes

$$\int_{1/x}^{\infty} f'(t)\, dt = f(\infty) - f\left(\frac{1}{x}\right) = O(x) \qquad (x \to \infty),$$

whence

$$f(x) = O\left(\frac{1}{x}\right) \qquad (x \to 0+),$$

so that (4) also holds when $k = 0$. Consequently we may apply Corollary 11a to the function $f(x) - f(\infty)$. Since

$$L_{k,t}[f(x) - f(\infty)] = L_{k,t}[f(x)] \qquad (k = 1, 2, \cdots),$$

our theorem is established.

## 12. Determining Function of Bounded Variation

We shall now discuss the question of what functions can be expressed as Laplace integrals, the determining function being of some prescribed class. In the present section we shall obtain necessary and sufficient conditions that

$$f(x) = \int_0^\infty e^{-xt}\, d\alpha(t),$$

for all positive $x$, the function $\alpha(t)$ being of bounded variation in $(0, \infty)$. We introduce the conditions in the form of a definition.

**DEFINITION 12.** *A function $f(x)$ satisfies Conditions A if it has derivatives of all orders in $0 < x < \infty$ and if there exists a constant $M$ such that*

(1) $\quad \displaystyle\int_0^\infty |L_{k,t}[f(x)]|\, dt = \int_0^\infty |f^{(k)}(u)|\, \frac{u^{k-1}}{(k-1)!}\, du < M \quad (k = 1, 2, \cdots).$

As an example take $f(x) = e^{-x}$. Then

$$\int_0^\infty |L_{k,t}[f(x)]|\, dt = \int_0^\infty e^{-u}\, \frac{u^{k-1}}{(k-1)!}\, du,$$

and this integral has the value unity regardless of $k$. Hence $e^{-x}$ satisfies Conditions $A$.

**THEOREM 12a.** *Conditions A are necessary and sufficient that*

(2) $\qquad\qquad f(x) = \displaystyle\int_0^\infty e^{-xt}\, d\alpha(t) \qquad\qquad (x > 0),$

*where $\alpha(t)$ is of bounded variation in $(0 \leq t < \infty)$.*

First, let us suppose $f(x)$ has the representation (2) and prove (1). We have

$$\int_0^\infty |f^{(k)}(u)|\, \frac{u^{k-1}}{(k-1)!}\, du \leq \int_0^\infty \frac{u^{k-1}}{(k-1)!}\, du \int_0^\infty e^{-ut} t^k\, |d\alpha(t)|.$$

By Theorem 15c of Chapter I, we may reverse the order of integration in the iterated integral and obtain

$$\int_0^\infty |f^{(k)}(u)|\, \frac{u^{k-1}}{(k-1)!}\, du \leq \int_0^\infty t^k\, |d\alpha(t)| \int_0^\infty e^{-ut}\, \frac{u^{k-1}}{(k-1)!}\, du$$

$$= \int_0^\infty |d\alpha(t)| = M.$$

This number is finite since $\alpha(t)$ was assumed of bounded variation on the infinite interval $(0 \leq t < \infty)$. Hence (1) is established.

Conversely, assume that $f(x)$ satisfies Conditions $A$. These conditions clearly imply the hypotheses of Theorem 11$b$, so that

(3) $$\lim_{k \to \infty} \int_{0+}^{\infty} e^{-xt} L_{k,t}[f(x)] \, dt = f(x) - f(\infty) \qquad (0 < x < \infty).$$

Since $e^{-xt}$ is bounded, it is clear from (1) that the integrals (3) converge absolutely, so that the $0+$ of the lower limit of integration may be replaced by 0.

Set

$$\alpha_k(t) = \int_0^t L_{k,u}[f(x)] \, du \qquad (0 \leq t < \infty).$$

The total variation of $\alpha_k(t)$ on $0 \leq t < \infty$ is less than $M$ by (1) for each $k = 1, 2, \cdots$. Hence the sequence is of uniformly bounded variation, and we may employ Helly's theorem, Theorem 16.3 of Chapter I. Hence we can find a sequence of integers $n_1, n_2, \cdots$, and a function $\alpha(t)$ of bounded variation on $(0 \leq t < \infty)$ such that

$$\lim_{i \to \infty} \alpha_{n_i}(t) = \alpha(t) \qquad (0 \leq t < \infty).$$

Hence

$$\lim_{i \to \infty} \int_0^{\infty} e^{-xt} L_{n_i,t}[f(x)] \, dt = f(x) - f(\infty),$$

or

$$\lim_{i \to \infty} \int_0^{\infty} e^{-xt} \, d\alpha_{n_i}(t) = f(x) - f(\infty).$$

By (1), integration by parts is possible for $x > 0$ and

$$\lim_{i \to \infty} x \int_0^{\infty} e^{-xt} \alpha_{n_i}(t) \, dt = f(x) - f(\infty).$$

Since

$$|\alpha_{n_i}(t)| < M \qquad (0 \leq t < \infty, i = 1, 2, \cdots),$$

we may apply Lebesgue's limit theorem and obtain

(4) $$f(x) - f(\infty) = x \int_0^{\infty} e^{-xt} \alpha(t) \, dt = \int_0^{\infty} e^{-xt} \, d\alpha(t) \qquad (0 \leq x < \infty).$$

If now we redefine $\alpha(t)$ so as to increase its value by $f(\infty)$ for every $t > 0$, keeping its value zero at $t = 0$, the term $f(\infty)$ on the left-hand side of (4) disappears, and our theorem is established.

THEOREM 12b. *If $f(x)$ satisfies Conditions A, then the limit*

$$\lim_{k\to\infty} \int_0^\infty |L_{k,t}[f(x)]|\,dt$$

*exists.*

For, by Theorem 12a, $f(x)$ has the representation (2) with

$$\int_0^\infty |d\alpha(t)| < \infty.$$

By Corollary 10

$$\lim_{k\to\infty} \int_0^\infty |L_{k,t}[f(x)]|\,dt = \int_0^\infty |d\alpha(t)| - |f(\infty)|.$$

This result is easily verified for the function $f(x) = e^{-x}$ given above.

### 13. Modified Conditions for Determining Functions of Bounded Variation

We now introduce conditions which are equivalent to Conditions $A$, but which involve a summation instead of an integral.*

DEFINITION 13. *A function $f(x)$ satisfies Conditions $A'$ if it has derivatives of all orders in $(0 < x < \infty)$ and if there exists a constant $M'$ such that*

(1) $$\sum_{k=0}^\infty |f^{(k)}(x)|\frac{x^k}{k!} < M' \qquad (0 < x < \infty).$$

As an example, take $f(x) = e^{-x}$ when the series (1) becomes

$$e^{-x}\sum_{k=0}^\infty \frac{x^k}{k!} = 1.$$

Hence $e^{-x}$ satisfies Conditions $A'$.

THEOREM 13. *Conditions $A$ and $A'$ are equivalent.*

We prove first that $A'$ implies $A$. Let (1) hold, and let $a$ be an arbitrary positive constant. Then we can show that $f(x)$ is analytic for $x > 0$ and continuous for $0 \leq x < \infty$ if $f(0)$ is defined as $f(0+)$. For, by Taylor's expansion with remainder we have

$$f(x) = \sum_{n=0}^k f^{(n)}(a)\frac{(x-a)^n}{n!} + f^{(k+1)}(\xi)\frac{(x-a)^{k+1}}{(k+1)!} \qquad (0 < x < \xi < a).$$

By (1)

$$|f^{(k+1)}(x)| < \frac{M'(k+1)!}{x^{k+1}} \qquad (0 < x < \infty).$$

* See D. V. Widder [1936] p. 293.

so that
$$\left|f^{(k+1)}(\xi)\frac{(x-a)^{k+1}}{(k+1)!}\right| < M'\left(\frac{|x-a|}{\xi}\right)^{k+1}.$$

If $a/2 < x \leqq a$, this gives
$$\left|f^{(k+1)}(\xi)\frac{(x-a)^{k+1}}{(k+1)!}\right| < M'\left(\frac{2a-2x}{a}\right)^{k+1} = o(1) \quad (k \to \infty).$$

Hence the series
$$\sum_{k=0}^{\infty} f^{(k)}(a)\frac{(x-a)^k}{k!}$$

converges to $f(x)$ for $a/2 < x \leqq a$. Since $a$ is arbitrary $f(x)$ is analytic in $(0 < x < \infty)$. By Abel's theorem one sees that
$$f(0) = \sum_{k=0}^{\infty} \frac{(-a)^k}{k!} f^{(k)}(a) \qquad (a > 0).$$

The series converges by (1), so that $f(x)$ is continuous in $(0 \leqq x < \infty)$.
Computing the successive derivatives of $f(x)$, we have
$$f^{(k+1)}(u) = \sum_{n=0}^{\infty} f^{(k+n+1)}(a)\frac{(u-a)^n}{n!} \qquad (0 < u \leqq a),$$

so that
$$|f^{(k+1)}(u)| \leqq \sum_{n=0}^{\infty} |f^{(k+n+1)}(a)|\frac{(a-u)^n}{n!},$$

and if $0 < \epsilon < a$, we obtain
$$\int_\epsilon^a |f^{(k+1)}(u)|\frac{u^k}{k!} du \leqq \sum_{n=0}^{\infty} \frac{|f^{(k+n+1)}(a)|}{n!k!} \int_\epsilon^a (a-u)^n u^k du.$$

The inequality is strengthened if $\epsilon$ is replaced by zero on the right-hand side, and the series continues to converge since it becomes
$$\sum_{n=k+1}^{\infty} |f^{(n)}(a)|\frac{a^n}{n!},$$

and this converges by (1). Hence for each $a > 0$
$$\int_0^a \frac{u^k}{k!} |f^{(k+1)}(u)| du < M' \qquad (k = 0, 1, 2, \ldots)$$

from which it is evident that $f(x)$ satisfies Conditions $A$.

Conversely, suppose $f(x)$ satisfies Conditions $A$. Then the hypotheses of Theorem 11b are satisfied. In the proof of that theorem we showed that $f(\infty)$ exists and that

$$[f(x) - f(\infty)]^{(k)} = o(x^{-k}) \qquad (x \to \infty\,;\, k = 0, 1, 2, \cdots).$$

But this is sufficient to guarantee that

$$f(a) - f(\infty) = (-1)^{k+1} \int_a^\infty \frac{(t-a)^k}{k!} f^{(k+1)}(t)\, dt \qquad (k = 0, 1, 2, \cdots; a > 0).$$

Applying this formula to $f^{(p)}(x)$ for an arbitrary positive integer $p$, we have

$$f^{(p)}(a) = (-1)^{k+p+1} \int_a^\infty \frac{(t-a)^{k-p}}{(k-p)!} f^{(k+1)}(t)\, dt \qquad (p = 1, 2, \cdots, k).$$

Consequently

$$\frac{a^p}{p!} |f^{(p)}(a)| \leq \int_a^\infty \frac{(t-a)^{k-p} a^p}{(k-p)!\, p!} |f^{(k+1)}(t)|\, dt \qquad (p = 1, 2, \cdots, k),$$

and

$$\sum_{p=0}^k \frac{a^p}{p!} |f^{(p)}(a)| \leq \int_a^\infty \frac{t^k}{k!} |f^{(k+1)}(t)|\, dt + |f(\infty)|$$

$$< M + |f(\infty)|$$

by Conditions $A$. Since the right-hand side is independent of $k$, we have

$$\sum_{p=0}^\infty \frac{a^p}{p!} |f^{(p)}(a)| \leq M + |f(\infty)|,$$

and this is precisely (1) with $M' = M + |f(\infty)|$.

## 14. Determining Function Non-decreasing

We now discuss conditions on the generating function which will insure that the determining function should be non-decreasing. We have seen earlier, Theorem 12a of Chapter IV, that for this it is necessary and sufficient that $f(x)$ should be completely monotonic. We here obtain a new proof by use of Theorem 12a.

**Theorem 14a.** *A necessary and sufficient condition that $f(x)$ can be expressed in the form*

(1) $$f(x) = \int_0^\infty e^{-xt}\, d\alpha(t) \qquad (0 \leq x < \infty),$$

*where $\alpha(t)$ is non-decreasing and bounded in $(0 \leq t < \infty)$, is that $f(x)$ should be completely monotonic in $(0 \leq x < \infty)$.*

## §14] DETERMINING FUNCTION NON-DECREASING

If $f(x)$ has the form (1), then clearly

$$(-1)^k f^{(k)}(x) = \int_0^\infty e^{-xt} t^k \, d\alpha(t) \qquad (x > 0; k = 0, 1, 2, \cdots),$$

and

$$f(0+) = \int_0^\infty d\alpha(t) < \infty,$$

so that $f(x)$ is completely monotonic in $(0 \leq x < \infty)$.

Conversely, if $f(x)$ is completely monotonic in $(0 \leq x < \infty)$, then $f(x)$ is analytic for $(0 < x < \infty)$ by Theorem 3a of Chapter IV. For any positive number $a$

$$(2) \qquad f(x) = \sum_{k=0}^\infty f^{(k)}(a) \frac{(x-a)^k}{k!} \qquad (0 < x < 2a).$$

Since each term of the series is positive when $x < a$, we have

$$f^{(k)}(a) \frac{(x-a)^k}{k!} \leq f(x) \leq f(0+) \qquad (0 < x < a).$$

Allowing $x$ to approach zero this becomes

$$f^{(k)}(a) \frac{(-a)^k}{k!} \leq f(0+) \qquad (0 < a < \infty).$$

Hence

$$f^{(k)}(x) = O(x^{-k}) \qquad (x \to \infty, x \to 0+; k = 0, 1, 2, \cdots).$$

Since $f(\infty)$ and $f(0+)$ both exist under the conditions assumed, we have by Theorem 4.4 of Chapter V

$$f^{(k)}(x) = o(x^{-k}) \qquad (x \to \infty, x \to 0+; k = 1, 2, \cdots).$$

Hence

$$\int_0^\infty |L_{k,t}[f(x)]| \, dt = (-1)^k \int_0^\infty f^{(k)}(t) \frac{t^{k-1}}{(k-1)!} dt = f(\infty) - f(0+)$$

as one sees by successive integration by parts. Hence Conditions $A$ are satisfied by $f(x)$.

Alternatively, we could show that Conditions $A'$ are satisfied. Equation (2) shows that

$$(3) \qquad f(0+) = \sum_{k=0}^\infty (-1)^k f^{(k)}(a) \frac{a^k}{k!},$$

the series converging. For, since the coefficients $(-1)^k f^{(k)}(a)$ of the series (2) are all non-negative, it is clear that the divergence of (3) would imply $f(0+) = \infty$, contrary to assumption. But (3) shows at once that $f(x)$ satisfies Conditions $A'$.

Hence by Theorem 12a or Theorem 13 we have equation (1) with $\alpha(t)$ of bounded variation in $(0, \infty)$. We may assume $\alpha(t)$ normalized. By Theorem 7a or Theorem 7c we see that $\alpha(t)$ is the limit of a sequence of non-decreasing functions, and is consequently non-decreasing. Finally, we note that $\alpha(t)$ must be bounded. For, otherwise we should have

$$f(0+) = \int_0^\infty d\alpha(t) = \infty,$$

contrary to hypothesis. This completes the proof of the theorem.

As an example, take $f(x) = e^{-x}$. It is completely monotonic for all $x$ and has the representation (1) with $\alpha(t)$ zero for $t < 1$ and unity for $t \geq 1$. The theorem does not apply, for example, to the function $f(x) = x^{-1}$ which is completely monotonic in $(0 < x < \infty)$ but not in $(0 \leq x < \infty)$. To cover such functions we have the following result, proved in Chapter IV as Theorem 12b, restated here for completeness.

THEOREM 14b. *A necessary and sufficient condition that $f(x)$ should have the representation (1) for $x > 0$, with $\alpha(t)$ non-decreasing in $(0 \leq t < \infty)$, is that $f(x)$ should be completely monotonic in $(0 < x < \infty)$.*

If $f(x) = x^{-1}$, then (1) holds with $\alpha(t) = t$, a non-decreasing, unbounded function.

## 15. The Class $L^p$, $p > 1$

We next discuss the case in which the determining function $\alpha(t)$ is an integral of a function $\phi(t)$ of class $L^p$ in $(0 \leq t < \infty)$. We first introduce a further definition.

DEFINITION 15. *A function $f(x)$ satisfies Conditions B if it has derivatives of all orders in $(0 < x < \infty)$, if it vanishes at infinity, and if there exist constants $M$ and $p$ $(p > 1)$ such that*

(1) $$\int_0^\infty |L_{k,t}[f(x)]|^p \, dt < M \qquad (k = 1, 2, \cdots).$$

From the definition of the operator $L_{k,t}[f(x)]$ we see that (1) is equivalent to

$$\frac{k}{(k!)^p} \int_0^\infty |f^{(k)}(t)|^p t^{kp+p-2} \, dt < M \qquad (k = 1, 2, \cdots).$$

As an example, take $f(x) = (x + 1)^{-1}$. Then

$$\int_0^\infty |L_{k,t}[f(x)]|^p \, dt = \frac{k}{kp + p - 1} \qquad (k = 1, 2, \cdots),$$

and it is clear that Conditions B are satisfied. On the other hand the functions $f(x) = x^{-1}$ and $f(x) = e^{-x}$ do not satisfy Conditions B. In the latter case

$$\int_0^\infty |L_{k,t}[f(x)]|^p \, dt = \frac{k\,\Gamma(kp+p-1)}{(k!)^p\, p^{kp+p-1}},$$

and this becomes infinite with $k$.

**THEOREM 15a.** *Conditions B are necessary and sufficient that*

(2) $$f(x) = \int_0^\infty e^{-xt} \phi(t) \, dt \qquad (x > 0),$$

*where $\phi(t)$ belongs to $L^p$ in $(0 \leq t < \infty)$:*

(3) $$\int_0^\infty |\phi(t)|^p \, dt < \infty.$$

First assume the representation (2). Then (2) converges absolutely for $x > 0$. For, by Hölder's inequality

$$\int_0^\infty e^{-xt} |\phi(t)| \, dt \leq \left[\int_0^\infty |\phi(t)|^p \, dt\right]^{1/p} [xq]^{-1/q},$$

where $x > 0$ and $p^{-1} + q^{-1} = 1$. The integral on the right exists by the assumption (3).

Another application of Hölder's inequality gives

$$|L_{k,t}[f(x)]|^p = \left|\int_0^\infty e^{-ku/t} \frac{u^k}{k!}\left(\frac{k}{t}\right)^{k+1} \phi(u)\, du\right|^p$$

$$\leq \int_0^\infty e^{-ku/t} \frac{u^k}{k!}\left(\frac{k}{t}\right)^{k+1} |\phi(u)|^p \, du \left[\int_0^\infty e^{-ku/t} \frac{u^k}{k!}\left(\frac{k}{t}\right)^{k+1} du\right]^{p/q}.$$

This last factor is equal to unity so that

(4) $$\int_0^\infty |L_{k,t}[f(x)]|^p \, dt \leq \int_0^\infty dt \int_0^\infty e^{-ku/t} \frac{u^k}{k!}\left(\frac{k}{t}\right)^{k+1} |\phi(u)|^p \, du.$$

By use of Fubini's theorem we may invert the order of integration in the iterated integral, since the resulting integral

(5) $$\int_0^\infty |\phi(u)|^p u^k \, du \int_0^\infty e^{-ku/t} \frac{1}{k!}\left(\frac{k}{t}\right)^{k+1} dt$$

converges and has the value

(6) $$\int_0^\infty |\phi(u)|^p \, du = M.$$

Thus (1) is satisfied. Since
$$f(\infty) = \lim_{t \to 0+} \int_0^t \phi(u)\, du = 0,$$
we see that Conditions $B$ are satisfied by $f(x)$.

Conversely, suppose Conditions $B$ hold. Then by Hölder's inequality
$$\int_0^t |L_{k,u}[f(x)]|\, du \leq M^{1/p} t^{(p-1)/p} \qquad (k = 1, 2, \cdots).$$

Hence for each positive integer
$$\int_0^t L_{k,u}[f(x)]\, du = O(t) \qquad (t \to \infty),$$

so that the hypotheses of Theorem 11b are satisfied. Consequently $f(\infty)$ exists and
$$\lim_{k \to \infty} \int_0^\infty e^{-xt} L_{k,t}[f(x)]\, dt = f(x) - f(\infty) \qquad (0 < x < \infty).$$

But $f(\infty) = 0$ by hypothesis. But by Theorem 17a of Chapter I, there exists a sequence $\{k_i\}_0^\infty$ of the positive integers and a function $\phi(t)$ of class $L^p$ in $(0 \leq t < \infty)$ such that
$$\lim_{i \to \infty} \int_0^\infty e^{-xt} L_{k_i,t}[f(x)]\, dt = \int_0^\infty e^{-xt} \phi(t)\, dt.$$

Hence
$$f(x) = \int_0^\infty e^{-xt} \phi(t)\, dt,$$

and the theorem is proved.

THEOREM 15b. *If $f(x)$ has the representation (2) with $\phi(t)$ belonging to $L^p$ in $(0 \leq t < \infty)$, then*
$$\lim_{k \to \infty} \int_0^\infty |L_{k,t}[f(x)]|^p\, dt = \int_0^\infty |\phi(t)|^p\, dt.$$

For, by (4), (5) and (6) we see that
$$\overline{\lim_{k \to \infty}} \int_0^\infty |L_{k,t}[f(x)]|^p\, dt \leq \int_0^\infty |\phi(t)|^p\, dt.$$

On the other hand, by Fatou's lemma,*

---

* Compare Theorem 17a of Chapter I; for Fatou's lemma see E. C. Titchmarsh [1932] p. 346.

$$\int_0^\infty |\phi(t)|^p \, dt \leq \lim_{k\to\infty} \int_0^\infty |L_{k,t}[f(x)]|^p \, dt,$$

so that the result is proved.

## 16. Determining Function the Integral of a Bounded Function

If we let $p$ become infinite in Conditions $B$ we are led to the case in which $L_{k,t}[f(x)]$ is bounded uniformly in $t$ and $k$. As one would expect from Theorem 15a, this case is that in which $f(x)$ has the representation §15 (2) with $\phi(t)$ uniformly bounded in $(0, \infty)$. We introduce

DEFINITION 16. *A function $f(x)$ satisfies Conditions $C$ if it has derivatives of all orders in $(0 < x < \infty)$ and if there exists a constant $M$ such that for $(0 < x < \infty)$*

(1) $\qquad |L_{k,t}[f(x)]| < M \qquad (k = 1, 2, \cdots)$

(2) $\qquad |xf(x)| < M.$

We observe that inequalities (1) and (2) are equivalent to

(3) $\qquad |f^{(k)}(x)| < \dfrac{Mk!}{x^{k+1}} \qquad (k = 0, 1, 2, \cdots; 0 < x < \infty).$

Since for $x \geq a > 0$, inequality (3) becomes

$$|f^{(k)}(x)| < \frac{Mk!}{a^{k+1}},$$

we see that $f(x)$ is analytic at any point of the complex plane within a distance $a$ from that part of the real axis between $x = a$ and $x = +\infty$.

As an example, take $f(x) = (x+1)^{-2}$. Then

$$\frac{x^{k+1}|f^{(k)}(x)|}{k!} = \frac{(k+1)x^{k+1}}{(x+1)^{k+2}},$$

and it is easily seen that the maximum value of the function on the right is at $x = k + 1$. But

$$\lim_{k\to\infty} \left(\frac{k+1}{k+2}\right)^{k+2} = \frac{1}{e},$$

so that (3) holds for any $M$ greater than $e^{-1}$.

THEOREM 16a. *Conditions $C$ are necessary and sufficient that*

(4) $\qquad f(x) = \displaystyle\int_0^\infty e^{-xt}\phi(t) \, dt,$

*where $\phi(t)$ is bounded in $(0 < t < \infty)$.*

For the necessity we have

(5) $$\left|\frac{f^{(k)}(x)x^{k+1}}{k!}\right| \leq x^{k+1}\int_0^\infty e^{-xt}\frac{t^k}{k!}|\phi(t)|\,dt$$

$$(k = 0, 1, \cdots; 0 < x < \infty)$$

$$\leq \underset{0 \leq t < \infty}{\text{u.b.}} |\phi(t)|,$$

so that inequalities (3) are satisfied.

Conversely, if (1) and (2) hold, then $f(\infty) = 0$ and

$$\int_0^x L_{k,t}[f(x)]\,dt = O(x) \qquad (x \to \infty).$$

Hence by Theorem 11b

$$f(x) = \lim_{k\to\infty}\int_0^\infty e^{-xt}L_{k,t}[f(x)]\,dt \qquad (0 < x < \infty).$$

By Theorem 17b of Chapter I, inequalities (1) and (2) imply the existence of a subset $\{k_i\}$ of the positive integers and a bounded function $\phi(t)$ such that

$$\lim_{i\to\infty}\int_0^\infty e^{-xt}L_{k_i,t}[f(x)]\,dt = \int_0^\infty e^{-xt}\phi(t)\,dt.$$

Hence

$$f(x) = \int_0^\infty e^{-xt}\phi(t)\,dt,$$

so that the theorem is proved.

**THEOREM 16b.** *If $f(x)$ has the representation (4) with $\phi(t)$ bounded in $(0 < t < \infty)$, then*

$$\lim_{k\to\infty}\underset{0 < t < \infty}{\text{u.b.}} |L_{k,t}[f(x)]| = \underset{0 < t < \infty}{\text{true max}} |\phi(t)|.$$

For, by Theorem 17b of Chapter I there exists in every infinite sequence of positive integers a subsequence $\{k_i\}_{i=0}^\infty$ such that

$$\underline{\lim_{i\to\infty}}\,\underset{0 \leq t < \infty}{\text{u.b.}} |L_{k_i,t}[f(x)]| \geq \underset{0 \leq t < \infty}{\text{true max}} |\psi(t)|$$

for some bounded function $\psi(t)$. For each of these sequences the function $\psi(t)$ must be $\phi(t)$ almost everywhere by the uniqueness theorem. But by (5)

$$\overline{\lim_{k\to\infty}}\,\underset{0 \leq t < \infty}{\text{u.b.}} |L_{k,t}[f(x)]| \leq \underset{0 \leq t < \infty}{\text{true max}} |\phi(t)|$$

so that the theorem is established.

As an example take the function $f(x) = (x + 1)^{-2}$ considered above. For this function

$$\underset{(0 \leq t < \infty)}{\text{u.b.}} L_{k,t}[f(x)] = \left(\frac{k+1}{k+2}\right)^{k+2}$$

and this tends to $e^{-1}$ as $k$ becomes infinite. But for this generating function $\phi(t) = te^{-t}$, and the maximum value of this function is $e^{-1}$.

## 17. The Class L

The developments of Section 15 are not valid when $p = 1$. For, when $p = 1$ Conditions B become Conditions A. For Theorem 15a to be valid for $p = 1$ the class of functions of bounded variation would have to be identical, by Theorem 12a, with the class of absolutely continuous functions, which is certainly not the case. We introduce a new set of conditions in

DEFINITION 17. *A function $f(x)$ satisfies Conditions D if it has derivaives of all orders in $(0 < x < \infty)$, vanishes at infinity, if*

$$\int_0^\infty |L_{k,t}[f(x)]| \, dt < \infty \qquad (k = 1, 2, \cdots),$$

*and if*

(1) $$\lim_{\substack{j \to \infty \\ k \to \infty}} \int_0^\infty |L_{k,t}[f(x)] - L_{j,t}[f(x)]| \, dt = 0.$$

For example, the function $f(x) = (x + 1)^{-1}$ can be shown to satisfy Conditions D. Equations (1) mean that $L_{k,t}[f(x)]$ converges in the mean (exponent unity). In discussing the class of functions satisfying these conditions we shall need the following lemma.

LEMMA 17. *If $\phi(t)$ belongs to L in $(0 \leq t < \infty)$, then*

(2) $$g(u) = \int_0^\infty |\phi(tu) - \phi(t)| \, dt = o(1) \qquad (u \to 1)$$

(3) $$g(u) = O(1) \qquad (u \to \infty)$$

(4) $$= O(u^{-1}) \qquad (u \to 0+).$$

By a change of variable

$$g(e^y) = \int_{-\infty}^\infty |\phi(e^{x+y}) - \phi(e^x)| e^x \, dx.$$

If $\psi(x) = \phi(e^x)e^x$, then

$$\int_{-\infty}^\infty |\psi(x)| \, dx = \int_0^\infty |\phi(t)| \, dt < \infty.$$

Then

$$|g(e^y)| \leq \int_{-\infty}^{\infty} \{|\psi(x+y)e^{-y} - \psi(x)e^{-y}| + |\psi(x)e^{-y} - \psi(x)|\}\, dx$$

$$= e^{-y} \int_{-\infty}^{\infty} |\psi(x+y) - \psi(x)|\, dx + |e^{-y} - 1| \int_{-\infty}^{\infty} |\psi(x)|\, dx.$$

The right-hand side of this inequality approaches zero as $y$ approaches zero,* so that (2) is established.

Moreover,

$$|g(u)| \leq \int_0^{\infty} |\phi(ut)|\, dt + \int_0^{\infty} |\phi(t)|\, dt = (1 + u^{-1}) \int_0^{\infty} |\phi(t)|\, dt,$$

so that (3) and (4) are also proved.

THEOREM 17a. *Conditions D are necessary and sufficient that*

(5) $$f(x) = \int_0^{\infty} e^{-xt} \phi(t)\, dt,$$

*where*

(6) $$\int_0^{\infty} |\phi(t)|\, dt < \infty.$$

Suppose first that (5) and (6) hold. Then

$$|L_{k,t}[f(x)] - \phi(t)| \leq \frac{1}{k!}\left(\frac{k}{t}\right)^{k+1} \int_0^{\infty} e^{-ku/t} u^k |\phi(u) - \phi(t)|\, du$$

$$= \frac{k^{k+1}}{k!} \int_0^{\infty} e^{-ku} u^k |\phi(tu) - \phi(t)|\, du.$$

Hence

(7) $$\int_0^{\infty} |L_{k,t}[f(x)] - \phi(t)|\, dt \leq \frac{k^{k+1}}{k!} \int_0^{\infty} dt \int_0^{\infty} e^{-ku} u^k |\phi(tu) - \phi(t)|\, du$$

$$= \frac{k^{k+1}}{k!} \int_0^{\infty} e^{-ku} u^k\, du \int_0^{\infty} |\phi(tu) - \phi(t)|\, dt.$$

In inverting the order of integration in the iterated integral we have used Fubini's theorem, observing that the last integral certainly converges for $k \geq 1$ by virtue of (3) and (4). But by Corollary 3c.1 this integral approaches $g(1) = 0$ as $k$ becomes infinite, if $g(x)$ is defined as in Lemma 17. That is, $L_{k,t}[f(x)]$ converges in mean to $\phi(t)$ on

---

* See, for example, N. Wiener [1933] p. 14.

$0 \leq t < \infty$, so that (1) is established. It is clear from (7) that each function $L_{k,t}[f(x)]$ belongs to $L$ on $0 \leq t < \infty$. Finally, equation (5) implies that

$$f(\infty) = \lim_{t \to 0+} \int_0^t \phi(u)\,du = 0.$$

Conversely, if Conditions $D$ hold, the general theory of mean convergence shows the existence of a function $\phi(t)$ of class $L$ in $(0 \leq t < \infty)$ such that

$$\lim_{k \to \infty} \int_0^\infty |L_{k,t}[f(x)] - \phi(t)|\,dt = 0$$

and

(8) $$\lim_{k \to \infty} \int_0^\infty |L_{k,t}[f(x)]|\,dt = \int_0^\infty |\phi(t)|\,dt.$$

This latter equation implies the existence of an integer $k_0$ such that

$$\int_0^\infty |L_{k,t}[f(x)]|\,dt \leq 1 + \int_0^\infty |\phi(t)|\,dt = M \qquad (k \geq k_0)$$

Hence it is clear that $f(x)$ satisfies Conditions $A$, and, by Theorem 12a, that

$$f(x) = \int_0^\infty e^{-xt}\,d\alpha(t),$$

where $\alpha(t)$ is a normalized function of bounded variation in $(0, \infty)$. But $\alpha(t)$ must be an integral. For, by Theorem 7a

$$\lim_{k \to \infty} \int_0^t L_{k,u}[f(x)]\,du = \alpha(t) - \alpha(0+) = \alpha(t).$$

Clearly $\alpha(0+)$ is zero since $f(\infty) = 0$ by hypothesis. Since $L_{k,t}[f(x)]$ converges in mean to $\phi(t)$ we have

$$\lim_{k \to \infty} \int_0^t L_{k,u}[f(x)]\,du = \int_0^t \phi(u)\,du = \alpha(t) \qquad (0 \leq t < \infty).$$

Hence

$$f(x) = \int_0^\infty e^{-xt}\phi(t)\,dt,$$

and the proof is complete.

THEOREM 17b. *If*

$$f(x) = \int_0^\infty e^{-xt} \phi(t)\,dt$$

$$\int_0^\infty |\phi(t)|\,dt < \infty,$$

*then*

$$\lim_{k\to\infty} \int_0^\infty |L_{k,t}[f(x)]|\,dt = \int_0^\infty |\phi(t)|\,dt.$$

This follows from Theorem 12b. In the present case $f(\infty) = 0$ and the total variation of the function

$$\alpha(x) = \int_0^x \phi(t)\,dt$$

is the right-hand side of (8). One could also obtain the result from the general theory of mean convergence.

THEOREM 17c. *If $f(x)$ satisfies conditions D, and if*

$$\lim_{k\to\infty} \int_0^\infty |L_{k,t}[f(x)] - \psi(t)|\,dt = 0,$$

*then*

(9) $$\lim_{k\to\infty} L_{k,t}[f(x)] = \psi(t)$$

*for almost all positive values of $t$.*

For, by Theorem 17a, $f(x)$ must have the representation (5) (6). Hence (9) follows by Corollary 6a.1 and the uniqueness theorem.

## 18. The General Laplace-Stieltjes Integral

To discuss the general Laplace-Stieltjes integral we need the following preliminary result.

THEOREM 18a. *A necessary and sufficient condition that*

(1) $$f(x) = x \int_0^\infty e^{-xt} \phi(t)\,dt \qquad (0 < x < \infty),$$

*where $\phi(t)$ is bounded in $(0 \leqq t < \infty)$ and is such that the limit*

$$\lim_{t\to\infty} \frac{1}{t} \int_0^t \phi(u)\,du$$

exists, is that for some constant $M$

(2) $$\left| \int_x^\infty \frac{u^k}{k!} f^{(k+1)}(u)\, du \right| < M \qquad (0 < x < \infty; k = 0, 1, \cdots).$$

We first establish the necessity of the condition. If

$$\lim_{t \to \infty} \frac{1}{t} \int_0^t \phi(u)\, du = A,$$

then by Corollary 1a of Chapter V we see that $f(\infty) = A$. Set $F(x) = f(x)/x$. Then by Theorem 16a

(3) $$|F^{(k)}(x)| < Nk!/x^{k+1} \qquad (0 < x < \infty; k = 0, 1, \cdots),$$

where $N$ is an upper bound of $|\phi(t)|$. This, by use of Theorem 4.4 of Chapter V, gives

$$F^{(k)}(x) \sim (-1)^k \frac{f(\infty)k!}{x^{k+1}} \qquad (x \to \infty).$$

But

(4) $$x^k f^{(k+1)}(x) = x^k [xF(x)]^{(k+1)} = \frac{d}{dx}[F^{(k)}(x) x^{k+1}].$$

Hence for $0 < x < \infty$ and $k = 0, 1, 2, \cdots$ we have

(5) $$\int_x^\infty \frac{u^k}{k!} f^{(k+1)}(u)\, du = (-1)^k f(\infty) - F^{(k)}(x) \frac{x^{k+1}}{k!},$$

so that by (5) and (3) we have (2) with $M = |f(\infty)| + N$.

Conversely, if (2) holds it is clear from the existence of the integral (2) when $k = 0$ that $f(\infty)$ exists. By (4) and (2) it is evident that

$$F^{(k)}(x) x^{k+1} = O(1) \qquad (x \to \infty).$$

Then as before (5) holds. But relations (5) and (2) give us (3) with $N = |f(\infty)| + M$. By Theorem 16a

$$xF(x) = f(x) = x \int_0^\infty e^{-xt} \phi(t)\, dt,$$

where $\phi(t)$ is bounded in $(0 \leq t < \infty)$. Finally, to prove that

$$\lim_{t \to \infty} \frac{1}{t} \int_0^t \phi(u)\, du$$

exists we make use of the fact that $\phi(t)$ is bounded and apply a Tauberian theorem, Theorem 4.6 of Chapter V.

THEOREM 18b. *A set of necessary and sufficient conditions that*

$$f(x) = \int_0^\infty e^{-xt}\, d\alpha(t),$$

*the integral converging for $x > 0$ is that for each positive $\epsilon$ there should exist a constant $M_\epsilon$ such that*

(6) $$\left| \int_x^\infty \frac{u^k}{k!} f^{(k+1)}(u)\, du \right| < M_\epsilon \left( \frac{x}{x-\epsilon} \right)^{k+1} \qquad (x > \epsilon;\ k = 0, 1, \cdots)$$

*and that for each positive $R$ there should exist a constant $N_R$ such that*

(7) $$\int_{(k+1)/R}^\infty \frac{u^k}{k!} |f^{(k+1)}(u)|\, du < N_R \qquad (k = 0, 1, \cdots).$$

We prove first the necessity of the conditions. Set $F(x) = f(x)/x$. Then

$$F(x) = \int_0^\infty e^{-xt} \alpha(t)\, dt \qquad (0 < x < \infty),$$

and by Theorem 2.2a of Chapter II for each positive $\epsilon$ there exists a constant $M_\epsilon^*$ such that

$$|\alpha(t)| < M_\epsilon^* e^{\epsilon t} \qquad (0 \leqq t < \infty),$$

whence

(8) $$|F^{(k)}(x)| < \frac{M_\epsilon^* k!}{(x-\epsilon)^{k+1}} \qquad (x > \epsilon).$$

From this it is clear that

$$F^{(k)}(x) = O(x^{-k-1}) \qquad (x \to \infty)$$

and that

$$F(x) \sim f(\infty) x^{-1} = \alpha(0+) x^{-1} \qquad (x \to \infty).$$

As in the proof of Theorem 18a this implies that

$$(-1)^k F^{(k)}(x) \sim \alpha(0+) k! x^{-k-1} \qquad (x \to \infty)$$

$$\int_x^\infty \frac{u^k}{k!} f^{(k+1)}(u)\, du = \int_x^\infty \left[ F^{(k)}(u) \frac{u^{k+1}}{k!} \right]'\, du$$

$$= (-1)^k \alpha(0+) - F^{(k)}(x) \frac{x^{k+1}}{k!}.$$

By (8) we now have

$$\left| \int_x^\infty f^{(k+1)}(u) \frac{u^k}{k!} du \right| < |\alpha(0+)| + M_\epsilon^* \left( \frac{x}{x-\epsilon} \right)^{k+1},$$

so that (6) is established if $M_\epsilon = |\alpha(0+)| + M_\epsilon^*$.
By Theorem 10

$$\lim_{k \to \infty} \int_0^R |L_{k+1,t}[f(x)]| \, dt = \lim_{k \to \infty} \int_{(k+1)/R}^\infty |f^{(k+1)}(u)| \frac{u^k}{k!} du$$
$$= \int_0^R |d\alpha(t)| - |f(\infty)|,$$

from which (7) is evident.

Conversely, if (6) holds, then as in the proof of Theorem 18a we have

(9) $$\int_x^\infty \frac{u^k}{k!} f^{(k+1)}(u) \, du = (-1)^k f(\infty) - F^{(k)}(x) \frac{x^{k+1}}{k!}.$$

Hence (6) becomes

$$\left| F^{(k)}(x) \frac{x^{k+1}}{k!} \right| \leq |f(\infty)| + M_\epsilon \left( \frac{x}{x-\epsilon} \right)^{k+1} \qquad (x > \epsilon)$$

$$|F^{(k)}(x+\epsilon)| \leq \frac{k!}{x^{k+1}} (M_\epsilon + |f(\infty)|) \qquad (x > 0),$$

so that by Theorem 16a

$$F(x+\epsilon) = \int_0^\infty e^{-xt} \psi_\epsilon(t) \, dt$$

$$F(x) = \frac{f(x)}{x} = \int_0^\infty e^{-xt} \phi(t) \, dt \qquad (x > \epsilon),$$

where $\psi_\epsilon(t)$ is bounded in $(0 \leq t < \infty)$ and where

$$\phi(t) = e^{\epsilon t} \psi_\epsilon(t)$$
(10) $$\phi(t) = O(e^{\epsilon t}) \qquad (t \to \infty).$$

The function $\phi(t)$ is independent of $\epsilon$ by virtue of the uniqueness theorem.
Set

$$\alpha_k(t) = (-1)^{k+1} \int_{k/t}^\infty f^{(k+1)}(u) \frac{u^k}{k!} du + f(\infty) \qquad (0 < t < \infty)$$

$$\alpha_k(0) = 0.$$

324    INVERSION OF LAPLACE TRANSFORM    [CH. VII

Then
$$\int_0^R |d\alpha_k(t)| = |f(\infty)| + \int_{k/R}^\infty |f^{(k+1)}(u)| \frac{u^k}{k!} du$$
$$\leq |f(\infty)| + \int_{(k+1)/2R}^\infty |f^{(k+1)}(u)| \frac{u^k}{k!} du$$
$$\leq |f(\infty)| + N_{2R} \qquad (k = 1, 2, \cdots).$$

That is, the set of functions $\alpha_k(t)$ is of uniformly bounded variation in $(0 \leq t \leq R)$. From equation (9) we have for $(0 < t < \infty)$

$$\frac{(-1)^k}{k!} F^{(k)}\left(\frac{k}{t}\right)\left(\frac{k}{t}\right)^{k+1} = f(\infty) + (-1)^{k+1} \int_{k/t}^\infty f^{(k+1)}(u) \frac{u^k}{k!} du = \alpha_k(t).$$

By Corollary 6a.1
$$\lim_{k \to \infty} \alpha_k(t) = \phi(t)$$

on a set $E$ of almost all positive $t$. Since the set of functions $\{\alpha_k(t)\}_0^\infty$ is of uniform bounded variation in $(0 \leq t \leq R)$, the limit function $\phi(t)$ is of bounded variation on that part of $E$ which is in $(0, R)$. Then $\phi(t)$ coincides* on $E$ with a function $\alpha(t)$ which is of bounded variation on $(0, R)$ for every positive $R$. Hence if we redefine $\phi(t)$ to coincide with $\alpha(t)$ everywhere, a process which may be performed without changing $f(x)$, then

(11) $$f(x) = x \int_0^\infty e^{-xt} \alpha(t) dt.$$

We may evidently assume that $\alpha(0) = 0$. Then by (10) we have on integrating (11) by parts

$$f(x) = \int_0^\infty e^{-xt} d\alpha(t) \qquad (0 < x < \infty),$$

and our result is established.

* See, for example, S. Saks [1933] p. 149, Theorem 1.

# CHAPTER VIII

# THE STIELTJES TRANSFORM

## 1. Introduction

The Stieltjes transform arises naturally as an iteration of the Laplace transform. For, if

$$f(x) = \int_0^\infty e^{-xt} \varphi(t)\, dt,$$

where

$$\varphi(x) = \int_0^\infty e^{-xt} \psi(t)\, dt,$$

then we have formally

$$f(x) = \int_0^\infty e^{-xu}\, du \int_0^\infty e^{-ut} \psi(t)\, dt$$

$$f(x) = \int_0^\infty \psi(t)\, dt \int_0^\infty e^{-u(x+t)}\, du$$

(1) $$f(x) = \int_0^\infty \frac{\psi(t)}{x+t}\, dt.$$

This last equation we refer to as the Stieltjes transform, or from another point of view as the Stieltjes integral equation. However, we shall usually be concerned with the more general case in which (1) is replaced by a Stieltjes integral

(2) $$f(x) = \int_0^\infty \frac{d\alpha(t)}{x+t}.$$

In fact it was in this form that the equation was originally considered by Stieltjes [1894] in connection with his work on continued fractions. It is clear from the above formal work that the properties of the integral (2) will be intimately related to those of the Laplace integral.*

## 2. Elementary Properties of the Transform

Let $\alpha(t)$ be of bounded variation in $0 \leq t \leq R$ for every positive $R$. Then we set

(1) $$f(s) = \int_0^\infty \frac{d\alpha(t)}{s+t} = \lim_{R \to \infty} \int_0^R \frac{d\alpha(t)}{s+t}$$

* See D. V. Widder [1937].

for any complex value of $s$, $s = \sigma + i\tau$, which makes the limit exist. The corresponding improper integral (1) is then said to converge.

It will also be convenient to assume in some of our work that $\alpha(t)$ is of bounded variation in $\epsilon \leq t \leq R$ for every pair of positive numbers $\epsilon$, $R$ ($\epsilon < R$). Then if the limits

$$\lim_{\epsilon \to 0+} \int_\epsilon^1 \frac{d\alpha(t)}{s+t}$$

$$\lim_{R \to \infty} \int_1^R \frac{d\alpha(t)}{s+t}$$

both exist, we write

(2) $$f(s) = \int_{0+}^\infty \frac{d\alpha(t)}{s+t}.$$

The difference in notation in (1) and (2) will be sufficient to distinguish between the difference in the assumptions about $\alpha(t)$ in the two cases. For example, if $\alpha(t) = t \sin(1/t)$ when $0 < t < 1$, and if $\alpha(t) = 0$ for all other $t$, we would write our transform in the form (2). On the other hand, the function $1/s$ can be represented in the form (1) though not in the form (2). In the next two theorems we restrict ourselves to integrals (1).

**THEOREM 2a.** *If the integral* (1) *converges for a point* $s = s_0$ *not on the negative real axis*, $\tau = 0$, $\sigma \leq 0$, *then it converges for every such point.*

For, set

$$\beta(t) = \int_0^t \frac{d\alpha(u)}{s_0 + u} \qquad (0 \leq t < \infty).$$

Then for any $s$ not on the negative real axis and for any positive $R$

$$\int_0^R \frac{d\alpha(t)}{s+t} = \int_0^R \frac{s_0+t}{s+t} d\beta(t) = \beta(R)\frac{s_0+R}{s+R} + (s_0 - s)\int_0^R \frac{\beta(t)}{(s+t)^2} dt.$$

Since $\beta(R)$ approaches $f(s_0)$ when $R$ becomes infinite it is clear that

$$\int_0^\infty \frac{\beta(t)}{(s+t)^2} dt$$

converges absolutely. Hence (1) converges and

(3) $$f(s) = \int_0^\infty \frac{d\alpha(t)}{s+t} = f(s_0) + (s_0 - s)\int_0^\infty \frac{\beta(t)}{(s+t)^2} dt.$$

This completes the proof. We observe that (1) may converge without converging absolutely as the example

$$f(s) = \sum_{n=0}^{\infty} \frac{(-1)^n}{s+n}$$

shows. However, equation (3) enables us to replace (1) by an absolutely convergent integral. We point out a contrast with the Laplace integral. For the latter integral a simple integration by parts replaces a conditionally convergent integral by an absolutely convergent one. That the same is not true for the integral (1) is shown by the example

$$\alpha(0) = 0$$

$$\alpha(t) = (-1)^n \frac{t+2}{\log(t+2)} \qquad (n < t < n+1; n = 0, 1, \cdots).$$

Here

$$\int_0^\infty \frac{|\alpha(t)|}{(t+2)^2} dt = \int_0^\infty \frac{dt}{(t+2)\log(t+2)},$$

and this integral clearly diverges. But

$$\int_0^\infty \frac{d\alpha(t)}{t+2} = \int_0^\infty \frac{\alpha(t)}{(t+2)^2} dt = \sum_{n=0}^{\infty} (-1)^n \log \frac{\log(n+3)}{\log(n+2)},$$

and this latter series converges.

THEOREM 2b. *If the integral* (1) *converges, it converges uniformly in any closed bounded region not containing a point of the negative real axis.*

For, using the notation of the previous section and defining $M$ as the maximum distance from the origin to a point $s$ of the region described in the theorem, we have for any $R$ greater than $M$

$$\int_R^\infty \frac{d\alpha(t)}{s+t} = f(s_0) - \beta(R)\frac{s_0+R}{s+R} + (s_0-s)\int_R^\infty \frac{\beta(t)}{(s+t)^2} dt$$

$$\left|\int_R^\infty \frac{d\alpha(t)}{s+t}\right| \leq |f(s_0) - \beta(R)| + |\beta(R)|\frac{|s-s_0|}{|s+R|}$$

$$+ |s_0-s|\int_R^\infty \frac{|\beta(t)|}{|s+t|^2} dt$$

$$\leq |f(s_0) - \beta(R)| + |\beta(R)|\frac{(M+|s_0|)}{R-M}$$

$$+ (M+|s_0|)\int_R^\infty \frac{|\beta(t)|}{(t-M)^2} dt.$$

Since the right-hand side is independent of $s$ and tends to zero as $R$ becomes infinite our result is proved.

**COROLLARY 2b.1.** *If the integral (1) converges it represents an analytic single-valued function in the complex plane cut along the negative real axis.*

**COROLLARY 2b.2.** *If the integral (1) converges to $f(s)$, then for any $s$ not on the negative real axis*

$$f^{(k)}(s) = (-1)^k k! \int_0^\infty \frac{d\alpha(t)}{(s+t)^{k+1}} \qquad (k = 0, 1, 2, \cdots).$$

We observe that all results of the present section apply equally well to the integral (2). For,

$$(4) \qquad \int_{0+}^\infty \frac{d\alpha(t)}{s+t} = \int_1^\infty \frac{d\alpha(t)}{s+t} - \frac{1}{s} \int_1^\infty \frac{t\, d\alpha(1/t)}{(1/s)+t},$$

and hence (2) can be expressed in terms of two integrals of type (1), in one of which $s$ has been replaced by its reciprocal. This is sufficient to prove our contention.

We may extend Theorem 2b to the case in which the kernel $(s+t)^{-1}$ is replaced by $(s+t)^{-\rho}$ with $\rho > 0$. We restrict ourselves to real values of $s$, since this was the only case needed in Chapter V.

**THEOREM 2c.** *If the integral*

$$(5) \qquad f(s) = \int_0^\infty \frac{d\alpha(t)}{(s+t)^\rho} \qquad (\rho > 0)$$

*converges for a positive value of $s$, it converges uniformly in any closed interval of the positive $s$-axis not including the origin.*

For, if the integral (5) converges at $s = s_0 > 0$ set

$$\gamma(t) = \int_0^t \frac{d\alpha(u)}{(s_0+u)^\rho} \qquad (0 \le t < \infty),$$

so that $\gamma(\infty) = f(s_0)$. Then

$$\int_0^\infty \frac{d\alpha(t)}{(s+t)^\rho} = f(s_0) + \rho(s_0 - s) \int_0^\infty \frac{(s_0+t)^{\rho-1}}{(s+t)^{\rho+1}} \gamma(t)\, dt,$$

so that (5) converges for all positive $s$. If $R > 0$, then

$$\int_R^\infty \frac{d\alpha(t)}{(s+t)^\rho} = f(s_0) - \gamma(R) \left(\frac{s_0+R}{s+R}\right)^\rho + \rho(s_0 - s) \int_R^\infty \frac{(s_0+t)^{\rho-1}}{(s+t)^{\rho+1}} \gamma(t)\, dt.$$

It is clear that $(s_0+R)^\rho (s+R)^{-\rho}$ tends uniformly to unity in an interval $0 < a \le s \le b$. Moreover, in that interval

$$\left| \rho(s_0 - s) \int_R^\infty \frac{(s_0+t)^{\rho-1}}{(s+t)^{\rho+1}} \gamma(t)\, dt \right| \le \rho(s_0+b) \int_R^\infty \frac{(s_0+t)^{\rho-1}}{(t-b)^{\rho+1}} |\gamma(t)|\, dt$$

## 3. Asymptotic Properties of Stieltjes Transforms

We first consider what necessary conditions are imposed on $\alpha(t)$ by the convergence of the integrals §2 (1) or §2 (2).

**THEOREM 3a.** *If either of the integrals*

(1) $$\int_0^\infty \frac{d\alpha(t)}{s+t}$$

(2) $$\int_{0+}^\infty \frac{d\alpha(t)}{s+t}$$

*converges, then $\alpha(0+)$ exists and*

(3) $$\alpha(t) = o(t) \qquad (t \to \infty).$$

Let (1) or (2) converge at $s = 1$ (and hence everywhere). Set

$$\beta(t) = \int_1^t \frac{d\alpha(u)}{u+1} \qquad (1 \leq t < \infty).$$

Then

$$\alpha(t) - \alpha(1) = \int_1^t d\alpha(u) = \int_1^t (u+1)\, d\beta(u) = \beta(t)(t+1) - \int_1^t \beta(u)\, du$$

$$\frac{\alpha(t)}{t} = \frac{\alpha(1)}{t} + \beta(t)\left(1 + \frac{1}{t}\right) - \frac{1}{t}\int_1^t \beta(u)\, du$$

$$\frac{\alpha(t)}{t} \sim \beta(\infty) - \beta(\infty) \qquad (t \to \infty),$$

so that (3) is established.

If (1) exists, the existence of $\alpha(0+)$ is trivial. If (2) exists set

$$\gamma(t) = \int_t^1 \frac{d\alpha(u)}{u+1} \qquad (0 < t \leq 1),$$

so that

(4) $$\alpha(t) - \alpha(1) = -\int_t^1 d\alpha(u) = \int_t^1 (u+1)\, d\gamma(u)$$
$$= -(t+1)\gamma(t) - \int_t^1 \gamma(u)\, du.$$

Since $\gamma(0+)$ exists by hypothesis it is clear from equation (4) that $\alpha(0+)$ exists.

COROLLARY 3a.1. *If the integral* (1) *converges, then*

$$\int_0^\infty \frac{d\alpha(t)}{s+t} = -\frac{\alpha(0)}{s} + \int_0^\infty \frac{\alpha(t)}{(s+t)^2}\,dt.$$

COROLLARY 3a2. *If the integral* (2) *converges, then*

$$\int_{0+}^\infty \frac{d\alpha(t)}{s+t} = -\frac{\alpha(0+)}{s} + \int_0^\infty \frac{\alpha(t)}{(s+t)^2}\,dt.$$

COROLLARY 3a.3. *If the integral*

$$\int_0^\infty \frac{d\alpha(t)}{(s+t)^\rho} \qquad (\rho > 0)$$

*converges, then*

$$\alpha(t) = o(t^\rho) \qquad (t \to \infty).$$

This follows easily from the equation

$$\alpha(t) - \alpha(1) = \gamma(t)(t+1)^\rho - \rho \int_1^t (u+1)^{\rho-1} \gamma(u)\,du,$$

where

$$\gamma(t) = \int_1^t \frac{d\alpha(u)}{(u+1)^\rho} \qquad (1 \leq t < \infty).$$

It is important to observe that the conditions on $\alpha(t)$ of Theorem 3a, necessary for convergence of (1) or (2) are by no means sufficient. For example, if

$$\alpha(t) = \int_0^t \frac{du}{\log(u+2)} \qquad (0 \leq t < \infty),$$

then $\alpha(0+) = 0$ and $\alpha(t) = o(t)$ as $t$ becomes infinite. Yet the integral

$$\int_0^\infty \frac{dt}{(s+t)\log(t+2)}$$

clearly diverges at $s = 2$ (and hence everywhere).

The following sufficient condition for convergence is useful.

THEOREM 3b. *If for some positive* $\delta$

(5) $$\alpha(t) = O(t^{1-\delta}) \qquad (t \to \infty),$$

*then* (1) *converges. If in addition* $\alpha(0+)$ *exists, then* (2) *converges.*

According to our convention for the integral (1), $\alpha(t)$ is of bounded variation in $0 \leq t \leq R$ for every $R > 0$. Hence

$$\int_0^R \frac{d\alpha(t)}{s+t} = \frac{\alpha(R)}{s+R} - \frac{\alpha(0)}{s} + \int_0^R \frac{\alpha(t)}{(s+t)^2}\,dt,$$

and it is clear by virtue of (5) that this approaches a limit as $R$ becomes infinite. A similar proof holds for the integral (2).

We wish to study next the asymptotic behavior of the Stieltjes transform $f(s)$ as $s$ becomes infinite or approaches zero.

**THEOREM 3c.** *If $f(s)$ is defined by the convergent integral (1) with $\alpha(0) = 0$, then*

(6) $\quad f^{(n)}(s) \sim (-1)^n n!\alpha(0+)s^{-n-1} \quad (s \to 0+; n = 0, 1, 2, \cdots)$

(7) $\quad f^{(n)}(s) = o(s^{-n}) \quad\quad\quad\quad (s \to \infty; n = 0, 1, 2, \cdots),$

*where $s$ is real and positive.*

First suppose that $n \geqq 1$. Then

$$f^{(n)}(s) = (-1)^n n! \int_0^\infty \frac{d\alpha(t)}{(s+t)^{n+1}}$$

(8) $$= (-1)^n (n+1)! \int_0^\infty \frac{\alpha(t)}{(s+t)^{n+2}} dt.$$

The integral (8) converges absolutely by Theorem 3a. Then

$$I = \frac{(-1)^n}{n!} f^{(n)}(s) s^{n+1} - \alpha(0+) = (n+1)s^{n+1} \int_0^\infty \frac{\alpha(t) - \alpha(0+)}{(s+t)^{n+2}} dt.$$

Given $\epsilon > 0$, we choose $\delta$ so small that for $(0 \leqq t \leqq \delta)$

$$|\alpha(t) - \alpha(0+)| < \epsilon.$$

Then

$$|I| \leqq (n+1)s^{n+1} \int_0^\delta \frac{\epsilon}{(s+t)^{n+2}} dt + (n+1)s^{n+1} \int_\delta^\infty \frac{M}{t^{n+1}} dt,$$

where

$$M = \operatorname*{u.b.}_{\delta \leqq t < \infty} |\alpha(t) - \alpha(0+)| t^{-1}.$$

This is a finite number by Theorem 3a. Hence

$$|I| \leqq \epsilon + 2s^{n+1}\delta^{-n} M$$

$$\overline{\lim_{s \to 0+}} |I| \leqq \epsilon$$

$$\lim_{s \to 0+} I = 0,$$

and (6) is established for $n \geqq 1$.

Again using Theorem 3a we determine $R$ so large that for $R \leqq t < \infty$

$$|\alpha(t)| < \epsilon t.$$

From (8) we have for $n \geqq 1$

$$|s^n f^{(n)}(s)| \leqq s^n(n+1)! \int_0^R \frac{|\alpha(t)|}{(s+t)^{n+2}} dt + s^n \epsilon(n+1)! \int_R^\infty \frac{t}{(s+t)^{n+2}} dt$$

$$\leqq \frac{(n+1)!}{s^2} \int_0^R |\alpha(t)| dt + \epsilon \frac{(n+1)!}{n}.$$

Allowing $s$ to become infinite, we have

$$\varlimsup_{s \to \infty} |s^n f^{(n)}(s)| \leqq \epsilon \frac{(n+1)!}{n}$$

$$\lim_{s \to \infty} s^n f^{(n)}(s) = 0.$$

That is, we have proved (7) for $n \geqq 1$.

The case $n = 0$ must be treated differently since certain of the integrals involved in the above calculation do not converge absolutely in this case. We now make use of §2 (3) with $s_0 = 1$. Then

(9) $$f(s) = f(1) + (1-s) \int_0^\infty \frac{\beta(t)}{(s+t)^2} dt$$

$$\beta(t) = \int_0^t \frac{d\alpha(u)}{1+u} \qquad (0 \leqq t < \infty).$$

To complete the proof of (6) it will evidently be enough to prove that

$$\lim_{s \to 0+} s \int_0^\infty \frac{\beta(t)}{(s+t)^2} dt = \beta(0+) = \alpha(0+).$$

But

$$\left| s \int_0^\infty \frac{\beta(t)}{(s+t)^2} dt - \beta(0+) \right| \leqq s \int_0^\delta \frac{|\beta(t) - \beta(0+)|}{(s+t)^2} dt$$

$$+ s \int_\delta^\infty \frac{|\beta(t) - \beta(0+)|}{(s+t)^2} dt.$$

Given $\epsilon$, we choose $\delta$ so small that for $0 \leqq t \leqq \delta$

$$|\beta(t) - \beta(0+)| < \epsilon.$$

Let $M$ be an upper bound of $|\beta(t) - \beta(0+)|$ for all positive $t$. Then

$$\left| s \int_0^\infty \frac{\beta(t)}{(s+t)^2} dt - \beta(0+) \right| \leqq \epsilon + \frac{Ms}{\delta}$$

$$\varlimsup_{s \to 0+} \left| s \int_0^\infty \frac{\beta(t)}{(s+t)^2} dt - \beta(0+) \right| \leqq \epsilon,$$

from which (6) follows for $n = 0$.

§3] ASYMPTOTIC PROPERTIES 333

Finally, to prove (7) when $n = 0$ we again use equation (9). It will be sufficient to show that

$$\lim_{s \to \infty} s \int_0^\infty \frac{\beta(t)}{(s+t)^2} \, dt = f(1) = \beta(\infty).$$

But

$$s \int_0^\infty \frac{\beta(t)}{(s+t)^2} \, dt - \beta(\infty) = s \int_0^\infty \frac{\beta(t) - \beta(\infty)}{(s+t)^2} \, dt.$$

This integral converges absolutely, so that the proof may be completed in an obvious way.

THEOREM 3d. *If $f(s)$ is defined by the convergent integral (2), then*

(10) $\qquad f^{(n)}(s) = o(s^{-n-1}) \qquad (s \to 0+;\ n = 0, 1, 2, \cdots)$

(11) $\qquad f^{(n)}(s) = o(s^{-n}) \qquad (s \to \infty;\ n = 0, 1, 2, \cdots),$

*where $s$ is real and positive.*

For, by procedure similar to that used in obtaining (9) we have

$$f(s) = f(1) + (1 - s) \int_0^\infty \frac{\beta(t)}{(s+t)^2} \, dt$$

$$\beta(t) = \int_{0+}^t \frac{d\alpha(u)}{1+u}.$$

But we showed, merely on the assumption that $\beta(0+)$ exists, that

$$\lim_{s \to 0+} s \int_0^\infty \frac{\beta(t)}{(s+t)^2} \, dt = \beta(0+).$$

In the present case $\beta(0+) = 0$, so that

$$f(s) = o(s^{-1}) \qquad (s \to 0+).$$

By Corollary 3a.2, we see that

(12) $\qquad f(s) = -\dfrac{\alpha(0+)}{s} + \displaystyle\int_0^\infty \dfrac{\alpha(t)}{(s+t)^2} \, dt.$

But we showed in the proof of Theorem 3c that

(13) $\qquad \dfrac{d^n}{ds^n} \displaystyle\int_0^\infty \dfrac{\alpha(t)}{(s+t)^2} \, dt \sim \dfrac{(-1)^n \alpha(0+) n!}{s^{n+1}} \qquad (s \to 0+)$

for all positive integers. In the present case we do not know that $\alpha(t)$ is of bounded variation in any interval including the origin. However, we do know that $\alpha(0+)$ exists, and this was the essential hypothesis in the above proof. Hence equations (12) and (13) give

$$f^{(n)}(s) = o(s^{-n-1}) \qquad (s \to 0+; n = 1, 2, \cdots),$$

so that (10) is established. In a similar way one proves (11) by use of (12) and the results of Theorem 3c.

### 4. Relation to the Laplace Transform

We saw in Section 1 that the Stieltjes transform may be regarded as the result of iterating the Laplace transform. We make this result more precise in the present section.

**THEOREM 4a.** *If the integral*

(1) $$f(s) = \int_0^\infty \frac{d\alpha(t)}{s+t}$$

*converges, then*

(2) $$f(s) = \int_{0+}^\infty e^{-st} \varphi(t)\, dt \qquad (\sigma > 0),$$

*where*

(3) $$\varphi(t) = \int_0^\infty e^{-tu}\, d\alpha(u) \qquad (t > 0).$$

For, by Theorem 3a the convergence of (1) implies that

$$\alpha(t) = o(t) \qquad (t \to \infty).$$

Hence (3) converges for $t > 0$ and converges uniformly in $\epsilon \leq t \leq R$ for arbitrary positive numbers $\epsilon$ and $R$. Hence

$$\int_\epsilon^R e^{-st}\, dt \int_0^\infty e^{-tu}\, d\alpha(u) = \int_0^\infty \frac{e^{-(s+u)\epsilon} - e^{-(s+u)R}}{s+u}\, d\alpha(u).$$

If $s$ is any point not on the negative real axis, the integral

$$\int_0^\infty \frac{e^{-\epsilon u}}{s+u}\, d\alpha(u)$$

clearly converges, so that

(4) $$\int_\epsilon^R e^{-st} \varphi(t)\, dt = e^{-s\epsilon} \int_0^\infty \frac{e^{-\epsilon u}}{s+u}\, d\alpha(u) - e^{-sR} \int_0^\infty \frac{e^{-uR}}{s+u}\, d\alpha(u).$$

The first Laplace integral on the right-hand side converges uniformly ($s$ being fixed) in the interval $0 \leq \epsilon \leq 1$, and hence approaches $f(s)$ as $\epsilon$ approaches zero. Moreover, by Corollary 1a of Chapter V we have

$$\lim_{R\to\infty}\int_0^\infty \frac{e^{-uR}}{s+u}\,d\alpha(u) = \lim_{t\to 0+}\int_0^t \frac{d\alpha(u)}{s+u} = \frac{\alpha(0+) - \alpha(0)}{s}$$

so that the last term of equation (4) approaches zero as $R$ becomes infinite. Letting $\epsilon$ and $R^{-1}$ approach zero in (4) thus gives (2), and the theorem is proved.

THEOREM 4b. *If the integral*

$$f(s) = \int_{0+}^\infty \frac{d\alpha(t)}{s+t}$$

*converges, then $f(s)$ has the representation (2) with*

$$\varphi(t) = \int_{0+}^\infty e^{-tu}\,d\alpha(u) \qquad (t > 0).$$

The proof of this result is similar to that of Theorem 4a and is omitted.

We observe that the converse of Theorem 4a is not true. That is, the integrals (2) and (3) may converge in the regions indicated without having (1) hold. For example, take

$$\varphi(t) = \frac{1}{(1+e^{-t})^2} = \sum_{n=0}^\infty (-1)^n (n+1) e^{-nt} \qquad (t > 0)$$

$$f(s) = \int_0^\infty \frac{e^{-st}}{(1+e^{-t})^2}\,dt \qquad (\sigma > 0).$$

But (1) becomes

$$f(s) = \sum_{n=0}^\infty (-1)^n \frac{n+1}{s+n},$$

a series which diverges for all $s$.

THEOREM 4c. *If the integral (3) converges for $t \geq 0$, and if $f(s)$ is defined by (2) for $\sigma > 0$, then $f(s)$ also has the representation (1) for all $s$ not on the negative real axis.*

For, under the present hypotheses, $\alpha(\infty)$ clearly exists. Hence by Theorem 3b the integral (1) converges. Hence we may apply Theorem 4a to obtain our result.

THEOREM 4d. *If $\alpha(u)$ is non-decreasing and is such that the integrals (2) and (3) converge for $\sigma > 0$ and $t > 0$ respectively, then the function $f(s)$ defined by (2) also has the representation (1) for any $s$ not on the negative real axis.*

To prove this, substitute the integral (3) in the integral (2). In the resulting iterated integral reverse the order of integration to obtain (1). This is permissible by Theorem 15c of Chapter I.

## 5. Uniqueness

It is important to observe that a function $f(s)$ which is a Stieltjes transform is such a transform in essentially only one way.

**THEOREM 5a.** *If $\alpha(t)$ is a normalized function of bounded variation in every finite interval $0 \leq t \leq R$ and is such that the integral*

$$(1) \qquad f(s) = \int_0^\infty \frac{d\alpha(t)}{s+t}$$

*converges, and if $f(s)$ vanishes at a set of points in arithmetic progression,*

$$(2) \qquad f(s_0 + nl) = 0 \qquad (l > 0,\ n = 0,\ 1,\ 2,\ \cdots),$$

*then $\alpha(t)$ is identically zero.*

For, by Theorem 4a, $f(s)$ is the Laplace integral §4 (2). Then by Corollary 6.2a of Chapter II we see that $\varphi(t)$ is identically zero. Then by Theorem 6.2 of Chapter II $\alpha(t)$ is identically zero. Hence our theorem is proved.

**COROLLARY 5a.** *Equations (1) and (2) imply that $f(s)$ is identically zero.*

**THEOREM 5b.** *If $\alpha(t)$ and $\beta(t)$ are normalized functions of bounded variation in every finite interval $0 \leq t \leq R$ such that*

$$\int_0^\infty \frac{d\alpha(t)}{s+t} = \int_0^\infty \frac{d\beta(t)}{s+t},$$

*both integrals converging. Then $\alpha(t)$ is identically equal to $\beta(t)$.*

This follows in an obvious way from Theorem 5a.

## 6. The Stieltjes Transform Singular at the Origin

We have noted that a function $f(s)$ which is a Stieltjes transform is analytic in the entire $s$-plane with the negative real axis removed. However, the points of this axis need not be singular points. For example if $\alpha(t)$ is unity except at $t = 0$ where it is zero, then $f(s) = s^{-1}$, which is analytic on the whole negative real axis. The origin itself need not be a singular point of $f(s)$ as the example $f(s) = (s+1)^{-1}$ shows. On the other hand the negative real axis may be a cut for the function $f(s)$. For example, $f(s)$ may reduce to the series

$$f(s) = \sum_{n=0}^\infty \frac{\beta_n}{s - a_n},$$

where the points $a_n$ are dense on the negative real axis and

$$\sum_{n=0}^\infty |\beta_n| < \infty.$$

E. Goursat* has shown that the line on which the $a_n$ are densely distributed is a cut for the function $f(s)$.

It is useful to obtain a sufficient condition that $f(s)$ should have a singularity at the origin.

THEOREM 6. *If $\alpha(t)$ is a real non-decreasing function for which the point $t = 0$ is a point of increase and for which the integral*

$$f(s) = \int_0^\infty \frac{d\alpha(t)}{s+t}$$

*converges, then $f(s)$ has a singularity at $s = 0$.*

For, if this were not the case the series

$$f(s) = \sum_{n=0}^\infty f^{(n)}(1) \frac{(s-1)^n}{n!}$$

would converge for $s$ some real negative number, $-\epsilon$, and

$$f(-\epsilon) = \sum_{n=0}^\infty (-1)^n f^{(n)}(1) \frac{(\epsilon+1)^n}{n!}$$

We may assume without loss of generality that $\alpha(0) = 0$. Then by Corollary 2b.2 and Theorem 3a we have

$$f(-\epsilon) = \sum_{n=0}^\infty \int_0^\infty \frac{(\epsilon+1)^n}{(t+1)^{n+1}} d\alpha(t)$$

$$= \sum_{n=0}^\infty (n+1) \int_0^\infty \frac{(\epsilon+1)^n}{(t+1)^{n+2}} \alpha(t) \, dt.$$

This series dominates the series

(1) $$\sum_{n=0}^\infty (n+1) \int_\epsilon^\infty \frac{(\epsilon+1)^n}{(t+1)^{n+2}} \alpha(t) \, dt,$$

so that the latter also converges. Since the integrand is non-negative, and since the series

$$\sum_{n=0}^\infty (n+1) \frac{(\epsilon+1)^n}{(t+1)^{n+2}} = \frac{1}{(t-\epsilon)^2}$$

converges for $t > \epsilon$, we may interchange integral and summation symbols in (1) to obtain the convergent integral

(2) $$\int_\epsilon^\infty \frac{\alpha(t)}{(t-\epsilon)^2} dt.$$

* See E. Borel [1917] p. 37.

But since our hypothesis asserts that $t = 0$ is a point of increase of $\alpha(t)$, it follows that $\alpha(\epsilon+) > 0$ and

$$\lim_{t \to \epsilon+} \frac{\alpha(t)}{t - \epsilon} = +\infty.$$

Hence (2) cannot converge, and our assumption that $f(s)$ is regular at $s = 0$ must have been a false one. This proves the theorem.

## 7. Complex Inversion Formula

An inversion formula for the integral §2 (1) was obtained by Stieltjes [1894] in connection with his work on continued fractions. This formula applies, however, only to the case in which $\alpha(t)$ is a real non-decreasing function. One may easily treat the general case by similar methods. We need certain preliminary results.

**LEMMA 7.1.** *If $\alpha(t)$ belongs to class $L$ in $0 \leq t \leq R$ and if $\alpha(0+)$ exists, then*

(1) $$\lim_{\eta \to 0+} \frac{1}{\pi} \int_0^R \frac{\alpha(t)\eta}{t^2 + \eta^2} dt = \frac{\alpha(0+)}{2}.$$

Since for any positive number $\eta$

$$\int_0^R \frac{\eta}{t^2 + \eta^2} dt = \tan^{-1} \frac{R}{\eta} \to \frac{\pi}{2} \qquad (R \to \infty),$$

it will clearly be sufficient to assume $\alpha(0+) = 0$. Given $\epsilon > 0$, we determine $\delta < R$ so that $|\alpha(t)| < \epsilon$ for $0 \leq t \leq \delta$. Then

$$\left| \int_0^R \frac{\alpha(t)\eta}{t^2 + \eta^2} dt \right| \leq \epsilon \int_0^\delta \frac{\eta}{t^2 + \eta^2} dt + \eta \int_\delta^R \frac{|\alpha(t)|}{t^2} dt,$$

$$\overline{\lim_{\eta \to 0+}} \left| \int_0^R \frac{\alpha(t)\eta}{t^2 + \eta^2} dt \right| \leq \epsilon \frac{\pi}{2}.$$

This gives us (1) at once.

**LEMMA 7.2.** *If $\alpha(t)$ belongs to $L$ in $0 \leq t \leq R$, if $0 < \xi < R$, and if $\alpha(\xi+)$ and $\alpha(\xi-)$ exist, then*

$$\lim_{\eta \to 0+} \frac{1}{\pi} \int_0^R \frac{\alpha(t)\eta}{(t - \xi)^2 + \eta^2} dt = \frac{\alpha(\xi+) + \alpha(\xi-)}{2}.$$

This is proved by breaking the integral into two parts corresponding to the intervals $(0, \xi)$ and $(\xi, R)$ and applying Lemma 7.1 to each part. The integral is known as Poisson's integral for the half-plane or as Cauchy's singular integral.*

---

*See, for example, E. C. Titchmarsh [1937] p. 30.

THEOREM 7a. *If the integral*

$$f(s) = \int_0^\infty \frac{d\alpha(t)}{s+t}$$

*converges, then for any positive number* $\xi$

(2)
$$\lim_{\eta \to 0+} \frac{1}{2\pi i} \int_0^\xi [f(-\sigma - i\eta) - f(-\sigma + i\eta)] d\sigma$$
$$= \frac{\alpha(\xi+) + \alpha(\xi-)}{2} - \frac{\alpha(0+) + \alpha(0)}{2}.$$

Let us assume that $\alpha(0) = 0$. This is no restriction. Then

$$f(s) = \int_0^\infty \frac{\alpha(t)}{(s+t)^2} dt.$$

Since this integral converges uniformly along the line segments $\tau = \pm \eta$, $0 \leq \sigma \leq \xi$, we have

$$I_\eta = \frac{1}{2\pi i} \int_0^\xi [f(-\sigma - i\eta) - f(-\sigma + i\eta)] d\sigma$$

$$= \frac{1}{2\pi i} \int_0^\infty \alpha(t) \left\{ \frac{1}{t - \xi - i\eta} - \frac{1}{t - i\eta} + \frac{1}{t + i\eta} - \frac{1}{t - \xi + i\eta} \right\} dt$$

$$= \frac{1}{\pi} \int_0^\infty \alpha(t) \eta \left\{ \frac{1}{(t-\xi)^2 + \eta^2} - \frac{1}{t^2 + \eta^2} \right\} dt.$$

Choose a number $R > \xi$. The integral

$$\int_R^\infty \frac{|\alpha(t)|}{(t-\xi)^2 t} dt$$

converges by §3 (3). Then

$$\left| I_\eta - \frac{1}{\pi} \int_0^R \alpha(t) \eta \left\{ \frac{1}{(t-\xi)^2 + \eta^2} - \frac{1}{t^2 + \eta^2} \right\} dt \right|$$
$$\leq \frac{\xi \eta}{\pi} \int_R^\infty \frac{|\alpha(t)| |2t - \xi| dt}{[(t-\xi)^2 + \eta^2](t^2 + \eta^2)} < \frac{2\eta \xi}{\pi} \int_R^\infty \frac{|\alpha(t)|}{(t-\xi)^2 t} dt.$$

Letting $\eta$ approach zero we see that

$$\lim_{\eta \to 0+} I_\eta = \lim_{\eta \to 0+} \frac{1}{\pi} \int_0^R \frac{\alpha(t) \eta}{(t-\xi)^2 + \eta^2} dt - \lim_{\eta \to 0+} \frac{1}{\pi} \int_0^R \frac{\alpha(t) \eta}{t^2 + \eta^2} dt,$$

and by Lemmas 7.1 and 7.2 we have

$$\lim_{\eta \to 0+} I_\eta = \frac{\alpha(\xi+) + \alpha(\xi-)}{2} - \frac{\alpha(0+)}{2},$$

and the proof is completed by replacing $\alpha(t)$ by $\alpha(t) - \alpha(0)$.

We observe that the left-hand side of equation (2) can be written

$$\lim_{\eta \to 0+} \left\{ \frac{1}{2\pi i} \int_{-\xi-i\eta}^{-i\eta} f(s)\, ds + \frac{1}{2\pi i} \int_{i\eta}^{-\xi+i\eta} f(s)\, ds \right\},$$

so that we are integrating $f(s)$ along opposite sides of the negative real axis in opposite directions and at infinitesimal distance from the axis. In particular if $\alpha(t)$ is real then the conjugate of $f(-\sigma - i\eta)$ is $f(-\sigma + i\eta)$ and $f(-\sigma - i\eta) - f(-\sigma + i\eta)$ reduces to $2i$ multiplied by the imaginary part of $f(-\sigma - i\eta)$. We thus obtain the result of Stieltjes.

COROLLARY 7a. *If $\alpha(t)$ is real, then the real part of*

$$\frac{1}{\pi i} \int_{-\xi-i\eta}^{-i\eta} f(s)\, ds$$

*approaches*

$$\frac{\alpha(\xi+) + \alpha(\xi-)}{2} - \frac{\alpha(0+) + \alpha(0)}{2}$$

*as $\eta$ approaches zero through positive values.*

As an example take $f(s) = s^{-1}$ with $\alpha(t) = 1$ for $0 < t < \infty$ and $\alpha(0) = 0$. Our inversion formula (2) gives for $t > 0$

$$\alpha(t) - \frac{\alpha(0+)}{2} = \lim_{\eta \to 0+} \frac{1}{\pi} \tan^{-1} \frac{t}{\eta} = \frac{1}{2},$$

and this is the correct result.

We turn next to a similar inversion formula for the case when $\alpha(t)$ is an integral,

$$f(s) = \int_0^\infty \frac{\varphi(t)}{s+t}\, dt.$$

We prove

THEOREM 7b. *If $\varphi(t)$ belongs to $L$ in $(0 \leq t \leq R)$ for every positive $R$ and is such that the integral*

$$f(s) = \int_0^\infty \frac{\varphi(t)}{s+t}\, dt$$

*converges, then*

(3) $$\lim_{\eta \to 0+} \frac{f(-\xi - i\eta) - f(-\xi + i\eta)}{2\pi i} = \frac{\varphi(\xi+) + \varphi(\xi-)}{2}$$

*for any positive $\xi$ at which $\varphi(\xi+)$ and $\varphi(\xi-)$ exist.*

For, simple computation gives

(4) $$\frac{f(-\xi - i\eta) - f(-\xi + i\eta)}{2\pi i} = \frac{1}{\pi} \int_0^\infty \frac{\eta \varphi(t)}{(t-\xi)^2 + \eta^2}\, dt.$$

Now choose $R$ greater than $\xi$ and express the integral (4) as the sum of two integrals $I_1$ and $I_2$ corresponding to the intervals $(0, R)$ and $(R, \infty)$ respectively. Then by Lemma 7.2

$$\lim_{\eta \to 0+} I_1 = \frac{\varphi(\xi+) + \varphi(\xi-)}{2}.$$

The integral $I_2$ need not converge absolutely. On this account we introduce the function

$$\alpha(t) = \int_0^t \varphi(u) \, du \qquad (0 \leq t < \infty).$$

Then integration by parts gives

$$I_2 = \frac{-\eta \alpha(R)}{\pi[(R-\xi)^2 + \eta^2]} + \frac{2\eta}{\pi} \int_R^\infty \frac{\alpha(t)(t-\xi)}{[(t-\xi)^2 + \eta^2]^2} \, dt.$$

The first term on the right clearly tends to zero with $\eta$. Denote the second by $I_3$. By Theorem 3a there exists a constant $M$ such that

$$|\alpha(t)| < Mt \qquad (0 \leq t < \infty).$$

Hence

$$|I_3| < \frac{2\eta M}{\pi} \int_R^\infty \frac{t}{(t-\xi)^3} \, dt.$$

The dominant integral converges and $I_3$ approaches zero as $\eta$ approaches zero. Hence

$$\lim_{\eta \to 0+} I_2 = 0,$$

and our theorem is established.

We observe for later reference that the result may be written symbolically as

(5) $$\frac{f(xe^{-i\pi}) - f(xe^{i\pi})}{2\pi i} = \varphi(x).$$

This* of course is to be understood as equivalent to equation (3).

## 8. A Singular Integral

We wish to obtain next a real inversion operator involving the successive derivatives of $f(s)$ and analogous to that obtained in Section 6

---

* Compare E. C. Titchmarsh [1937] p. 318, equation (11.8.4). There our equation (3) is proved for the case in which $\varphi(t)$ belongs to $L^2$.

of Chapter VII. Preliminary to the derivation of such a formula we prove:

**THEOREM 8a.** *If*

1. $\varphi(x) \; \varepsilon \; L \; (0 < t \leqq x \leqq R)$ *for a fixed $t$ and every larger $R$;*

2. $\int_t^\infty x^c \varphi(x)\, dx$ *converges for a fixed real constant $c$;*

3. $\qquad \int_t^x [\varphi(u) - \varphi(t)]\, du = o(x - t) \qquad (x \to t+);$

*then*

(1) $\qquad \lim_{k \to \infty} \frac{(2k-1)!}{k!(k-2)!} \int_t^\infty \frac{t^{k-1} u^k}{(t+u)^{2k}} \varphi(u)\, du = \frac{\varphi(t)}{2}$

To save writing let us set

$$d_k = \frac{(2k-1)!}{k!(k-2)!}.$$

Let $\delta$ be a positive number, and set

$$I_k' = d_k \int_{t+\delta}^\infty \frac{t^{k-1} u^k}{(t+u)^{2k}} \varphi(u)\, du$$

$$I_k'' = d_k \int_t^{t+\delta} \frac{t^{k-1} u^k}{(t+u)^{2k}} \varphi(u)\, du.$$

Set

$$\alpha(x) = \int_t^x u^c \varphi(u)\, du \qquad (t \leqq x < \infty),$$

so that there exists a constant $M$ for which

$$|\alpha(x)| < M \qquad (t \leqq x < \infty).$$

Integration by parts gives

$$I_k' = -d_k t^{k-1} \frac{(t+\delta)^{k-c}}{(2t+\delta)^{2k}} \alpha(t+\delta) - d_k \int_{t+\delta}^\infty t^{k-1} \alpha(u)\, d\left\{ \frac{u^{k-c}}{(t+u)^{2k}} \right\}$$

provided $k > -c$. We observe that in practice $c$ will usually be a negative constant. The function $u^{k-c}(t+u)^{-2k}$ considered as a function of $u$ has its only maximum at $u = t(k-c)/(k+c)$, a number which is less than $t + \delta$ for $k$ sufficiently large, say for $k > k_0$. That is, the function is decreasing in $(t + \delta \leqq u < \infty)$ when $k > k_0$. Hence

$$|I_k'| \leqq M d_k t^{k-1} \frac{(t+\delta)^{k-c}}{(2t+\delta)^{2k}} + M d_k t^{k-1} \frac{(t+\delta)^{k-c}}{(2t+\delta)^{2k}} = l_k.$$

But
$$\frac{l_{k+1}}{l_k} = \frac{(2k+1)(2k)}{(k+1)(k-1)} \frac{t(t+\delta)}{(2t+\delta)^2} \to \frac{4t(t+\delta)}{(2t+\delta)^2} \qquad (k \to \infty).$$

This limit is clearly less than unity, so that $I'_k$ tends to zero as $k$ becomes infinite.

To $I''_k$ we apply Theorem 2b of Chapter VII, choosing
$$h(x) = \log x - 2\log(x+t),$$
$a = t$ and $b = t + \delta$. Then
$$h''(t) = \frac{-1}{2t^2},$$
so that
$$I''_k \sim \frac{(2k-1)!}{k!(k-2)!} \varphi(t) t^{k-1} \frac{t^k}{(2t)^{2k}} \left(\frac{\pi t^2}{k}\right)^{1/2} \qquad (k \to \infty).$$

By use of Stirling's formula this becomes
$$I''_k \sim \frac{\varphi(t)}{2}.$$

This completes the proof of the theorem. A result of a similar nature is contained in:

THEOREM 8b. *If*

1. $\varphi(x) \, \varepsilon \, L \; (0 < \epsilon \leq x \leq t)$ *for a fixed $t$ and every smaller positive $\epsilon$;*

2. $\int_{0+}^{t} x^c \varphi(x) \, dx$ *converges for a fixed real constant $c$;*

3. $\displaystyle\int_t^x [\varphi(u) - \varphi(t)] \, du = o(t - x) \qquad (x \to t-);$

*then*

(2) $$\lim_{k \to \infty} \frac{(2k-1)!}{k!(k-2)!} \int_{0+}^{t} \frac{t^{k-1} u^k}{(t+u)^{2k}} \varphi(u) \, du = \frac{\varphi(t)}{2}.$$

We prove first that the hypothesis implies
$$\int_{1/t}^{x} \left[\frac{\varphi(u^{-1})}{u^2} - t^2 \varphi(t)\right] du = o\left(x - \frac{1}{t}\right) \qquad \left(x \to \frac{1}{t}+\right),$$
or that
$$\lim_{x \to t-} \frac{xt}{x-t} \int_t^x \left[\varphi(u) - \frac{t^2}{u^2} \varphi(t)\right] du = 0.$$

By hypothesis 3. this will follow if

$$\lim_{x \to t-} \frac{1}{x-t} \int_t^x \left[ \varphi(t) - \frac{t^2}{u^2} \varphi(t) \right] du = 0.$$

This is seen to be true by performing the indicated integration.

After this preliminary remark, we may easily reduce our present problem to that of Theorem 8a. For by a change of variable we have

$$\int_{0+}^t \frac{t^{k-1} u^k}{(t+u)^{2k}} \varphi(u)\, du = \frac{1}{t^2} \int_{1/t}^\infty \frac{(1/t)^{k-1} u^k}{(u+t^{-1})^{2k}} \frac{\varphi(u^{-1})}{u^2} du.$$

To the latter integral we may apply Theorem 8a, replacing the function $\varphi(u)$ of that theorem by $\varphi(u^{-1})u^{-2}$. Using (1) we clearly obtain (2), and our result is established.

**THEOREM 8c.** *If*

1. $\varphi(x) \, \varepsilon \, L \, (R^{-1} \leq x \leq R)$ *for every* $R > 1$;

2. $\displaystyle\int_1^\infty \varphi(x) x^c\, dx$ *converges for a fixed real constant* $c$;

3. $\displaystyle\int_{0+}^1 \varphi(x) x^{c'}\, dx$ *converges for a fixed real constant* $c'$;

4. $\displaystyle\int_t^x [\varphi(u) - \varphi(t)]\, du = o(|x-t|) \qquad (x \to t);$

*then*

(3) $$\lim_{k \to \infty} \frac{(2k-1)!}{k!(k-2)!} \int_{0+}^\infty \frac{t^{k-1} u^k}{(t+u)^{2k}} \varphi(u)\, du = \varphi(t).$$

The proof is obtained by combining (1) and (2).

**COROLLARY 8c.1.** *If hypothesis 4. is replaced by the existence of* $\varphi(t+)$ *and* $\varphi(t-)$, *then*

$$\lim_{k \to \infty} \frac{(2k-1)!}{k!(k-2)!} \int_{0+}^\infty \frac{t^{k-1} u^k}{(t+u)^{2k}} \varphi(u)\, du = \frac{\varphi(t+) + \varphi(t-)}{2}.$$

For then hypothesis 3. of Theorem 8a holds with $\varphi(t)$ replaced by $\varphi(t+)/2$, and hypothesis 3. of Theorem 8b holds with $\varphi(t)$ replaced by $\varphi(t-)/2$.

**COROLLARY 8c.2.** *If hypothesis 4. is omitted, equation* (3) *holds for almost all positive* $t$.

For hypothesis 1. implies hypothesis 4. for almost all positive $t$.

## 9. The Inversion Operator for the Stieltjes Transform with $\alpha(t)$ an Integral

We now define the following differential operator, which will serve to invert the Stieltjes transform

DEFINITION 9. *An operator $L_{k,t}[f(x)]$ is defined for any real positive number $t$ by the equations*

$$L_{k,t}[f(x)] = \frac{(-t)^{k-1}}{k!(k-2)!} \frac{d^{2k-1}}{dt^{2k-1}} [t^k f(t)] \quad (k = 2, 3, \cdots),$$

$$L_{0,t}[f(x)] = f(t),$$

$$L_{1,t}[f(x)] = \frac{d}{dt}[tf(t)].$$

We assume, of course, that a function $f(x)$ to which we are applying the operator has derivatives of all orders less than $2k$. In practice the functions employed will have derivatives of all orders.

As an example take $f(x) = (x + a)^{-1}$ where $a > 0$. Then

$$L_{k,t}[f(x)] = \frac{(2k-1)!}{k!(k-2)!} \frac{t^{k-1} a^k}{(t+a)^{2k}} \quad (t > 0; k = 2, 3, \cdots).$$

THEOREM 9. *If $\varphi(t)$ belongs to $L$ in $0 \leq t \leq R$ for every positive $R$ and is such that the integral*

(1) $$f(x) = \int_0^\infty \frac{\varphi(t)}{x + t} dt$$

*converges, then*

(2) $$\lim_{k \to \infty} L_{k,t}[f(x)] = \varphi(t),$$

*at all points $t$ of the Lebesgue set for the function $\varphi(t)$.*

That is, we wish to prove (2) for all $t$ for which

(3) $$\int_t^x |\varphi(u) - \varphi(t)| du = o(|x - t|) \quad (x \to t).$$

Since

$$\frac{t^k}{t+u} = \frac{t^k - (-u)^k}{t+u} + \frac{(-u)^k}{t+u} = t^{k-1} - t^{k-2}u + \cdots \pm u^{k-1} + \frac{(-u)^k}{t+u},$$

we see that

$$L_{k,t}[f(x)] = \frac{(2k-1)!}{k!(k-2)!} \int_0^\infty \frac{t^{k-1} u^k}{(t+u)^{2k}} \varphi(u) du.$$

If $t$ is a point satisfying (3) we are in a position to apply Theorem 8c. We may take $c = -1$. For, since (1) converges we know that the function

$$\alpha(x) = \int_1^x \frac{\varphi(t)}{t+1} dt \qquad (1 \leq x < \infty)$$

approaches a limit as $x$ becomes infinite. Then

$$\int_1^\infty \frac{\varphi(t)}{t} dt = \int_1^\infty \frac{t+1}{t} d\alpha(t) = \alpha(\infty) + \int_1^\infty \frac{\alpha(t)}{t^2} dt,$$

provided the latter integral converges. But it clearly converges absolutely, so that it is sufficient to take $c = -1$. We may take $c' = 0$ since $\varphi(t) \, \varepsilon \, L$ in $(0 \leq t \leq 1)$. Hypothesis 4. follows *a fortiori* from (3). Then §8 (3) gives §9 (2), and our theorem is proved.

COROLLARY 9.1. *Equation (2) holds for almost all positive t.*
COROLLARY 9.2. *Equation (2) holds for all t where $\varphi(u)$ is continuous.*
COROLLARY 9.3. *At all points t in a neighborhood of which $\varphi(u)$ is of bounded variation.*

$$\lim_{k \to \infty} L_{k,t}[f(x)] = \frac{\varphi(t+) + \varphi(t-)}{2}.$$

As an example of the theorem take $f(x) = x^{-\delta}$ where $\delta$ is a positive number less than unity. By use of the familiar formula

$$(4) \qquad \int_0^\infty \frac{x^{m-1}}{(1+x)^{m+n}} dx = \frac{\Gamma(m)\Gamma(n)}{\Gamma(m+n)}$$

we see that

$$x^{-\delta} = \frac{1}{\Gamma(1-\delta)\Gamma(\delta)} \int_0^\infty \frac{t^{-\delta}}{x+t} dt.$$

Simple computation gives

$$L_{k,t}[x^{-\delta}] = \frac{\Gamma(k-\delta+1)\Gamma(k+\delta-1)}{\Gamma(k+1)\Gamma(k-1)} \frac{t^{-\delta}}{\Gamma(1-\delta)\Gamma(\delta)}$$

But

$$\frac{\Gamma(k+\alpha)}{\Gamma(k)} \sim k^\alpha \qquad (\alpha > 0, k \to \infty),$$

so that we have

$$\lim_{k \to \infty} L_{k,t}[x^{-\delta}] = \frac{t^{-\delta}}{\Gamma(1-\delta)\Gamma(\delta)},$$

as predicted by the theorem.

## 10. The Inversion Operator for the Stieltjes Transform in the General Case

We now obtain an inversion formula for the general convergent integral

$$(1) \qquad f(x) = \int_0^\infty \frac{d\alpha(t)}{x+t}.$$

We shall need the following preliminary result.

**LEMMA 10a.** *If* (1) *converges, then*

$$\lim_{x \to 0+} \frac{(2k-1)!}{k!(k-2)!} x^k \int_0^\infty \frac{t^{k-1}}{(x+t)^{2k}} \alpha(t)\, dt = \alpha(0+) \frac{k-1}{k} \qquad (k = 2, 3, \cdots).$$

For, given an arbitrary positive $\epsilon$, we determine $\delta$ such that

$$|\alpha(t) - \alpha(0+)| < \epsilon \qquad (0 \leq t \leq \delta).$$

Then defining $d_k$ as in Section 8 we have

$$|I(x)| = \left| d_k x^k \int_0^\infty \frac{t^{k-1}[\alpha(t) - \alpha(0+)]}{(x+t)^{2k}}\, dt \right|$$

$$\leq \epsilon d_k x^k \int_0^\delta \frac{t^{k-1}}{(x+t)^{2k}}\, dt + \left| d_k x^k \int_\delta^\infty \frac{t^{k-1}}{(x+t)^{2k}} [\alpha(t) - \alpha(0+)]\, dt \right|.$$

Since the last term is not greater than

$$d_k x^k \int_\delta^\infty \frac{|\alpha(t) - \alpha(0+)|}{t^{k+1}}\, dt,$$

the integral converging by §3 (3), it tends to zero with $x$. Moreover

$$d_k x^k \int_0^\delta \frac{t^{k-1}}{(x+t)^{2k}}\, dt < d_k x^k \int_0^\infty \frac{t^{k-1}}{(x+t)^{2k}}\, dt,$$

and by use of formula §9 (4)

$$d_k x^k \int_0^\infty \frac{t^{k-1}}{(x+t)^{2k}}\, dt = \frac{k-1}{k} < 1.$$

Hence

$$\overline{\lim_{x \to 0+}} |I(x)| \leq \epsilon,$$

so that

$$\lim_{x \to 0+} d_k x^k \int_0^\infty \frac{t^{k-1} \alpha(t)}{(x+t)^{2k}}\, dt = \lim_{x \to 0+} \alpha(0+) d_k x^k \int_0^\infty \frac{t^{k-1}}{(x+t)^{2k}}\, dt$$
$$= \alpha(0+) \frac{k-1}{k}.$$

THEOREM 10a. *If $\alpha(t)$ is a normalized function of bounded variation in $0 \leq t \leq R$ for every positive $R$, and if the integral*

(1) $$f(x) = \int_0^\infty \frac{d\alpha(t)}{x+t}$$

*converges, then*

(2) $$\lim_{k \to \infty} \int_0^t L_{k,u}[f(x)]\, du = \alpha(t) - \alpha(0+).$$

For, the computations of the previous section show that

$$L_{k,u}[f(x)] = d_k \int_0^\infty \frac{u^{k-1} y^k}{(u+y)^{2k}}\, d\alpha(y) \quad (k = 2, 3, \cdots).$$

Since $\alpha(0) = 0$ and $\alpha(y) = o(y)$ as $y$ becomes infinite, we have after integration by parts

(3) $$L_{k,u}[f(x)] = -d_k \int_0^\infty \alpha(y) \frac{\partial}{\partial y}\left[\frac{u^{k-1} y^k}{(u+y)^{2k}}\right] dy.$$

But by Euler's theorem regarding homogeneous functions

$$\frac{\partial}{\partial u}\left[\frac{u^k y^{k-1}}{(u+y)^{2k}}\right] = -\frac{\partial}{\partial y}\left[\frac{u^{k-1} y^k}{(u+y)^{2k}}\right].$$

Hence

$$L_{k,u}[f(x)] = d_k \frac{\partial}{\partial u} \int_0^\infty \frac{u^k y^{k-1}}{(u+y)^{2k}} \alpha(y)\, dy.$$

Consequently if $0 < \epsilon < t$, we have

(4) $$\int_\epsilon^t L_{k,u}[f(x)]\, du = d_k \int_0^\infty \frac{t^k y^{k-1}}{(t+y)^{2k}} \alpha(y)\, dy - d_k \int_0^\infty \frac{\epsilon^k y^{k-1}}{(\epsilon+y)^{2k}} \alpha(y)\, dy.$$

Allowing $\epsilon$ to approach zero and making use of Lemma 10a we have

(5) $$\int_{0+}^t L_{k,u}[f(x)]\, du = d_k \int_0^\infty \frac{t^k y^{k-1}}{(t+y)^{2k}} \alpha(y)\, dy - \alpha(0+)\frac{k-1}{k}.$$

To the integral on the right we now apply Corollary 8c.1, taking $\varphi(u) = \alpha(u)/u$. We thus obtain

$$\lim_{k \to \infty} \int_{0+}^t L_{k,u}[f(x)]\, du = \alpha(t) - \alpha(0+).$$

Set $f(x) = f_1(x) + f_2(x)$, where $f_1(x)$ is the integral (1) from 0 to 1. By a change in order of integration

$$\int_0^1 |L_{k,u}[f_1(x)]|\, du \leq \frac{k-1}{k} \int_0^1 |d\alpha(t)|.$$

By §3 (3) there is an upper bound $M$ for $|\alpha(y)|y^{-1}$ in $(1 \leq y < \infty)$, so that

$$L_{k,u}[f_2(x)] = -\alpha(1)d_k \frac{u^k}{(1+u)^{2k}} + kd_k \int_1^\infty \alpha(y)(y-u) \frac{u^{k-1}y^{k-1}}{(u+y)^{2k+1}} dy$$

$$|L_{k,u}[f_2(x)]| \leq |\alpha(1)|d_k + kd_k M \int_1^\infty y^{-k} dy \qquad (0 \leq u < 1).$$

Hence the integral of $|L_{k,u}[f]|$ from 0 to 1 exists, so that

$$\int_{0+}^t L_{k,u}[f(x)] du = \int_0^t L_{k,u}[f(x)] du,$$

and our theorem is proved.

COROLLARY 10a.1. *If $\alpha(t)$ is continuous in $a \leq t \leq b$, then (2) holds uniformly in $a' \leq t \leq b'$, where $a < a' < b' < b$ if $a > 0$ and $a' = 0$ if $a = 0$.*

To prove this one would use the fact that

$$\alpha(ty) - \alpha(t) = o(1) \qquad\qquad (y \to 1)$$

uniformly for $t$ in $(a' \leq t \leq b')$.

COROLLARY 10a.2. *A function $f(x)$ cannot have two different representations in the form (1).*

For, if it had the representation (1) and also the representation

$$f(x) = \int_0^\infty \frac{d\beta(t)}{x+t},$$

where $\alpha(t)$ and $\beta(t)$ are both normalized, then we should have

$$0 = \int_0^\infty \frac{d[\alpha(t) - \beta(t)]}{x+t}.$$

Then by Theorem 3c

$$\alpha(0+) - \beta(0+) = 0$$

and by Theorem 10a

$$\alpha(t) - \beta(t) = 0 \qquad (0 < t < \infty).$$

We thus have a new proof of Theorem 5b.

COROLLARY 10a.3. *If $\alpha(\infty)$ exists, then*

$$\lim_{k \to \infty} \int_0^\infty L_{k,t}[f(x)] dt = \alpha(\infty) - \alpha(0+).$$

For, we have from Lemma 10a after replacing $x$ by $x^{-1}$ and $t$ by $t^{-1}$

$$\lim_{x\to\infty} \frac{(2k-1)!}{k!(k-2)!} x^k \int_0^\infty \frac{t^{k-1}}{(x+t)^{2k}} \alpha\left(\frac{1}{t}\right) dt = \alpha(0+)\frac{k-1}{k},$$

or

$$\lim_{x\to\infty} \frac{(2k-1)!}{k!(k-2)!} x^k \int_0^\infty \frac{t^{k-1}}{(x+t)^{2k}} \alpha(t)\, dt = \alpha(\infty)\frac{k-1}{k}.$$

If we allow $t$ to become infinite in equation (5) we obtain

$$\int_0^\infty L_{k,u}[f(x)]\, du = \frac{k-1}{k}[\alpha(\infty) - \alpha(0+)].$$

Now allowing $k$ to become infinite we obtain the desired result.

We now give another equivalent inversion formula for the integral (1).

**THEOREM 10b.** *Under the conditions of Theorem 10a*

$$\lim_{k\to\infty} \frac{(-1)^{k-1}}{k!(k-2)!}[t^{2k-1}f^{(k-1)}(t)]^{(k-1)} = \alpha(t) \quad (0 < t < \infty).$$

For, by use of the equation

$$f(x) = \int_0^\infty \frac{\alpha(y)}{(x+y)^2}\, dy$$

we have by direct computation

(6) $$\frac{(-1)^{k-1}}{k!(k-2)!}[t^{2k-1}f^{(k-1)}(t)]^{(k-1)} = \frac{(2k-1)!}{k!(k-2)!}\int_0^\infty \frac{t^k y^{k-1}}{(t+y)^{2k}} \alpha(y)\, dy.$$

But this integral is precisely the integral on the right-hand side of (5), and we have seen that it tends to $\alpha(t)$ as $k$ becomes infinite.

Comparing equations (5) and (6) we are led to conjecture the following result.

**THEOREM 10c.** *If $f(x)$ has a derivative of order $(2k-1)$, then*

$$L_{k,t}[f(x)] = \frac{(-1)^{k-1}}{k!(k-2)!}[t^{2k-1}f^{(k-1)}(t)]^{(k)}.$$

This is certainly true if $f(x)$ has the representation (1). For then (5) and (6) give

$$\int_0^t L_{k,u}[f(x)]\, du = \frac{(-1)^{k-1}}{k!(k-2)!}[t^{2k-1}f^{(k-1)}(t)]^{(k-1)} - \alpha(0+)\frac{k-1}{k}.$$

On differentiating both sides of this equation with respect to $t$ we obtain the desired result. However, the result is true for any function having the requisite number of derivatives. Evidently we must show that

(7) $$x^{k-1}[x^k f(x)]^{(2k-1)} = [x^{2k-1}f^{(k-1)}(x)]^{(k)}.$$

The proof consists merely in computing both sides of the equation by Leibniz's rule. In each case we obtain

$$\sum_{p=0}^{k} \frac{(2k-1)!k!}{(2k-p-1)!p!(k-p)!} f^{(2k-p-1)}(x) x^{2k-p-1}.$$

Or we could prove the result without computation by observing that both sides of (7) are zero if $f(x)$ is any one of the $(2k-1)$ linearly independent functions

$$x^n \qquad (n = -k, -k+1, \cdots, k-2).$$

The coefficient of $f^{(2k-1)}(x)$ in the expanded form of each side of (7) is clearly $x^{2k-1}$

To illustrate Theorems 10a and 10b take $f(x) = x^{-1}$. As we have just observed, $L_{k,t}[x^{-1}] = 0$. On the other hand $\alpha(t) = 1$ for $0 < t < \infty$, $\alpha(0) = 0$. Since

$$\lim_{k \to \infty} \int_0^t L_{k,u}[x^{-1}] du = 0 = \alpha(t) - \alpha(0+),$$

Theorem 10a is verified in this special case. For Theorem 10b we have

$$\frac{(-1)^{k-1}}{k!(k-2)!} [t^{2k-1} f^{(k-1)}(t)]^{(k-1)} = \frac{k-1}{k},$$

and this tends to unity for all positive $t$ as $k$ becomes infinite.

## 11. The Jump Operator

We now define the operator which serves to compute the saltus of $\alpha(t)$ at a given point in terms of $f(x)$.

DEFINITION 11. *An operator $l_{k,t}[f(x)]$ is defined by the equation*

$$l_{k,t}[f(x)] = 2t\pi^{1/2} k^{-1/2} L_{k,t}[f(x)].$$

For example, if $l_{k,t}[f(x)] = (1 + x)^{-1}$,

$$l_{k,t}[f(x)] = 2\pi^{1/2} k^{-1/2} d_k t^k (t+1)^{-2k},$$

where $d_k$ is the constant defined in Section 8.

THEOREM 11. *If the integral*

$$f(x) = \int_0^\infty \frac{d\alpha(t)}{x+t}$$

*converges, then*

(1) $$\lim_{k \to \infty} l_{k,t}[f(x)] = \alpha(t+) - \alpha(t-) \qquad (0 < t < \infty).$$

For, direct computation gives

$$l_{k,t}[f(x)] = 2t^k \left(\frac{\pi}{k}\right)^{1/2} d_k \int_0^\infty \frac{u^k}{(u+t)^{2k}} d\alpha(u).$$

By the change of variable $u = vt$, this becomes (assuming $\alpha(0) = 0$)

$$l_{k,t}[f(x)] = 2\left(\frac{\pi}{k}\right)^{1/2} d_k \int_0^\infty \frac{v^k}{(v+1)^{2k}} d\alpha(vt)$$

$$= -2\left(\frac{\pi}{k}\right)^{1/2} d_k \int_0^\infty \alpha(vt) \frac{d}{dv} \frac{v^k}{(v+1)^{2k}} dv.$$

Let $\delta$ be a positive number less than unity, and break the latter integral into the sum of four others: $I_1(k)$, $I_2(k)$, $I_3(k)$, and $I_4(k)$, corresponding to the intervals $(0, 1 - \delta)$, $(1 - \delta, 1)$, $(1, 1 + \delta)$, and $(1 + \delta, \infty)$, respectively. To integrals $I_2(k)$ and $I_3(k)$ we apply Theorems 8a and 8b of Chapter VII. For example

$$I_3(k) = -2\left(\frac{\pi}{k}\right)^{1/2} d_k \int_1^{1+\delta} \alpha(vt) k \frac{v^{k-1}(1-v)}{(v+1)^{2k+1}} dv.$$

Taking the function $h(x)$ of Theorem 8a equal to $\log[x(x+1)^{-2}]$ and $\varphi(x)$ equal to $\alpha(xt)x^{-1}(x+1)^{-1}$ we obtain

$$\int_1^{1+\delta} (v-1) \frac{v^k}{(v+1)^{2k}} \frac{\alpha(vt)}{v(v+1)} dv \sim \frac{\alpha(t+)}{2k} \frac{2}{2^{2k}} \quad (k \to \infty),$$

since $h''(1) = -\frac{1}{2}$. As in the proof of Theorem 9 we see by use of Stirling's formula that

(2) $$d_k \sim 2^{2k-1} k^{1/2} \pi^{-1/2} \quad (k \to \infty).$$

Combining these results we have

(3) $$\lim_{k \to \infty} I_3(k) = \alpha(t+).$$

In a similar way

(4) $$\lim_{k \to \infty} I_2(k) = -\alpha(t-).$$

Since $v^k(v+1)^{-2k}$ is increasing and $\alpha(vt)$ is bounded in $0 \leq v \leq 1 - \delta$, we have by use of (2)

$$|I_1(k)| \leq A 2^{2k-1} \int_0^{1-\delta} d\left[\frac{v^k}{(v+1)^{2k}}\right] = A 2^{2k-1} \frac{(1-\delta)^k}{(2-\delta)^{2k}},$$

where $A$ is a suitable positive constant. Since

$$\frac{4(1-\delta)}{(2-\delta)^2} < 1,$$

it is clear that $I_1(k)$ tends to zero with $1/k$. By §3 (3) we have for a suitable positive constant $B$,

$$|I_4(k)| \leqq -B2^{2k-1}\int_{1+\delta}^{\infty} v\, d\left[\frac{v^k}{(v+1)^{2k}}\right]$$

$$= B2^{2k-1}\frac{(1+\delta)^{k+1}}{(2+\delta)^{2k}} + B2^{2k-1}\int_{1+\delta}^{\infty}\frac{v^k}{(v+1)^{2k}}\, dv.$$

It is easily seen that each term on the right approaches zero with $1/k$ since

$$\frac{4(1+\delta)}{(2+\delta)^2} < 1.$$

That is, $I_4(k)$ also tends to zero. Hence taking account of (3) and (4) we have (1).

To illustrate the theorem take $f(x) = (x+1)^{-1}$, so that $\alpha(t) = 1$ for $t > 1$ and $\alpha(t) = 0$ for $t < 1$. Then by use of (2) we have

$$l_{k,t}[(x+1)^{-1}] \sim \frac{(4t)^k}{(t+1)^{2k}} \qquad (k \to \infty).$$

The right-hand side clearly approaches zero with $1/k$ for any positive $t$ not unity since

$$4t < (t+1)^2 \qquad (t \neq 1);$$

but for $t = 1$ it is equal to unity for all $k$. Since $\alpha(t)$ is continuous except at $t = 1$ where it has a unit jump, the theorem is verified in this special case.

## 12. The Variation of $\alpha(t)$

We now wish to obtain a formula which will express the variation of $\alpha(t)$ in terms of $f(x)$ and its derivatives. The result is expressed in the following theorem.

THEOREM 12. *If $\alpha(t)$ has variation $V(R)$ in the interval $0 \leqq t \leqq R$ and if the integral*

$$f(x) = \int_0^{\infty} \frac{d\alpha(t)}{x+t}$$

*converges then*

$$\lim_{k \to \infty} \int_0^R |L_{k,t}[f(x)]|\, dt = V(R) - V(0+).$$

It is sufficient to prove the theorem with $\alpha(0+) = \alpha(0) = V(0+) = 0$. Let $S$ and $R$ be any two positive numbers, $R < S$. Set

$$f_1(x) = \int_0^S \frac{d\alpha(t)}{x+t}$$

$$f_2(x) = \int_S^\infty \frac{d\alpha(t)}{x+t}.$$

Then

$$\int_0^R |L_{k,t}[f(x)]|\,dt \leq \int_0^R |L_{k,t}[f_1(x)]|\,dt + \int_0^R |L_{k,t}[f_2(x)]|\,dt$$

$$= I_1(k) + I_2(k).$$

We see at once that

$$I_1(k) \leq \int_0^R L_{k,t}\left[\int_0^S \frac{dV(u)}{x+u}\right] dt.$$

By Theorem 10a the right-hand side of this inequality approaches $V(R)$ as $k$ becomes infinite, so that

$$\varlimsup_{k\to\infty} I_1(k) \leq V(R).$$

On the other hand by equation §10 (3)

$$L_{k,u}[f_2(x)] = -d_k \int_S^\infty [\alpha(y) - \alpha(S)] \frac{\partial}{\partial y}\left[\frac{u^{k-1} y^k}{(u+y)^{2k}}\right] dy.$$

If $0 \leq u \leq R$, the maximum of the function $y^k(u+y)^{-2k}$, considered as a function of $y$ occurs outside the interval $S \leq y < \infty$, and its derivative with respect to $y$ is negative throughout that interval. Hence

$$|L_{k,u}[f_2(x)]| \leq -d_k \int_S^\infty |\alpha(y) - \alpha(S)| \frac{\partial}{\partial y}\left[\frac{u^{k-1} y^k}{(u+y)^{2k}}\right] dy.$$

Then proceeding as in Section 10 we have

$$I_2(k) \leq d_k \int_S^\infty |\alpha(y) - \alpha(S)| \frac{R^k y^{k-1}}{(R+y)^{2k}}\,dy.$$

But since $R < S$, the right-hand member clearly tends to zero as $k$ becomes infinite. Hence

(1) $$\varlimsup_{k\to\infty} \int_0^R |L_{k,t}[f(x)]|\,dt \leq V(R).$$

Now set

$$\alpha_k(t) = \int_0^t L_{k,u}[f(x)]\,du \qquad (0 \leq t < \infty).$$

If $t_0, t_1, \cdots, t_n$ form a subdivision of $(0 \leq t \leq R)$ we have

$$\sum_{i=0}^{n-1} |\alpha_k(t_{i+1}) - \alpha_k(t_i)| \leq \int_0^R |L_{k,u}[f(x)]| \, du,$$

and allowing $k$ to become infinite and using Theorem 10a

$$\sum_{i=0}^{n-1} |\alpha(t_{i+1}) - \alpha(t_i)| \leq \lim_{k \to \infty} \int_0^R |L_{k,u}[f(x)]| \, du.$$

The left-hand side can be brought as near to $V(R)$ as desired by choice of the points $t_i$. Hence

(2) $$V(R) \leq \lim_{k \to \infty} \int_0^R |L_{k,u}[f(x)]| \, du.$$

Combining (1) and (2), the theorem is established.

COROLLARY 12. *If $V(\infty) < \infty$, then*

$$V(\infty) - V(0+) = \lim_{k \to \infty} \int_0^\infty |L_{k,t}[f(x)]| \, dt.$$

The proof is the same as that of Corollary 10, Chapter VII, and is omitted.

## 13. A General Representation Theorem

Following the analogy with Section 11 of the previous chapter we should expect that for a very large class of functions $f(x)$ we should have

$$\lim_{k \to \infty} \int_0^\infty \frac{L_{k,t}[f(x)]}{x+t} \, dt = f(x).$$

This is in fact the case. It is sufficient that $f(x)$ should have derivatives of all orders and that these derivatives behave in a specified way at $x = 0$ and at $x = \infty$.

THEOREM 13. *If $f(x)$ has derivatives of all orders in $0 < x < \infty$ which satisfy the conditions*

(1) $\qquad f^{(k)}(x) = o(x^{-k-1}) \qquad (x \to 0+, k = 0, 1, 2, \cdots),$

$\qquad\qquad\quad = o(x^{-k}) \qquad\quad (x \to \infty, k = 0, 1, 2, \cdots),$

*then*

$$\lim_{k \to \infty} \int_{0+}^\infty \frac{L_{k,t}[f(x)]}{x+t} \, dt = f(x) \qquad (0 < x < \infty).$$

Set $g(t) = (x+t)^{-1}$. Then for a fixed positive $x$

(2)   $g^{(k)}(t) = O(1)$    $(t \to 0+, k = 0, 1, \cdots)$
            $= O(t^{-k-1})$   $(t \to \infty, k = 0, 1, \cdots)$.

Consequently

(3)   $\int_{0+}^{\infty} [t^{2k-1}f^{(k-1)}(t)]^{(k)} g(t)\, dt = (-1)^k \int_{0}^{\infty} t^{2k-1}f^{(k-1)}(t) g^{(k)}(t)\, dt,$

if either integral exists. For, when we integrate by parts the integrated part is of the form

$[t^{2k-1}f^{(k-1)}(t)]^{(k-p)} g^{(p-1)}(t)$

$= g^{(p-1)}(t) \sum_{r=0}^{k-p} a_r t^{2k-r-1} f^{(2k-p-r-1)}(t) \quad (p = 1, 2, \cdots, k),$

where the $a_r$ are constants. By (1) each term in the summation is $o(t^{p-1})$ as $t \to 0+$ and is $o(t^p)$ as $t \to \infty$. Hence by (2)

$[t^{2k-1}f^{(k-1)}(t)]^{(k-p)} g^{(p-1)}(t) = o(1) \qquad (t \to 0+, t \to \infty).$

This establishes (3) on the assumption that either integral exists. Now integrate the right-hand member of (3) by parts. The integrated part has the form

$[t^{2k-1}g^{(k)}(t)]^{(p)} f^{(k-2-p)}(t)$

$= f^{(k-2-p)}(t) \sum_{r=0}^{p} b_r t^{2k-r-1} g^{(k+p-r)}(t) \quad (p = 0, 1, \cdots, k-2),$

where the $b_r$ are constants. Again using (1) and (2) we see that

$[t^{2k-1}g^{(k)}(t)]^{(p)} f^{(k-2-p)}(t) = o(1) \qquad (t \to \infty)$
$= o(t^k) \qquad (t \to 0+).$

Hence

$\int_0^{\infty} t^{2k-1} f^{(k-1)}(t) g^{(k)}(t)\, dt = (-1)^{k-1} \int_0^{\infty} [t^{2k-1} g^{(k)}(t)]^{(k-1)} f(t)\, dt$

if either integral exists. But the integral on the right-hand side is

$-(2k-1)! \int_0^{\infty} \frac{t^k x^{k-1}}{(x+t)^{2k}} f(t)\, dt,$

and it is clear from (1) that this integral converges absolutely for large $k$. Hence

(4)   $\int_{0+}^{\infty} \frac{L_{k,t}[f(x)]}{x+t}\, dt = \frac{(2k-1)!}{k!(k-2)!} \int_0^{\infty} \frac{x^{k-1} t^k}{(x+t)^{2k}} f(t)\, dt.$

By Theorem 8c the right-hand side of this equation tends to $f(x)$ as $k$ becomes infinite, so that the proof is complete.

COROLLARY 13. *Under the hypotheses of Theorem* 13

$$-f'(x) = \lim_{k \to \infty} \int_{0+}^{\infty} \frac{L_{k,t}[f(x)]}{(x+t)^2} dt.$$

For, take $g(x) = (x+t)^{-2}$. Then relations (2) are satisfied *a fortiori* so that

$$\int_{0+}^{\infty} [t^{2k-1} f^{(k-1)}(t)]^{(k)} g(t)\, dt = (-1)^{(k)} \int_{0}^{\infty} [t^{2k-1} g^{(k)}(t)]^{(k-2)} f'(t)\, dt.$$

This equation is obtained from the above calculations by one less integration by parts. But by Euler's theorem employed earlier

$$\frac{\partial}{\partial t} \frac{t^{2k-1}}{x^{k-2}(x+t)^{k+2}} = -\frac{\partial}{\partial x} \frac{t^{2k-2}}{x^{k-3}(x+t)^{k+2}}$$

$$\frac{\partial^{k-2}}{\partial t^{k-2}} \frac{t^{2k-1}}{x^{k-2}(x+t)^{k+2}} = (-1)^{k-2} \frac{\partial^{k-2}}{\partial x^{k-2}} \frac{t^{k+1}}{(x+t)^{k+2}}$$

$$= \frac{(2k-1)!}{(k+1)!} \frac{t^{k+1}}{(x+t)^{2k}}.$$

Hence

$$\int_{0+}^{\infty} \frac{L_{k,t}[f(x)]}{(x+t)^2} dt = -\frac{(2k-1)!}{k!(k-2)!} \int_{0}^{\infty} \frac{x^{k-2} t^{k+1}}{(x+t)^{2k}} f'(t)\, dt.$$

By Theorem 8c the right-hand integral approaches $-f'(x)$ as $k$ becomes infinite, and the corollary is established.

## 14. Order Conditions

It will sometimes be convenient to replace condition 13 (1) by others of somewhat different type. We introduce the latter in the present section. We assume throughout that $f(x)$ is a function which has derivatives of all orders in the interval $(0 < x < \infty)$.

LEMMA 14. *If for a fixed positive integer* $k$

(1) $$f^{(k-1)}(t) = O(t^{-k}) \qquad (t \to 0+),$$

*then*

$$\int_{0+}^{t} u^{2k+1} f^{(k)}(u)\, du = O(t^{k+1}) \qquad (t \to 0+).$$

For, if $0 < \epsilon < t$, integration by parts gives

$$\int_\epsilon^t u^{2k+1} f^{(k)}(u)\, du = t^{2k+1} f^{(k-1)}(t) - \epsilon^{2k+1} f^{(k-1)}(\epsilon).$$

$$- (2k+1) \int_\epsilon^t u^{2k} f^{(k-1)}(u)\, du.$$

By (1) the second term on the right approaches zero with $\epsilon$ and the integral

$$\int_0^t u^{2k} f^{(k-1)}(u)\, du.$$

exists. Hence

$$\int_{0+}^t u^{2k+1} f^{(k)}(u)\, du = t^{2k+1} f^{(k-1)}(t) - (2k+1) \int_0^t u^{2k} f^{(k-1)}(u)\, du.$$

Again using (1) we see that both terms on the right are $O(t^{k+1})$ as $t$ approaches zero, so that the desired result is proved.

THEOREM 14a. *If the integrals*

(2) $$\int_{0+}^1 [u^{2k-1} f^{(k-1)}(u)]^{(k)}\, du \qquad (k = 1, 2, \cdots)$$

*all exist, then there is a constant $A$ such that*

$$(-1)^k f^{(k)}(x) \sim \frac{A k!}{x^{k+1}} \qquad (x \to 0+;\, k = 0, 1, 2, \cdots).$$

Taking $k = 1$ in (2) we see that as $x$ approaches zero the function $xf(x)$ approaches a limit which we denote by $A$. In particular $f(x) = O(x^{-1})$ as $x$ approaches zero. We proceed by induction and assume that

(3) $\qquad f^{(p)}(x) = O(x^{-p-1}) \qquad (x \to 0+;\, p = 0, 1, \cdots, k-1)$

and seek to prove (3) for $p = k$. By (2) with $k$ replaced by $k+1$ it is clear that

$$[x^{2k+1} f^{(k)}(x)]^{(k)} = O(1) \qquad (x \to 0+).$$

Hence

$$\int_0^x [u^{2k+1} f^{(k)}(u)]^{(k)}\, du = O(x) \qquad (x \to 0+).$$

Also $[x^{2k+1} f^{(k)}(x)]^{(k-1)}$ approaches a limit $c_1$ as $x$ approaches zero. Hence

$$[x^{2k+1} f^{(k)}(x)]^{(k-1)} - c_1 = O(x) \qquad (x \to 0+).$$

By successive integrations we obtain

$$\int_0^x u^{2k+1} f^{(k)}(u)\, du = P_k(x) + O(x^{k+1}) \qquad (x \to 0+),$$

where $P_k(x)$ is a polynomial of degree $k$ at most. By Lemma 14 this integral is itself $O(x^{k+1})$ so that

$$P_k(x) = O(x^{k+1}) \qquad (x \to 0+).$$

But this is impossible unless $P_k(x)$ is identically zero. But in the course of the above integrations we had at one stage

$$x^{2k+1} f^{(k)}(x) = P_k'(x) + O(x^k) \qquad (x \to 0+).$$

Hence

$$x^{2k+1} f^{(k)}(x) = O(x^k)$$

$$f^{(k)}(x) = O(x^{-k-1}) \qquad (x \to 0+),$$

and the induction is complete. By Theorem 4.4 of Chapter V we now have

$$(-1)^k f^{(k)}(x) \sim \frac{Ak!}{x^{k+1}} \qquad (x \to 0+),$$

and the theorem is established.

This result enables us to prove:

THEOREM 14b. *If*

(4) $\qquad \int_{0+}^x L_{k,t}[f(x)]\, dt = O(x) \qquad (x \to \infty; k = 1, 2, \cdots),$

*then*

(5) $\qquad (-1)^k f^{(k)}(x) \sim \dfrac{Ak!}{x^{k+1}} \qquad (x \to 0+; k = 0, 1, 2, \cdots)$

(6) $\qquad f^{(k)}(x) = O(x^{-k}) \qquad (x \to \infty; k = 0, 1, 2, \cdots),$

*where*

$$A = \lim_{x \to 0+} xf(x).$$

Hypothesis (4) implies the existence of the integrals (2) so that (5) follows from Theorem 14a. On the other hand the relation

$$\int_{0+}^x [t^{2k-1} f^{(k-1)}(t)]^{(k)}\, dt = O(x) \qquad (x \to \infty)$$

implies that

(7) $$[x^{2k-1}f^{(k-1)}(x)]^{(k-1)} = O(x) \quad (x \to \infty)$$
$$x^{2k-1}f^{(k-1)}(x) = O(x^k) \quad (x \to \infty),$$

from which (6) is evident.

## 15. General Representation Theorems

By use of the results of the previous section we can now modify the form of Theorem 13 as follows.

**THEOREM 15.** *If*

(1) $$\int_{0+}^{x} L_{k,t}[f(x)] \, dt = O(x) \quad (x \to \infty \, ; k = 1, 2, \cdots),$$

*and if* $f(\infty) = 0$, *then*

(2) $$f(x) = \lim_{k \to \infty} \int_{0+}^{\infty} \frac{L_{k,t}[f(x)]}{x+t} \, dt + \frac{A}{x} \quad (x > 0),$$

*where*

$$A = \lim_{x \to 0+} xf(x).$$

The existence of the constant $A$ is assured by Theorem 14b. Set

$$g(x) = f(x) - \frac{A}{x}.$$

Then by Theorem 14b and the hypothesis $f(\infty) = 0$ we have

$$g^{(k)}(x) = o(x^{-k-1}) \quad (x \to 0+ \, ; k = 0, 1, \cdots)$$
$$= o(x^{-k}) \quad (x \to \infty \, ; \; k = 0, 1, \cdots).$$

Consequently we may apply Theorem 13 to $g(x)$. Since

$$L_{k,t}[f(x)] = L_{k,t}[g(x)],$$

we have (2) at once.

**COROLLARY 15.1.** *If*

(3) $$\underset{0 \leq t < \infty}{\text{u.b.}} |L_{k,t}[f(x)]| < \infty \quad (k = 1, 2, \cdots),$$

*and if* $f(\infty) = 0$, *then* (2) *holds.*

For, it is clear that (3) implies (1).

**COROLLARY 15.2.** *If for some number* $p \geq 1$

(4) $$\int_{0}^{\infty} |L_{k,t}[f(x)]|^p \, dt < \infty \quad (k = 1, 2, \cdots),$$

*then* (2) *holds.*

For, if $p = 1$, then
$$\int_0^\infty |[tf(t)]'| \, dt < \infty$$
so that $f(\infty) = 0$. Moreover for $p = 1$, relation (4) clearly implies (1). If $p > 1$ we have by Hölder's inequality
$$\int_0^x |L_{k,t}[f(x)]| \, dt \leq \left[ \int_0^\infty |L_{k,t}[f(x)]|^p \, dt \right]^{1/p} x^{(p-1)/p}.$$
This implies (1). In particular for $k = 1$ it gives
$$xf(x) = O(x^{(p-1)/p}) \qquad (x \to \infty),$$
whence $f(\infty) = 0$. Hence the result is established.

### 16. The Function $\alpha(t)$ of Bounded Variation

We are now in a position to study what functions are Stieltjes transforms. We treat first the case in which $\alpha(t)$ is of bounded variation in $(0, \infty)$. We introduce:

DEFINITION 16. *A function $f(x)$ satisfies Conditions A if it has derivatives of all orders in $(0 < x < \infty)$ and if there exists a constant $M$ such that*

(1) $$\int_0^\infty |L_{k,t}[f(x)]| \, dt < M \qquad (k = 1, 2, \cdots).$$

We observe that (1) is equivalent to
$$\int_0^\infty t^{k-1} |[t^k f(t)]^{(2k-1)}| \, dt < Mk!(k-2)! \qquad (k = 1, 2, \cdots),$$
or to
$$\int_0^\infty |[t^{2k-1} f^{(k-1)}(t)]^{(k)}| \, dt < Mk!(k-2)! \qquad (k = 1, 2, \cdots).$$

As an example, the function $f(x) = (x + 1)^{-1}$ satisfies Conditions $A$. For, in this case the left-hand side of (1) becomes $(k - 1)/k$ and $M$ may be taken equal to unity.

THEOREM 16. *A necessary and sufficient condition that*

(2) $$f(x) = \int_0^\infty \frac{d\alpha(t)}{x + t}$$

*with $\alpha(t)$ of bounded variation in $(0, \infty)$ is that $f(x)$ should satisfy Conditions $A$.*

We showed earlier that if (2) converges then
$$L_{k,t}[f(x)] = \frac{(2k-1)!}{k!(k-2)!} \int_0^\infty \frac{u^k t^{k-1}}{(t+u)^{2k}} \, d\alpha(u) \qquad (k = 1, 2, \cdots).$$

Hence

$$\int_0^\infty |L_{k,t}[f(x)]|\,dt \leq \frac{(2k-1)!}{k!(k-2)!} \int_0^\infty u^k |d\alpha(u)| \int_0^\infty \frac{t^{k-1}}{(t+u)^{2k}}\,dt$$

provided the iterated integral on the right exists. It clearly does under our hypotheses and has the value

$$\frac{k-1}{k} \int_0^\infty |d\alpha(u)|.$$

Hence we may take the total variation of $\alpha(t)$ as the constant $M$ of our theorem.

Conversely, if (1) holds then Corollary 15.2 with $p = 1$ gives

$$f(x) = \lim_{k \to \infty} \int_0^\infty \frac{L_{k,t}[f(x)]}{x+t}\,dt + \frac{A}{x} \qquad (x > 0)$$

$$A = \lim_{x \to 0+} xf(x).$$

Each function

(3) $$\alpha_k(t) = \int_0^t L_{k,u}[f(x)]\,du \qquad (k = 1, 2, \cdots)$$

clearly has total variation less than $M$ by (1). We may consequently apply Theorem 16.3 of Chapter I and obtain a sequence $\alpha_{k_i}(t)$ selected from the sequence (3) which approaches a function $\alpha(t)$ of bounded variation in $(0, \infty)$. Hence

$$f(x) = \lim_{i \to \infty} \int_0^\infty \frac{d\alpha_{k_i}(t)}{x+t} + \frac{A}{x} \qquad (x > 0).$$

Integrating by parts and applying Lebesgue's limit theorem we have after a second integration by parts

$$f(x) \doteq \int_0^\infty \frac{d\alpha(t)}{x+t} + \frac{A}{x}.$$

Since the term $A/x$ is itself an integral (2) with $\alpha(t)$ a step-function discontinuous only at the origin, our result is established.

COROLLARY 16. *If $f(x)$ satisfies Conditions A, then*

(4) $$\lim_{k \to \infty} \int_0^\infty |L_{k,t}[f(x)]|\,dt$$

*exists.*

For, by Theorem 16 equation (2) holds with the total variation of $\alpha(t)$ finite. Then by Corollary 12 we obtain the existence of the limit (4).

## 17. The Function $\alpha(t)$ Non-decreasing and Bounded

For $\alpha(t)$ to be non-decreasing the basic condition is that $L_{k,t}[f(x)] \geq 0$ as one would expect from analogy with the previous chapter. For $\alpha(t)$ to be also bounded we need a further condition on the behavior of $f(x)$ at infinity. We prove first two preliminary theorems.*

THEOREM 17a. *If*

(1) $\quad (-1)^{k-1}[x^{2k-1}f^{(k-1)}(x)]^{(k)} \geq 0 \qquad (k = 1, 2, \cdots; 0 < x < \infty),$

*and if*

(2) $\qquad\qquad\qquad \lim_{x \to \infty} xf(x) = B,$

*then*

(3) $\qquad\qquad f^{(k)}(x) \sim (-1)^k \dfrac{Bk!}{x^{k+1}} \quad (x \to \infty\, ; k = 0, 1, 2, \cdots).$

It is evident that inequalities (1) are equivalent to

$$L_{k,t}[f(x)] \geq 0 \qquad (k = 1, 2, \cdots\,; 0 < t < \infty).$$

Integrating (1) from 1 to $x$ we have

$$(-1)^{k-1}[x^{2k-1}f^{(k-1)}(x)]^{(k-1)} \geq O(1) \qquad (x \to \infty).$$

Repeating the operation sufficiently often gives us

$$(-1)^{k-1}[x^{2k-1}f^{(k-1)}(x)] \geq O(x^{k-1})$$

(4) $\qquad\qquad (-1)^{k-1}f^{(k-1)}(x) \geq O(x^{-k}) \qquad (x \to \infty).$

Take $k = 3$ in (4). This condition with (2) enables us to apply Corollary 4.4a of Chapter V. The conclusion is (3) with $k = 1$. Then step by step we obtain (3) generally.

THEOREM 17b. *If $f(x)$ is non-negative for positive $x$ and if*

(5) $\quad (-1)^{k-1}[x^k f(x)]^{(2k-1)} \geq 0 \qquad (0 < x < \infty\,; k = 1, 2, \cdots),$

*then $xf(x)$ has a limit $A$ as $x$ approaches zero and*

(6) $\qquad\qquad f(x) \sim (-1)^k \dfrac{Ak!}{x^{k+1}} \qquad (x \to 0+\,; k = 0, 1, 2, \cdots).$

Inequalities (5) are clearly equivalent to (1). By (5) with $k = 1$ we see that $xf(x)$ is non-decreasing. Since it is also non-negative the existence of $A$ is assured. It is no restriction to suppose that $A$ is zero.

---
* These two theorems are special cases of a much more general theorem of R. P. Boas [1937], Theorem 2, p. 643.

For, clearly the function $g(x) = f(x) - Ax^{-1}$ also satisfies (5). If we can show that

(7) $\qquad g^{(k)}(x) = o(x^{-k-1}) \qquad (x \to 0+; k = 0, 1, 2, \cdots),$

it will follow that (6) also holds.

We use induction. Suppose that (7) with $g(x)$ replaced by $f(x)$ holds for $k = 0, 1, 2, \cdots, p - 1$. Then

(8) $\qquad [x^{p+1}f(x)]^{(p-1)} = o(x) \qquad (x \to 0+).$

By (5)

$$(-1)^p[x^{p+1}f(x)]^{(2p+1)} \geqq 0 \qquad (0 < x < \infty).$$

That is, $(-1)^p[x^{p+1}f(x)]^{(2p)}$ is non-decreasing. It is consequently bounded above in $(0 < x \leqq 1)$, or $(-1)^{p-1}[x^{p+1}f(x)]^{(2p)}$ is bounded below. Repetition of this reasoning shows that $[x^{p+1}f(x)]^{(p+1)}$ is bounded below. That is,

$$[x^{p+1}f(x)]^{(p+1)} \geqq O(1) \qquad (x \to 0+).$$

Then *a fortiori*

(9) $\qquad [x^{p+1}f(x)]^{(p+1)} \geqq O(x^{-1}) \qquad (x \to 0+).$

Using (8) and (9) we may apply Theorem 4.4 of Chapter V and obtain

$$[x^{p+1}f(x)]^{(p)} = o(1) \qquad (x \to 0+).$$

If we expand by Leibniz's rule we obtain by use of (7) with $k < p$

$$x^{p+1}f^{(p)}(x) = o(1) \qquad (x \to 0+),$$

and the induction is complete.

We are now in a position to prove the main result of the section.

**THEOREM 17c.** *Necessary and sufficient conditions that*

(10) $\qquad f(x) = \int_0^\infty \dfrac{d\alpha(t)}{x + t}$

*with $\alpha(t)$ non-decreasing and bounded are that*

$$f(x) \geqq 0$$

(11) $\qquad (-1)^{k-1}[x^k f(x)]^{(2k-1)} \geqq 0 \qquad (0 < x < \infty; k = 1, 2, \cdots)$

*and that $xf(x)$ should approach a limit $B$ as $x$ becomes infinite.*

We prove first the necessity of these conditions. It is clear from (10) that $f(x)$ cannot be negative for positive $x$ and that if $\alpha(0) = 0$

$$B = \lim_{x \to \infty} \int_0^\infty \dfrac{x\,d\alpha(t)}{x + t} = \alpha(\infty).$$

Moreover,

$$(-1)^{k-1}[x^k f(x)]^{(2k-1)} = (2k-1)! \int_0^\infty \frac{t^k}{(x+t)^{2k}} d\alpha(t) \quad (k = 1, 2, \cdots),$$

so that (11) is also evident.

To prove the converse we employ Theorem 16. By (1) Conditions $A$ become

(12) $\qquad \dfrac{(-1)^{k-1}}{k!(k-2)!} \displaystyle\int_0^\infty [t^{2k-1} f^{(k-1)}(t)]^{(k)} dt < M \qquad (k = 1, 2, \cdots).$

But by use of Theorems 17a and 17b it is a simple matter to compute this integral. The left-hand side of (12) is seen to be

$$\frac{k-1}{k} [B - A],$$

where $A$ and $B$ are the constants defined in Theorems 17a and 17b. Taking $M = B - A$, we see that Conditions $A$ are satisfied. Hence $f(x)$ has representations (10) with $\alpha(t)$ of bounded variation in $(0, \infty)$. By Theorem 10a

$$\alpha(t) - \alpha(0+) = \lim_{k \to \infty} \int_0^t L_{k,u}[f(x)] \, du,$$

so that $\alpha(t)$ is the limit of non-decreasing functions and hence is itself non-decreasing. Finally, $\alpha(t)$ is bounded since it is of finite total variation. In fact

$$\alpha(\infty) - \alpha(0+) = \lim_{k \to \infty} \int_0^\infty L_{k,t}[f(x)] \, dt,$$

as we saw in Corollary 16.

## 18. The Function $\alpha(t)$ Non-decreasing and Unbounded

We wish to treat next the case in which $\alpha(t)$ may increase without limit. We show first that if $L_{k,t}[f(x)]$ is non-negative for all $k$ and $t$ then $f(x)$ necessarily approaches a limit as $x$ becomes infinite.

THEOREM 18a. *If*

$$(-1)^{k-1}[x^{2k-1} f^{(k-1)}(x)]^{(k)} \geq 0 \qquad (k = 1, 2, \cdots ; 0 < x < \infty),$$

*then $f(\infty)$ exists.*

Our hypothesis is the same as §17 (1). Since §17 (4) was proved without use of §17 (2) we see that §17 (4) holds under our present hypotheses. In particular

(1) $\qquad\qquad\qquad f'(x) < \dfrac{M}{x^2} \qquad\qquad (1 \leq x < \infty)$

for some constant $M$. If $1 < x < y$, we have by integrating (1)

$$f(y) - f(x) < \frac{M}{x} - \frac{M}{y}$$

and

$$\overline{\lim_{y \to \infty}} f(y) \leq f(x) + \frac{M}{x}.$$

Now letting $x$ become infinite

$$\overline{\lim_{y \to \infty}} f(y) \leq \underline{\lim_{x \to \infty}} f(x)$$

so that $f(\infty)$ either exists or is $-\infty$. The latter case is impossible since §17 (4) for $k = 1$ gives

$$f(x) \geq O\left(\frac{1}{x}\right) \qquad (x \to \infty),$$

whence

$$\underline{\lim_{x \to \infty}} f(x) \geq 0$$

This completes the proof of the theorem.

**THEOREM 18b.** *The conditions*

(2) $$f(x) \geq 0$$
(3) $$(-1)^{k-1}[x^k f(x)]^{(2k-1)} \geq 0 \qquad (0 < x < \infty ; k = 1, 2, \cdots)$$

*are necessary and sufficient that $f(x)$ should have the form*

$$f(x) = P + \int_0^\infty \frac{d\alpha(t)}{x+t},$$

*where $\alpha(t)$ is non-decreasing and $P$ is a non-negative constant.*

The necessity of the conditions is evident.
To prove the sufficiency we observe that by Theorem 17b there exists a constant $A$ such that

(4) $$f^{(k)}(x) \sim (-1)^k \frac{Ak!}{x^{k+1}} \qquad (x \to 0+ ; k = 0, 1, 2, \cdots).$$

By Theorem 18a we see that $f(\infty)$ exists. Set it equal to $E$. By §17 (4) and Theorem 4.4 of Chapter V we have

$$f^{(k)}(x) = o(x^{-k}) \qquad (x \to \infty ; k = 1, 2, \cdots).$$

Thus it is clear that the function

$$g(x) = f(x) - \frac{A}{x} - E$$

satisfies all the conditions of Theorem 13, so that

$$\lim_{k \to \infty} \int_0^\infty \frac{L_{k,t}[g(x)]}{x+t} \, dt = g(x) \qquad (0 < x < \infty),$$

or

(5) $$\lim_{k \to \infty} \int_0^\infty \frac{L_{k,t}[f(x)]}{x+t} \, dt = f(x) - \frac{A}{x} - E.$$

Now set

$$\beta_k(t) = \int_0^t \frac{L_{k,u}[f(x)]}{u+1} \, du.$$

By (2) and (3) the integrand is non-negative so that

$$\beta_k(t) \leq \int_0^\infty \frac{L_{k,u}[f(x)]}{u+1} \, du.$$

This integral converges by Theorem 13. But by §13 (4)

(6) $$\beta_k(t) \leq \frac{(2k-1)!}{k!(k-2)!} \int_0^\infty \frac{u^k}{(u+1)^{2k}} f(u) \, du.$$

By (4) the function $uf(u)$ is bounded in the neighborhood of the origin. And since $f(\infty)$ exists, $f(u)$ is bounded in the neighborhood of infinity. Hence there exists a positive constant $M$ such that

$$f(u) \leq M(u^{-1} + 1) \qquad (0 < u < \infty),$$

and making use of this in (6) we have

$$\beta_k(t) \leq M\left(1 + \frac{k-1}{k}\right) < 2M \qquad (0 \leq t < \infty).$$

Since the $\beta_k(t)$ form a uniformly bounded sequence of non-decreasing functions, we may pick a subsequence, $\beta_{k_i}(t)$, which approaches a non-decreasing function $\beta(t)$ by Helly's theorem.

Now consider the integral

(7) $$\int_0^\infty \frac{t+1}{x+t} \, d\beta_{k_i}(t) = \int_0^\infty \frac{L_{k_i,t}[f(x)]}{x+t} \, dt.$$

By (5) it approaches $f(x) - Ax^{-1} - E$ as $i$ becomes infinite. On the other hand, if we integrate by parts we have

$$\int_0^\infty \frac{t+1}{x+t} d\beta_{k_i}(t) = \int_0^\infty \frac{L_{k_i,u}[f(x)]}{u+1} du + (1-x) \int_0^\infty \frac{\beta_{k_i}(t)}{(x+t)^2} dt.$$

Since the functions $\beta_{k_i}(t)$ are uniformly bounded we may take the limit under the sign and obtain

(8)
$$f(x) - \frac{A}{x} - E = f(1) + (1-x) \int_0^\infty \frac{\beta(t)}{(x+t)^2} dt$$
$$= f(1) - \int_0^\infty \beta(t) \, d\left(\frac{t+1}{x+t}\right).$$

Now set

$$\alpha(t) = \int_0^t (u+1) \, d\beta(u) \qquad (0 \leq t < \infty),$$

so that $\alpha(t)$ is non-decreasing. Integrating (8) by parts gives

$$f(x) - \frac{A}{x} - E = f(1) - \beta(\infty) + \int_0^\infty \frac{t+1}{x+t} d\beta(t)$$
$$= f(1) - \beta(\infty) + \int_0^\infty \frac{d\alpha(t)}{x+t}.$$

Since $\beta_{k_i}(t)$ is a non-decreasing function it is clear that $\beta(\infty) \leq f(1)$. By (2) the constants $A$ and $E$ cannot be negative so that

$$f(x) = P + \frac{A}{x} + \int_0^\infty \frac{d\alpha(t)}{x+t},$$

where

$$P = f(1) - \beta(\infty) + E \geq 0.$$

By adding a step-function with a positive jump $A$ to $\alpha(t)$ the term $Ax^{-1}$ may be absorbed into the integral and our theorem is established.

## 19. The Class $L^p$, $p > 1$

We next treat the case of representation in the form

$$f(x) = \int_0^\infty \frac{\varphi(t)}{x+t} dt,$$

where $\varphi(t)$ is a function of $L^p$ $(p > 1)$ in $(0, \infty)$. That is,

$$\int_0^\infty |\varphi(t)|^p \, dt < \infty.$$

DEFINITION 19. *A function $f(x)$ satisfies Conditions B if it has derivatives of all orders in $(0 < x < \infty)$, if*

(1) $$\lim_{x \to 0+} xf(x) = 0,$$

*and if there exist constants $M$ and $p$ ($p > 1$) such that*

(2) $$\int_0^\infty |L_{k,t}[f(x)]|^p \, dt < M \qquad (k = 1, 2, \cdots).$$

The relations (2) are equivalent to

$$c_k^p \int_0^\infty t^{kp-p} |[t^k f(t)]^{(2k-1)}|^p \, dt < M \qquad (k = 1, 2, \cdots)$$

or to

$$c_{k'}^p \int_0^\infty |[t^{2k-1} f^{(k-1)}(t)]^{(k)}|^p \, dt < M \qquad (k = 1, 2, \cdots),$$

where

$$c_k = \frac{1}{k!(k-2)!} \qquad (k = 2, 3, \cdots)$$

$$c_1 = 1.$$

As an example, the function $\log(1 + x^{-1})$ satisfies Conditions $B$ for any $p$ greater than unity. This could be shown directly by computing the integral (2), but the computations are somewhat long. The fact will follow from

THEOREM 19a. *Conditions B are necessary and sufficient that*

(3) $$f(x) = \int_0^\infty \frac{\varphi(t)}{x+t} \, dt,$$

*where*

(4) $$\int_0^\infty |\varphi(t)|^p \, dt < \infty.$$

We first prove the necessity of the conditions. We assume then that relations (3) and (4) are true. But

$$\lim_{x \to 0+} xf(x) = \alpha(0+),$$

where

$$\alpha(t) = \int_0^t \varphi(u) \, du.$$

That is, we have established (1). Furthermore, direct computation gives

$$L_{k,t}[f(x)] = d_k t^{k-1} \int_0^\infty \frac{u^k}{(t+u)^{2k}} \varphi(u)\, du \qquad (k = 1, 2, \cdots),$$

where

$$d_k = (2k-1)!c_k.$$

These integrals converge absolutely since by Hölder's inequality

$$\int_0^\infty \frac{u^k}{(t+u)^{2k}} |\varphi(u)|\, du \leq \left[\int_0^\infty |\varphi(u)|^p\right]^{1/p} \left[\frac{\Gamma(qk+1)\Gamma(qk-1)}{t^{qk-1}\Gamma(2qk)}\right]^{1/q},$$

where

$$\frac{1}{p} + \frac{1}{q} = 1.$$

Furthermore, for $k = 2, 3, \cdots$, we have by Hölder's inequality

(5) $\quad |L_{k,t}[f(x)]|^p \leq d_k t^{k-1} \int_0^\infty \frac{u^k}{(t+u)^{2k}} |\varphi(u)|^p du \left[d_k \int_0^\infty \frac{t^{k-1} u^k}{(t+u)^{2k}} du\right]^{p/q},$

$$\int_0^\infty |L_{k,t}[f(x)]|^p\, dt \leq d_k \int_0^\infty t^{k-1}\, dt \int_0^\infty \frac{u^k}{(t+u)^{2k}} |\varphi(u)|^p\, du.$$

(6) $\quad \int_0^\infty |L_{k,t}[f(x)]|^p\, dt = \frac{k-1}{k} \int_0^\infty |\varphi(u)|^p\, du \leq \int_0^\infty |\varphi(u)|^p\, du.$

For $k = 1$ the above argument fails since in that case the second integral on the right-hand side of (5) diverges. In this case we make use of an inequality of Hilbert.* It states that if $K(t, u)$ is positive and homogeneous of degree $-1$, then

$$\int_0^\infty dt \left[\int_0^\infty K(t,u) |\varphi(u)|\, du\right]^p \leq \left[\int_0^\infty K(t,1) t^{(1-p)/p}\, dt\right]^p \int_0^\infty |\varphi(u)|^p\, du$$

if the integrals on the right exist. We have

$$|L_{1,t}[f(x)]| \leq \int_0^\infty \frac{u|\varphi(u)|}{(t+u)^2}\, du.$$

Take

$$K(t, u) = \frac{u}{(t+u)^2}.$$

* See, for example, G. H. Hardy, J. E. Littlewood and G. Pólya [1934] p. 229, Theorem 319.

Set
$$r = \int_0^\infty K(t, 1)t^{(1-p)/p} dt = \int_0^\infty \frac{t^{(1-p)/p}}{(t + 1)^2} dt = \frac{\Gamma\left(\frac{1}{p}\right)\Gamma\left(2 - \frac{1}{p}\right)}{\Gamma(2)}.$$

Then
$$\int_0^\infty |L_{1,t}[f(x)]|^p dt \leq r^p \int_0^\infty |\varphi(u)|^p du.$$

If $M$ is taken as the integral (4) or $r^p$ times that integral, whichever is larger, (2) clearly holds.

Conversely, if Conditions B hold we have

(7)
$$\int_0^x |L_{k,t}[f(x)]| dt \leq \left[\int_0^x |L_{k,t}[f(x)]|^p dt\right]^{1/p} x^{1/q}$$
$$\leq M^{1/p} x^{1/q}$$
$$\int_0^x L_{k,t}[f(x)] dt = O(x) \qquad (x \to \infty; k = 1, 2, \cdots).$$

Then by (1) and Theorem 14b

(8) $\qquad f^{(k)}(x) = o(x^{-k-1}) \qquad (x \to 0+; k = 0, 1, \cdots),$

(9) $\qquad f^{(k)}(x) = O(x^{-k}) \qquad (x \to \infty; k = 0, 1, \cdots).$

By (7) with $k = 1$ we have
$$\int_0^x [tf(t)]' dt = O(x^{1/q}) \qquad (x \to \infty),$$
whence
$$f(x) = O(x^{-1/p}) \qquad (x \to \infty),$$
$$f(\infty) = 0.$$

Hence (9) implies
$$f^{(k)}(x) = o(x^{-k}) \qquad (x \to \infty; k = 0, 1, \cdots).$$

By use of (8) and (9) we are now able to apply Theorem 13 to obtain

(10) $\qquad f(x) = \lim_{k \to \infty} \int_0^\infty \frac{L_{k,t}[f(x)]}{x + t} dt.$

Inequality (2) enables us to apply Theorem 17a of Chapter I. There exists by virtue of that theorem a set of positive integers $k_1, k_2, \cdots$ and a function $\varphi(t)$ of class $L^p$ in $(0, \infty)$ such that
$$\lim_{i \to \infty} \int_0^\infty \frac{L_{k_i,t}[f(x)]}{x + t} dt = \int_0^\infty \frac{\varphi(t)}{x + t} dt,$$

or by (10)

$$f(x) = \int_0^\infty \frac{\varphi(t)}{x+t} dt.$$

This completes the proof of the theorem.

We now see easily that $\log(1 + x^{-1})$ satisfies Conditions B. For,

$$\log\left(1 + \frac{1}{x}\right) = \int_0^1 \frac{dt}{x+t},$$

so that the corresponding function $\varphi(t)$ belongs to $L^p$ for any $p$.

THEOREM 19b. *If $f(x)$ has the representation* (3) (4), *then*

(11)  $$\lim_{k \to \infty} \int_0^\infty |L_{k,t}[f(x)]|^p dt = \int_0^\infty |\varphi(t)|^p dt.$$

For, by (6)

(12)  $$\varlimsup_{k \to \infty} \int_0^\infty |L_{k,t}[f(x)]|^p dt \leq \int_0^\infty |\varphi(t)|^p dt.$$

By Fatou's lemma*

(13)  $$\int_0^\infty |\varphi(t)|^p dt \leq \varliminf_{k \to \infty} \int_0^\infty |L_{k,t}[f(x)]|^p dt.$$

Or, we may use Theorem 17a of Chapter I. Inequalities (12) and (13) imply equation (11).

## 20. The Function $\varphi(t)$ Bounded

We now consider representation in the form §19 (3) with $\varphi(t)$ bounded.

DEFINITION 20. *A function $f(x)$ satisfies Conditions C if it has derivatives of all orders in $(0 < x < \infty)$, if*

(1)  $$\lim_{x \to 0} xf(x) = 0$$

(2)  $$\lim_{x \to \infty} f(x) = 0,$$

*and if there exists a constant $M$ such that*

(3)  $|L_{k,t}[f(x)]| < M$     $(0 < t < \infty ; k = 2, 3, \cdots).$

We shall prove that Conditions $C$ are necessary and sufficient for the representation in question. But first we wish to call attention to the contrast between them and Conditions $C$ of Chapter VII. The main difference is that $k = 1$ is excluded from the inequalities (3). Equation (2) might be thought of as replacing it. To see the reason for this

* See, for example, E. C. Titchmarsh [1932] p. 346.

change we look at what becomes of Conditions $B$ as $p$ becomes infinite. We had §19 (6) for $k = 2, 3, \cdots$ and

(4) $$\int_0^\infty |L_{1,t}[f(x)]|^p \, dt \leqq r^p \int_0^\infty |\varphi(u)|^p \, du$$

$$r = \Gamma\left(\frac{1}{p}\right)\Gamma\left(2 - \frac{1}{p}\right) \Big/ \Gamma(2).$$

If we allow $p$ to become infinite in §19 (6) we do indeed obtain (3) where $M$ is an upper bound of $|\varphi(u)|$. But this fails in (4) since $r$ clearly becomes infinite with $p$.

THEOREM 20a. *Conditions C are necessary and sufficient that*

(5) $$f(x) = \int_0^\infty \frac{\varphi(t)}{x+t} \, dt,$$

*with $\varphi(t)$ bounded in $(0, \infty)$.*

The conditions are necessary. For, if $|\varphi(t)| < M$ for $(0 < t < \infty)$, then

(6) $|L_{k,t}[f(x)]| \leqq \dfrac{(2k-1)!}{k!(k-2)!} \displaystyle\int_0^\infty \dfrac{u^k t^{k-1}}{(u+t)^{2k}} |\varphi(u)| \, du \quad (k = 2, 3, \cdots)$

$\qquad\qquad < M \qquad\qquad (0 < t < \infty\,;\, k = 2, 3, \cdots),$

so that (3) holds. Also

$$\lim_{x \to 0+} \int_0^\infty \frac{x\varphi(t)}{x+t} \, dt = \lim_{u \to 0+} \int_0^u \varphi(t) \, dt = 0$$

$$\lim_{x \to \infty} \int_0^\infty \frac{\varphi(t)}{x+t} \, dt = 0,$$

so that (1) and (2) also hold.

Conversely, (3) clearly implies

(7) $$\int_0^x L_{k,t}[f(x)] \, dt = O(x) \qquad (x \to \infty\,;\, k = 2, 3, \cdots).$$

From (1) and (2) we have for $k = 1$

$$\int_0^x L_{1,t}[f(x)] \, dt = \int_0^x [tf(t)]' \, dt = xf(x) = o(x) \qquad (x \to \infty),$$

so that (7) also holds for $k = 1$. Then by Theorem 14b, using (1) and (2) we have

$$f^{(k)}(x) = o(x^{-k-1}) \quad (x \to 0+\,;\, k = 0, 1, \cdots)$$
$$\qquad\quad = o(x^{-k}) \qquad (x \to \infty\,;\, k = 0, 1, \cdots).$$

Then by Corollary 13
$$\lim_{k \to \infty} \int_0^\infty \frac{L_{k,t}[f(x)]}{(x+t)^2} dt = -f'(x).$$

Furthermore, by Theorem 17b of Chapter I, we see that (3) guarantees the existence of a subset $k_i$ of the positive integers and a bounded function $\varphi(t)$ such that

$$-f'(x) = \lim_{i \to \infty} \int_0^\infty \frac{L_{k_i,t}[f(x)]}{(x+t)^2} dt = \int_0^\infty \frac{\varphi(t)}{(x+t)^2} dt.$$

Now let $0 < x < y$ and integrate this equation from $x$ to $y$. Since the integral on the right is uniformly convergent for $x \geqq \delta > 0$ we have

$$f(x) - f(y) = (y - x) \int_0^\infty \frac{\varphi(t)}{(x+t)(y+t)} dt.$$

Since $f(\infty) = 0$ by hypothesis, we have for any fixed $x$

$$\int_0^\infty \frac{\varphi(t)}{(x+t)(y+t)} dt \sim \frac{f(x)}{y} \qquad (y \to \infty).$$

The function $\varphi(t)$ is bounded so that for any fixed $x$

$$\frac{\varphi(t)}{x+t} = O\left(\frac{1}{t}\right) \qquad (t \to \infty).$$

Consequently we are in a position to apply Theorem 5a of Chapter V. The conclusion is that

$$f(x) = \int_0^\infty \frac{\varphi(t)}{x+t} dt,$$

and our result is established.

THEOREM 20b. *If $f(x)$ has the representation (5) with $\varphi(t)$ bounded, then*

$$\lim_{k \to \infty} \text{u.b.} \ |L_{k,t}[f(x)]| = \text{true max} \ |\varphi(t)|.$$
$$\phantom{\lim_{k \to \infty}} 0 < t < \infty \phantom{\ |L_{k,t}[f(x)]| = \text{true max} \ } 0 \leqq t < \infty$$

The proof is similar to that of Theorem 16b, Chapter VII.

### 21. The Class $L$

If $p$ is set equal to unity in Conditions $B$ they become essentially Conditions $A$. Hence it is evident that Theorem 19a is no longer valid when $p = 1$. To discuss representation in the form §19 (3) with $\varphi(t)$ of class $L$ we introduce:

DEFINITION 21. *A function $f(x)$ satisfies Conditions D if it has derivatives of all orders in $(0 < x < \infty)$, and if*

(1) $$\int_0^\infty |L_{k,t}[f(x)]|\,dt < \infty \qquad (k = 1, 2, \cdots),$$

(2) $$\lim_{k,l\to\infty} \int_0^\infty |L_{k,t}[f(x)] - L_{l,t}[f(x)]|\,dt = 0$$

(3) $$\lim_{x\to 0+} xf(x) = 0.$$

The essential meaning of these conditions is that $L_{k,t}[f(x)]$ converges in the mean (exponent unity) as $k$ becomes infinite.

THEOREM 21. *Conditions D are necessary and sufficient that*

(4) $$f(x) = \int_0^\infty \frac{\varphi(t)}{x+t}\,dt,$$

*where*

(5) $$\int_0^\infty |\varphi(t)|\,dt < \infty$$

For the necessity of the conditions, assume (4) and (5). As we have seen several times before

$$\lim_{x\to 0+} \int_0^\infty \frac{x\varphi(t)}{x+t}\,dt = \lim_{x\to 0+} \int_0^x \varphi(t)\,dt = 0,$$

so that (3) is established. Also

$$\int_0^\infty |L_{k,t}[f(x)]|\,dt \leq d_k \int_0^\infty dt \int_0^\infty \frac{t^{k-1}u^k}{(t+u)^{2k}}|\varphi(u)|\,du \qquad (k = 1, 2, \cdots)$$

where $d_1 = 1$ and

$$d_k = \frac{(2k-1)!}{k!(k-2)!} \qquad (k = 2, 3, \cdots).$$

Then by Fubini's theorem

$$\int_0^\infty |L_{k,t}[f(x)]|\,dt \leq d_k \int_0^\infty u^k |\varphi(u)|\,du \int_0^\infty \frac{t^{k-1}}{(t+u)^{2k}}\,dt$$

$$\leq \int_0^\infty |\varphi(u)|\,du \qquad (k = 1, 2, \cdots),$$

so that (1) is proved. Next we have

$$|L_{k,t}[f(x)] - \varphi(t)| \leq d_k \int_0^\infty \frac{t^{k-1}u^k}{(t+u)^{2k}} |\varphi(u) - \varphi(t)|\,du$$

$$\leq d_k \int_0^\infty \frac{u^k}{(u+1)^{2k}} |\varphi(tu) - \varphi(t)|\,du.$$

Hence

(6) $$\int_0^\infty |L_{k,t}[f(x)] - \varphi(t)|\, dt \leq d_k \int_0^\infty \frac{u^k}{(u+1)^{2k}} g(u)\, du,$$

where

$$g(u) = \int_0^\infty |\varphi(tu) - \varphi(t)|\, dt.$$

But we saw in Lemma 17 of Chapter VII that $g(u)$ tends to zero as $u$ approaches unity and that there exists a constant $M$ for which

$$|g(u)| < M(u^{-1} + 1) \qquad (0 < u < \infty).$$

Hence by Theorem 8c the right-hand side of (6) tends to zero with $1/k$. Therefore $L_{k,t}[f(x)]$ converges in the mean to $\varphi(t)$ on $(0, \infty)$, so that (2) is established.

Conversely, the assumption (2) implies the existence of a function $\varphi(t)$ of class $L$ such that

(7) $$\lim_{k \to \infty} \int_0^\infty |L_{k,t}[f(x)] - \varphi(t)|\, dt = 0$$

(8) $$\lim_{k \to \infty} \int_0^\infty |L_{k,t}[f(x)]|\, dt = \int_0^\infty |\varphi(t)|\, dt.$$

As in Section 17, Chapter VII, we see by use of Theorem 16 that

$$f(x) = \int_0^\infty \frac{d\alpha(t)}{x+t},$$

where $\alpha(t)$ is a normalized function of bounded variation in $(0, \infty)$. But

$$\left| \int_0^u L_{k,t}[f(x)]\, dt - \int_0^u \varphi(t)\, dt \right| \leq \int_0^\infty |L_{k,t}[f(x)] - \varphi(t)|\, dt$$

$$(0 < u < \infty; k = 1, 2, \cdots).$$

or by (7)

$$\lim_{k \to \infty} \int_0^u L_{k,t}[f(x)]\, dt = \int_0^u \varphi(t)\, dt.$$

But by Theorem 10a

$$\lim_{k \to \infty} \int_0^u L_{k,t}[f(x)]\, dt = \alpha(u) - \alpha(0+).$$

By (3) we see that $\alpha(0+) = 0$, so that
$$\alpha(u) = \int_0^u \varphi(t)\, dt,$$
from which (4) follows at once.

COROLLARY 21. *If $f(x)$ has the representation* (4) (5) *then*
$$\lim_{k \to \infty} \int_0^\infty |L_{k,t}[f(x)]|\, dt = \int_0^\infty |\varphi(t)|\, dt.$$

This follows by Fatou's lemma as in Section 19 or directly from (8)

## 22. The Function $\alpha(t)$ of Bounded Variation in Every Finite Interval

We wish to conclude our discussion of representation theory with an extension of Theorem 16 to the case in which $\alpha(t)$ is no longer restricted to be of bounded variation in the infinite interval. We introduce a new differential operator corresponding to the kernel $(x + t)^{-2}$.

DEFINITION 22. *The operators $M_{k,t}[f(x)]$ are defined by the equations*
$$M_{1,t}[f(x)] = tf(t)$$
$$M_{k,t}[f(x)] = \frac{(-1)^{k-1}}{k!(k-2)!}[t^{2k-1}f^{(k-1)}(t)]^{(k-1)} \qquad (k = 2, 3, \cdots)$$

It is easily seen that $M_{k,t}[f(x)]$ is a linear differential operator that annuls the $2k - 2$ linearly independent functions
$$x^{-p} \qquad (p = k, k - 1, \cdots, 3, 2)$$
$$x^p \qquad (p = 0, 1, \cdots, k - 2).$$

THEOREM 22a. *If*

(1) $\qquad\qquad\qquad f^{(n)}(t) = o(t^{-n-2}) \quad (t \to 0+;\ n = 0, 1, \cdots)$

(2) $\qquad\qquad\qquad\qquad\quad = o(t^{-n}) \quad (t \to \infty;\ n = 0, 1, \cdots),$

*then*

(3) $\qquad\qquad\qquad \displaystyle\lim_{k \to \infty} \int_0^\infty \frac{M_{k,t}[f(t)]}{(x+t)^2}\, dt = f(x)$

To prove this we return to equation §13 (3) and observe that in its derivation we used the full force of §13 (1) only in the first integration by parts. In the remaining ones (1) and (2) are clearly sufficient. Hence
$$\int_{0+}^\infty [t^{2k-1}f^{(k-1)}(t)]^{(k-1)} g'(t)\, dt = (-1)^k \int_0^\infty t^{2k-1}f^{(k-1)}(t) g^{(k)}(t)\, dt.$$

Consequently, instead of §13 (4) we have

$$\int_{0+}^{\infty} \frac{M_{k,t}[f(x)]}{(x+t)^2} dt = \frac{(2k-1)!}{k!(k-2)!} \int_0^{\infty} \frac{x^{k-1}t^k}{(x+t)^{2k}} f(t) dt,$$

and we saw in Section 13 that the integral on the right tends to $f(x)$ as $k$ becomes infinite, so that our result is established.

We prove next a representation theorem for the kernel $(x+t)^{-2}$.

THEOREM 22b. *A necessary and sufficient condition that*

(4) $$f(x) = \int_0^{\infty} \frac{\varphi(t)}{(x+t)^2} dt,$$

*where $\varphi(t)$ is bounded, is that there should exist a constant $N$ such that*

(5) $\qquad |M_{k,t}[f(x)]| < N \qquad (0 < t < \infty; k = 1, 2, \cdots).$

To prove the necessity we have, assuming (4),

$$M_{k,t}[f(x)] = \frac{(2k-1)!}{k!(k-2)!} \int_0^{\infty} \frac{x^k t^{k-1}}{(x+t)^{2k}} \varphi(t) dt \qquad (k = 2, 3, \cdots)$$

$$|M_{k,t}[f(x)]| \leq \underset{(0<t<\infty)}{\text{u.b.}} |\varphi(t)| \frac{k-1}{k} \qquad (k = 2, 3, \cdots)$$

$$|M_{1,t}[f(x)]| < \underset{(0<t<\infty)}{\text{u.b.}} |\varphi(t)|$$

Hence (5) is established, $N$ being any upper bound of $|\varphi(t)|$.

Conversely, if (5) holds, then it is clear that §14 (7) holds *a fortiori* in the present case, from which (2) is evident. It is also clear that

$$f(t) = O\left(\frac{1}{t^2}\right) \qquad (t \to 0+).$$

Use induction, assuming that

(6) $\qquad f^{(n)}(t) = O\left(\dfrac{1}{t^{n+2}}\right) \qquad (t \to 0+; n = 0, 1, 2, \cdots, 2k-2).$

Since

(7) $\qquad [t^{2k+1} f^{(k)}(t)]^{(k)} = O(1) \qquad (t \to 0+)$

it is clear by integration that we also have

(8) $\qquad [t^{2k+1} f^{(k)}(t)]^{(k-1)} = O(1) \qquad (t \to 0+).$

Then from (6) and (8) we see that (6) also holds for $n = 2k - 1$. Finally using (7) we see that (6) also holds for $n = 2k$, so that the relation (6) holds for all integers $n$. But clearly $t^2 f(t)$ approaches zero with $t$, so that (1) holds by Theorem 4.4 of Chapter V. Hence Theorem 22a is applicable and (3) holds. The proof of the theorem is now concluded in a familiar way by use of Theorem 17b of Chapter I. We call attention to the contrast between the present theorem and Theorem 20a by reason of the absence of any conditions of type §20 (1) and §20 (2). This difference results from the fact that the present kernel, $(x + t)^{-2}$, belongs to $L$ on $(0, \infty)$ as a function of $t$, whereas $(x + t)^{-1}$ does not.

We now prove a representation theorem in which $L_{k,t}[f(x)]$ rather than $M_{k,t}[f(x)]$ appears.

THEOREM 22c. *A necessary and sufficient condition that $f(x)$ should have the representation* (4), *with $\varphi(t)$ bounded and satisfying*

$$(9) \qquad \int_0^t \varphi(u)\, du \sim At \qquad (t \to 0+)$$

*for some constant $A$, is that*

$$(10) \qquad \left| \int_{0+}^t L_{k,u}[f(x)]\, du \right| < N \qquad (0 < t < \infty\,;\, k = 1, 2, \cdots)$$

*for some constant $N$.*

If $f(x)$ is given by (4), then

$$f^{(k)}(x) = (-1)^k (k+1)! \int_0^\infty \frac{\varphi(t)}{(x+t)^{k+2}}\, dt.$$

By Corollary 2a of Chapter V we see, using (9), that

$$f^{(k)}(x) \sim \frac{(-1)^k k! A}{x^{k+1}} \qquad (x \to 0+\,;\, k = 0, 1, 2, \cdots).$$

But this shows that

$$(11) \qquad \int_{0+}^t L_{k,u}[f(x)]\, du = M_{k,t}[f(x)] - (-1)^{k-1}\left(\frac{k-1}{k}\right) A$$

$$(k = 2, 3, \cdots)$$

$$(12) \qquad \int_{0+}^t L_{1,u}[f(x)]\, du = M_{1,t}[f(x)] - A.$$

By Theorem 22b the right-hand sides of equations (11) and (12) are bounded uniformly in $k$ and $t$, so that the necessity of conditions (10) is established.

On the other hand if (10) holds then the integrals §14 (2) exist, and by Theorem 14a there exists a constant $A$ such that

$$(-1)^k f^{(k)}(x) \sim \frac{Ak!}{x^{k+1}} \qquad (x \to 0+\,;\, k = 0, 1, 2, \cdots).$$

Hence (11) and (12) are again correct equations and (10) implies (5). By Theorem 22b, $f(x)$ has the representation (4) with $\varphi(t)$ bounded. It remains only to establish (9). This follows from

$$f(x) \sim \frac{A}{x} \qquad (x \to 0+)$$

and the boundedness of $\varphi(t)$ by use of Theorem 5b of Chapter V.

We can now prove the main result of this section.

**THEOREM 22d.** *A necessary and sufficient condition that*

$$f(x) = \int_0^\infty \frac{d\alpha(t)}{x+t},$$

*where $\alpha(t)$ is a normalized function of bounded variation in every finite interval and is bounded in the infinite interval, is that there should exist a constant $M$ and a positive function $N(t)$ such that*

(13) $\qquad \left| \int_0^R L_{k,t}[f(x)] \, dt \right| \leq M \qquad (R > 0; k = 1, 2, \cdots)$

(14) $\qquad \int_0^R | L_{k,t}[f(x)] | \, dt \leq N(R) \qquad (R > 0; k = 1, 2, \cdots).$

If $f(x)$ has the representation described, then

(15) $\qquad f(x) = \int_0^\infty \frac{\alpha(t)}{(x+t)^2} \, dt,$

$$\int_0^t \alpha(u) \, du \sim \alpha(0+)t \qquad (t \to 0+),$$

so that Theorem 22c is applicable and (10) or (13) follows. Also

(16) $\qquad \lim_{k \to \infty} \int_0^R | L_{k,t}[f(x)] | \, dt = V(R) - V(0+)$

by Theorem 12, so that the existence of $N(R)$ is assured.*

Conversely, by Theorem 22c (13) implies that $f(x)$ has the form (15) with $\alpha(t)$ bounded and

$$\int_0^t \alpha(u) \, du \sim A t \qquad (t \to 0+).$$

Set

$$\alpha_k(t) = \int_0^t L_{k,u}[f(x)] \, du = M_{k,t}[f(x)] - (-1)^{k-1} \frac{k-1}{k} A.$$

* The existence of the integral (16) for $k = 1$ is easily established.

Recall that

$$M_{k,t}[f(x)] = \frac{(2k-1)!}{k!(k-2)!} \int_0^\infty \frac{u^{k-1} t^k}{(t+u)^{2k}} \alpha(u)\, du.$$

By Theorem 8c this integral approaches $\alpha(t)$ except perhaps in a set $E$ of measure zero. But the variation of $\alpha_k(t)$ in $(0, R)$ is not greater than $N(R)$ by (14). Hence $\alpha(t)$ is a normalized function of bounded variation in $(0, R)$ if suitably redefined. This redefinition has no effect on $f(x)$ since $E$ is of measure zero. But

$$\int_0^\infty \frac{\alpha(t)}{(x+t)^2}\, dt = -\frac{\alpha(t)}{x+t}\Big|_0^\infty + \int_0^\infty \frac{d\alpha(t)}{x+t}.$$

The integrated term is zero since $\alpha(t)$ is bounded and since $\alpha(0) = 0$. Hence the theorem is established.

## 23. Operational Considerations

It is instructive to consider our inversion operator as a linear differential operator of infinite order. Since a fundamental system of solutions of the linear differential equation

(1) $$L_{k,x}[f(x)] = c_k(-x)^{k-1}[x^k f(x)]^{(2k-1)} = 0$$

is

(2) $$f(x) = x^p \qquad (p = -k, -k+1, \cdots, k-2),$$

we can easily set up a symbolic operator which will be equivalent to $L_{k,x}[f]$. Consider the operator

(3) $$-xD \prod_{p=-k+2}^{k}{}' \left(1 + \frac{xD}{p}\right) f(x),$$

where the prime indicates that there is no factor in the product corresponding to $p = 0$, and where the symbol $D$ indicates differentiation with respect to $x$. To apply the operator one applies the separate factors step by step. For example

$$\left(1 + \frac{xD}{p}\right) f(x) = f(x) + \frac{xf'(x)}{p}.$$

It is easily seen that the order of application is immaterial. But

$$\left(1 + \frac{xD}{p}\right) f(x)$$

is zero if $f(x) = x^{-p}$. That is, the operator (3) also annuls the functions (2). Since (3) is a linear differential operator of order $2k + 1$ it is the same as $L_{k,x}[f]$ if it has the same coefficient of the highest derivative. But this coefficient in each case is $(-1)^{k-1}/[k!(k-2)!]$. Hence the two operators are identical.

In our inversion of the Stieltjes transform we let $k$ become infinite, so that the inversion operator is symbolically

$$Lf(x) = -xD \prod_{p=-\infty}^{\infty}{}' \left(1 + \frac{xD}{p}\right) f(x),$$

or, making use of the infinite product expansion of the sine,

(4) $$Lf(x) = -\frac{\sin \pi xD}{\pi} f(x).$$

It is a familiar fact that an Euler differential equation such as (1) can be reduced to one with constant coefficients by means of the transformation $x = e^t$. Thus (4) becomes

$$Lf(x) = -\frac{\sin \pi D_t}{\pi} f(e^t) \qquad (x = e^t),$$

where $D_t$ indicates differentiation with respect to $t$.

Let us now use the power series expansion of the sine instead of the infinite product, first setting

$$-\frac{\sin \pi D_t}{\pi} = \frac{i}{2\pi} [e^{i\pi D_t} - e^{-i\pi D_t}].$$

But the exponential operator is identified as the translation operator as follows:

$$e^{aD_t} g(t) = g(t) + g'(t)a + g''(t)\frac{a^2}{2!} + \cdots = g(a + t).$$

Hence

$$\frac{i}{2\pi} [e^{i\pi D_t} - e^{-i\pi D_t}] f(e^t) = \frac{i}{2\pi} [f(e^{t+i\pi}) - f(e^{t-i\pi})],$$

and reverting to the variable $x$

(5) $$-\frac{\sin \pi xD}{\pi} = \frac{f(xe^{-i\pi}) - f(xe^{i\pi})}{2\pi i}.$$

But this is the complex inversion operator §7 (5) which we obtained earlier. Thus the infinite product expansion of the sine leads to the real inversion operator, the power series expansion to the complex.

The operator $L_{k,x}[f(x)]$ can be expressed as a product in still another way. By considerations similar to those above we have

$$L_{k,x}[f(x)] = g_k \frac{1}{\pi\sqrt{x}} \prod_{p=-k}^{k-2} \left(1 - \frac{2xD}{2p+1}\right)(\sqrt{x}f(x))$$

$$g_k = \frac{\pi}{2}\left(\frac{1}{2}\cdot\frac{3}{2}\cdot\frac{3}{4}\cdot\frac{5}{4}\cdots\frac{2k-5}{2k-4}\cdot\frac{2k-3}{2k-4}\right)\frac{2k-3}{2k-2}\cdot\frac{2k-1}{2k}.$$

By Wallis's product

$$\frac{2}{\pi} = \prod_{n=1}^{\infty}\left(\frac{2n-1}{2n}\cdot\frac{2n+1}{2n}\right),$$

we have

(6) $\qquad L[f(x)] = \lim_{k\to\infty} L_{k,x}[f(x)] = \frac{1}{\pi\sqrt{x}}(\cos \pi x D)(\sqrt{x}f(x)).$

The expansion of this operator in power series gives

$$\frac{1}{\pi\sqrt{x}}(\cos \pi x D)(\sqrt{x}f(x)) = \frac{e^{i\pi/2}f(xe^{i\pi}) + e^{-i\pi/2}f(xe^{-i\pi})}{2\pi},$$

and this is equivalent to (5). But the form (6) of the operator shows the relation of our work to certain results of Paley and Wiener.*

## 24. The Iterated Stieltjes Transform

We wish now to discuss briefly the work of R. P. Boas and the author on the iterated Stieltjes transform. If $f(x)$ has the representation

$$f(x) = \int_0^{\infty} \frac{g(y)}{x+y}\,dy,$$

where

$$g(y) = \int_0^{\infty} \frac{\varphi(t)}{t+y}\,dt,$$

then one has formally

$$f(x) = \int_0^{\infty} \frac{dy}{x+y}\int_0^{\infty} \frac{\varphi(t)}{t+y}\,dt = \int_0^{\infty} \varphi(t)\,dt \int_0^{\infty} \frac{dy}{(x+y)(t+y)}$$

$$= \int_0^{\infty} \frac{\log(x/t)\varphi(t)}{x-t}\,dt.$$

This is again an integral equation

$$f(x) = \int_0^{\infty} K(x, t)\varphi(t)\,dt$$

* N. Wiener [1934] pp. 41–44.

where the kernel now has the form

$$K(x, t) = [\log (x/t)][x - t]^{-1}.$$

Since the equation was obtained by iteration of the Stieltjes transform it is natural to suppose that two applications of the operator $L_{k,x}[f(x)]$ will solve or invert it. This is in fact the case. One can show* that

(1) $$L_{k,x}\{L_{k,x}[f(x)]\} = \int_0^\infty F_k(x, u)\varphi(u)\,du,$$

where

$$F_k(x, u) = d_k^2 x^{k-1} u^k \int_0^\infty \frac{t^{2k-1}\,dt}{(x + t)^{2k}(u + t)^{2k}},$$

$$d_k = \frac{(2k - 1)!}{k!(k - 2)!}.$$

It can then be proved by methods similar to those of section 9 that the integral (1) tends to $\varphi(x)$ as $k$ becomes infinite for almost all positive values of $x$. Starting from this fundamental inversion one may develop a complete theory of the iterated transform entirely analogous to the results of the present chapter.

## 25. Application to the Laplace Transform

We may use our inversion formula for the Stieltjes transform to obtain a new inversion of the Laplace transform. For, suppose that

$$f(x) = \int_0^\infty e^{-xt} \varphi(t)\,dt,$$

where for definiteness let $\varphi(t)$ belong to $L$ in $(0, \infty)$,

$$\int_0^\infty |\varphi(t)|\,dt < \infty$$

Then, by Fubini's theorem

$$F(y) = \int_0^\infty e^{-xy} f(x)\,dx = \int_0^\infty e^{-xy}\,dx \int_0^\infty e^{-xt}\varphi(t)\,dt.$$

(1) $$F(y) = \int_0^\infty \frac{\varphi(t)}{t + y}\,dt.$$

* R. P. Boas and D. V. Widder [1939] p. 18

If we now invert the integral (1) we obtain $\varphi(t)$. Explicitly, we shall show that

$$L_{k,y}[F(y)] = c_k(-1)^k y^{k-1} \int_0^\infty e^{-xy} x^{2k-1} f^{(k)}(x)\, dx.$$

To prove this we need

LEMMA 25. *If $f(x)$ is any function of class $C^k$, then*

$$\frac{\partial^k}{\partial x^k}\left[x^{k-1} f\left(\frac{y}{x}\right)\right] = \frac{(-y)^k}{x^{k+1}} f^{(k)}\left(\frac{y}{x}\right)$$

This follows by successive application of Euler's theorem for homogeneous functions. Thus

$$\frac{\partial}{\partial x}\frac{x^{k-1}}{y^k} f\left(\frac{y}{x}\right) = -\frac{\partial}{\partial y}\left[\frac{x^{k-2}}{y^{k-1}} f\left(\frac{y}{x}\right)\right]$$

since $(x/y)^{k-1} f(y/x)$ is homogeneous of order zero. Successive application of this result gives

$$\frac{\partial^k}{\partial x^k}\frac{x^{k-1}}{y^k} f\left(\frac{y}{x}\right) = (-1)^k \frac{\partial^k}{\partial y^k}\left[\frac{1}{x} f\left(\frac{y}{x}\right)\right].$$

and if we carry out the indicated differentiation on the right-hand side of this equation our result is established.

By use of this result we now prove:

THEOREM 25a. *If $\varphi(t)$ belongs to $L$ in $(0, \infty)$ and if*

$$f(x) = \int_0^\infty e^{-xt} \varphi(t)\, dt,$$

*then*

(2) $$\lim_{k\to\infty} \frac{(-1)^k y^{k-1}}{k!(k-2)!} \int_0^\infty e^{-xy} x^{2k-1} f^{(k)}(x)\, dx = \varphi(y)$$

*for almost all positive values of $y$.*

For, let us compute $L_{k,y}[F(y)]$, where $F(y)$ is defined as the Laplace transform of $f(x)$. We have

(3) $$L_{k,y}[F(y)] = (-1)^{k-1} c_k [y^{2k-1} F^{(k-1)}(y)]^{(k)}$$

But

$$(-1)^{k-1} F^{(k-1)}(y) = \int_0^\infty e^{-xy} x^{k-1} f(x)\, dx = \frac{1}{y}\int_0^\infty e^{-t}\left(\frac{t}{y}\right)^{k-1} f\left(\frac{t}{y}\right) dt,$$

where we have set $xy = t$. Using Lemma 25 we have

$$(-1)^{k-1}[y^{2k-1}F^{(k-1)}(y)]^{(k)} = \int_0^\infty e^{-t} t^{k-1} \frac{(-t)^k}{y^{k+1}} f^{(k)}\left(\frac{t}{y}\right) dt$$

$$= (-1)^k y^{k-1} \int_0^\infty e^{-xy} x^{2k-1} f^{(k)}(x) \, dx.$$

Here we have again set $xy = t$. Thus the integral (2) is precisely equal to (3). By by Theorem 9 we know that as $k$ becomes infinite $L_{k,y}[F(y)]$ approaches $\varphi(y)$ for almost all positive $y$. Observe that this formula depends on the values of all the derivatives of $f(x)$ over the whole range $(0, \infty)$.

Another form of the result is obtained by applying the operator $L_{k,y}$ to the Laplace integral directly. Thus

$$L_{k,y}[F(y)] = \int_0^\infty L_{k,y}[e^{-xy}] f(x) \, dx.$$

In applying the operator to $e^{-xy}$, $x$ is of course held constant. If we make the computation, we obtain

$$L_{k,y}[F(y)] = \int_0^\infty e^{-xy} P_{2k-1}(xy) f(x) \, dx,$$

where

$$P_{2k-1}(t) = \frac{(-1)^{k-1}(2k-1)!}{k!(k-2)!} \sum_{p=0}^k \binom{k}{p} \frac{(-t)^{2k-p-1}}{(2k-p-1)!}.$$

We can thus establish:

THEOREM 25b. *Under the hypotheses of Theorem 25a*

(4) $$\lim_{k \to \infty} \int_0^\infty e^{-xy} P_{2k-1}(xy) f(x) \, dx = \varphi(y)$$

*for almost all positive $y$*

The importance of this result lies in the fact that the left-hand side of (4) depends only on the values of $f(x)$ in $(0, \infty)$ and not on any of its derivatives. One could employ (4) to obtain a whole new series of representation theorems, in which the conditions involved would depend only on the values of $f(x)$. This has been done in a paper by R. P. Boas and the author [1940a].

## 26. Solution of an Integral Equation

We wish to solve the integral equation

(1) $$f(s) = \lambda \int_{0+}^{\infty} \frac{f(t)}{s+t} dt.$$

Various methods have been used.* We follow here the procedure of Hardy and Titchmarsh [1929], which solves the equation under the least restrictive conditions.

By Corollary 2b.1 any solution of (1) must be analytic in the $s$-plane cut along the negative real axis. Further necessary conditions are now obtained.

LEMMA 26a. *If $f(s)$ satisfies (1) then*

(2) $\quad\quad\quad\quad\quad\quad f(re^{i\theta}) = o(1) \quad\quad\quad\quad (r \to \infty)$

(3) $\quad\quad\quad\quad\quad\quad f(re^{i\theta}) = o(1/r) \quad\quad\quad (r \to 0+)$

*uniformly in* $-\pi/2 \leq \theta \leq \pi/2$.

Set

$$\beta(t) = \lambda \int_{0+}^{t} \frac{f(u)}{1+u} du \quad\quad (0 \leq t < \infty)$$

Since $\beta(\infty) = f(1)$ we have $|\beta(t)| < M$ for some constant $M$. By §2 (3)

$$f(s) = -s \int_{0}^{\infty} \frac{\beta(t) - \beta(\infty)}{(s+t)^2} dt + \int_{0}^{\infty} \frac{\beta(t)}{(s+t)^2} dt.$$

For $s = re^{i\theta}$, $-\pi/2 \leq \theta \leq \pi/2$, we have $|s+t|^2 \geq r^2 + t^2$. To an arbitrary positive $\epsilon$ we can determine $R$ so that

$$|\beta(t) - \beta(\infty)| < \epsilon \quad\quad (R \leq t < \infty)$$

Then

$$|f(s)| \leq r\epsilon \int_{R}^{\infty} \frac{dt}{r^2 + t^2} + 2Mr \int_{0}^{R} \frac{dt}{r^2} + M \int_{0}^{\infty} \frac{dt}{t^2 + r^2}$$

$$\leq \epsilon \frac{\pi}{2} + \frac{2MR}{r} + \frac{M\pi}{2r}$$

$$\varlimsup_{r \to \infty} |f(s)| \leq \frac{\epsilon\pi}{2},$$

so that (2) is proved. By §2 (4) we see that $s^{-1}f(s^{-1})$ is also a solution of (1), from which (3) follows.

* See J. Hyslop [1924] and E. C. Titchmarsh [1937], p. 310

**LEMMA 26b.** *If $f(s)$ is a solution of* (1), *then $f(is)$ and $f(-is)$ are also solutions.*

For, by use of (2) and (3) we see by contour integration that for any $s$ not on the negative real axis

$$f(\pm is) = \lambda \int_0^\infty \frac{f(\pm it)}{s+t}\, dt,$$

and the result is proved.

This shows by Corollary 2b.1 that $f(is)$ is analytic for $s$ on the imaginary axis, so that $f(s)$ is analytic on the negative real axis. That is, analytic continuation across this line is possible in either direction and $s = 0$ is the only possible singularity of $f(s)$. Another consequence is that (2) and (3) hold uniformly for $0 \leq \theta \leq 2\pi$.

**LEMMA 26c.** *If $f(s)$ is a solution of* (1) *then*

(4) $$f(re^{i\pi}) + 2\pi i \lambda f(r) - f(re^{-i\pi}) = 0 \qquad (0 < r < \infty).$$

This is an immediate consequence of Theorem 7b.

**THEOREM 26.** *For any real positive $\lambda \neq 1/\pi$ the only solutions of* (1) *are*

(5) $$f(s) = A s^{-a} + B s^{a-1},$$

*where $A$ and $B$ are arbitrary constants and $a$ is a solution of the equation*

$$\sin a\pi = \lambda \pi$$

*between 0 and 1 if $\lambda < 1/\pi$, and with real part $1/2$ if $\lambda > 1/\pi$. If $\lambda = 1/\pi$ the only solutions are*

(6) $$f(s) = \frac{1}{\sqrt{s}}[A + B \log s].$$

First suppose that $f(s)$ is a solution of (1) with $\lambda$ positive but unequal to $1/\pi$. Define $a$ is in the statement of the theorem. Set

(7) $$\varphi(s) = s^a[f(s) + e^{i\pi a}f(se^{-i\pi})]$$
(8) $$\psi(s) = s^{1-a}[f(s) - e^{-i\pi a}f(se^{-i\pi})].$$

By use of (4) one sees easily that $\varphi(re^{i\pi}) = \varphi(r)$ and $\psi(re^{i\pi}) = \psi(r)$. Hence $\varphi(s)$ and $\psi(s)$ are single-valued functions. By (2) and (3) we have

(9) $$\varphi(re^{i\theta}) = o(r^c) \qquad (r \to \infty)$$
(10) $$\varphi(re^{i\theta}) = o(r^{c-1}) \qquad (r \to 0+)$$

uniformly for $0 \leq \theta \leq 2\pi$, where $c$ is $a$ if $a$ is real and is $1/2$ if $a$ is complex. The function $\psi(s)$ satisfies a similar relation with $c$ replaced by $1 - c$.

It is now easy to prove that $\varphi(s)$ and $\psi(s)$ are constants. We treat with $\varphi(s)$ only. By (10) the function $s\varphi(s)$ is bounded in a neighborhood of the origin. By **Riemann's** theorem\* it can be defined so as to be analytic there. That is, $\varphi(s)$ has at worst a pole of the first order at $s = 0$. Since $c < 1$ this contradicts (10) unless $\varphi(s)$ is analytic at $s = 0$.

By (9) $\varphi(s) = O(|s|)$ as $|s|$ becomes infinite. Hence† $\varphi(s)$ is a polynomial of at most the first degree. But this contradicts (9) unless the degree is zero. Now replace $\varphi(s)$ and $\psi(s)$ by constants in (7) and (8) and solve these equations for $f(s)$. The result is (5).

Next suppose $\lambda = 1/\pi$. Set

(11)
$$i\pi\chi(s) = \sqrt{s}\,[if(se^{i\pi}) - f(s)]$$
$$\omega(s) = \sqrt{s}\,f(s) - \chi(s)\log s,$$

using the principal value of the logarithm. Again appealing to (4) with $\lambda = 1/\pi$ we have $\chi(re^{-i\pi}) = \chi(r)$ and $\omega(re^{-i\pi}) = \omega(r)$. It is clear by arguments similar to those employed above that $\omega(s)$ and $\chi(s)$ are constants. Then (11) is precisely (6)

It remains to show that (5) and (6) are in fact solutions of (1). If the real part of $a$ lies between zero and unity

(12)
$$\lambda \int_0^\infty \frac{t^{-a}}{s+t}\,dt = \lambda\Gamma(a)\Gamma(1-a)s^{-a} = \frac{\lambda\pi}{\sin \pi a}s^{-a} = s^{-a},$$

so that $s^{-a}$ is a solution. Equation (12) holds if $a$ is replaced by $1 - a$, so that $s^{a-1}$ is also a solution. To verify that $(\log s)/\sqrt{s}$ is a solution when $\lambda = 1/\pi$ we have

$$f(s) = \frac{1}{\pi}\int_0^\infty \frac{\log t}{\sqrt{t}(s+t)}\,dt = \frac{1}{\pi}\int_{-\infty}^\infty \frac{te^{t/2}}{s+e^t}\,dt$$

$$f(e^s)e^{s/2} = \frac{1}{2\pi}\int_{-\infty}^\infty \frac{t\,dt}{\cosh(s-t)/2} = \frac{1}{2\pi}\int_{-\infty}^\infty \frac{(s-t)}{\cosh(t/2)}\,dt$$

$$= \frac{s}{2\pi}\int_{-\infty}^\infty \frac{dt}{\cosh t/2}$$

By Theorem 17a of Chapter VI the right-hand side is equal to $s$, which establishes the result.

---
\* See W. F. Osgood [1923], p. 310.
† See E. C. Titchmarsh [1932] p. 87.

## 27. A Related Integral Equation

We may use the above result to solve the homogeneous Laplace integral equation

(1) $$f(s) = \lambda \int_{0+}^{\infty} e^{-st} f(t)\, dt.$$

Clearly, if $f(s)$ is a solution, then

$$f(\infty) = \lim_{t \to 0+} \int_{0+}^{t} f(u)\, du = 0.$$

Consequently the integral (1) converges for the real part of $s$ positive.

**LEMMA 27.** *If $f(s)$ is a solution of (1) it is also a solution of*

(2) $$f(s) = \lambda^2 \int_{0+}^{\infty} \frac{f(t)}{s+t}\, dt.$$

For if $0 < A < B$ we have by uniform convergence

$$\int_{A}^{B} e^{-sx} f(s)\, ds = \lambda \int_{0+}^{\infty} f(t) \frac{e^{-A(t+x)} - e^{-B(t+x)}}{t+x}\, dt.$$

Allowing $B$ to become infinite we have for every positive $x$

$$\int_{A}^{\infty} e^{-sx} f(s)\, ds = \lambda e^{-Ax} \int_{0+}^{\infty} \frac{f(t)}{t+x} e^{-At}\, dt.$$

Since

$$\frac{f(t)}{t+x} = o\left(\frac{1}{t}\right) \qquad\qquad (t \to \infty)$$

we may apply Theorem 3a of Chapter V to obtain

$$f(x) = \lambda \int_{0+}^{\infty} e^{-sx} f(s)\, ds = \lambda^2 \int_{0+}^{\infty} \frac{f(t)}{t+x}\, dt.$$

This completes the proof of the lemma.

**THEOREM 27.** *For any real $\lambda \neq \pm \pi^{-1/2}$ the only solutions of (1) are*

(3) $$f(s) = A\left[\sqrt{\Gamma(a)}\, s^{-a} \pm \sqrt{\Gamma(1-a)}\, s^{a-1}\right],$$

*where $A$ is an arbitrary constant and $a$ is a solution of the equation*

$$\sin \pi a = \lambda^2 \pi$$

*between zero and unity if $\lambda^2 < 1/\pi$ and with real part $1/2$ if $\lambda^2 > 1/\pi$. If $\lambda = \pi^{-1/2}$ the only solutions are*

(4) $$f(s) = A/\sqrt{s},$$

and if $\lambda = -\pi^{-1/2}$ the only solutions are

(5) $$f(s) = A\left[\frac{\Gamma'(1/2)}{\sqrt{\pi s}} - \frac{2\log s}{\sqrt{s}}\right].$$

By the lemma any solution of (1) must satisfy (2). By Theorem 26 it must then have the form §26 (5) or §26 (6). Thus the proof of the present result becomes merely a matter of verifying that §26 (5) and §26 (6) satisfy (1). We have

$$\lambda \int_0^\infty e^{-st}[At^{-a} + Bt^{a-1}]dt = \lambda A\Gamma(1-a)s^{a-1} + \lambda B\Gamma(a)s^{-a}.$$

Hence we must have

$$\lambda A\Gamma(1-a) = B \qquad \lambda B\Gamma(a) = A$$

$$\lambda^2\Gamma(a)\Gamma(1-a) = \frac{\lambda^2 \pi}{\sin \pi a} = 1.$$

Hence $f(s)$ is a solution of (1) if it has the form (3).

To treat the cases $\lambda = \pm\pi^{-1/2}$ we must substitute §26 (6) in (1). From the formula

$$\Gamma(x) = s^x \int_0^\infty e^{-su} u^{x-1} du$$

we have by differentiation

$$\Gamma'(\tfrac{1}{2}) = \sqrt{\pi}\log s + \sqrt{s}\int_0^\infty e^{-st}\frac{\log t}{\sqrt{t}}dt.$$

Hence

$$\lambda\int_0^\infty e^{-st}\left[\frac{A}{\sqrt{t}} + \frac{B}{\sqrt{t}}\log t\right]dt = \lambda\left[\frac{A\sqrt{\pi}}{\sqrt{s}} + \frac{B\Gamma'(1/2)}{\sqrt{s}} - B\sqrt{\pi}\frac{\log s}{\sqrt{s}}\right],$$

and this is to be equal to $As^{-1/2} + Bs^{-1/2}\log s$. That is,

$$B = -\lambda B\sqrt{\pi}$$
$$A = \lambda A\sqrt{\pi} + B\lambda\Gamma'(1/2).$$

If $\lambda = \pi^{-1/2}$ we must take $B = 0$, after which $A$ is arbitrary. This gives (4). If $\lambda = -\pi^{-1/2}$, $B$ is arbitrary and

$$2A = -B\Gamma'(1/2)\pi^{-1/2}.$$

That is, $f(s)$ must have the form (5). This completes the proof of the theorem.

# BIBLIOGRAPHY

ABEL, N. H.
1826. Untersuchungen über die Reihe $1 + \frac{m}{1}x + \frac{m(m-1)}{1.2}x^2 + \cdots$. Journal für die reine und angewandte Mathematik, vol. 1, pp. 311–339.
1828. Note sur le mémoire de Mr. L. Olivier No. 4 du second tome de ce journal, ayant pour titre "remarques sur les séries infinies et leur convergence." Journal für die reine und angewandte Mathematik, vol. 3, pp. 79–82.

BANACH, S.
1932. Théorie des opérations linéaires. Warsaw.

BERNSTEIN, S.
1914. Sur la définition et les propriétés des fonctions analytiques d'une variable réelle. Mathematische Annalen, vol. 75, pp. 449–468.
1928. Sur les fonctions absolument monotones. Acta Mathematica, vol. 51, pp. 1-66.

BIEBERBACH, L.
1931. Lehrbuch der Funktionentheorie, vol. 2. Leipzig and Berlin.

BLUMENTHAL, L. M.
1928. On definite algebraic quadratic forms. American Mathematical Monthly, vol. 35, pp. 551-554.

BOAS, R. P., JR.
1937a. Ph.D. Thesis, Harvard University. The iterated Stieltjes transform.
1937b. Asymptotic relations for derivatives. Duke Mathematical Journal, vol. 3, pp. 637–646.
1939. The Stieltjes moment problem for functions of bounded variation. Bulletin of the American Mathematical Society, vol. 45, pp. 399-404.
1941. Functions with positive derivatives. Duke Mathematical Journal, vol. 8, pp. 163-172.

BOAS, R. P., JR., AND WIDDER, D. V.
1939. The iterated Stieltjes transform. Transactions of the American Mathematical Society, vol. 45, pp. 1–72.
1940a. An inversion formula for the Laplace integral. Duke Mathematical Journal, vol. 6, pp. 1–26.
1940b. Functions with positive differences. Duke Mathematical Journal, vol. 7, pp. 496 503.

BOCHNER, S.
1932. Vorlesungen über Fouriersche Integrale. Leipzig.

BOHR, H.
1913. Darstellung der gleichmässigen Konvergenzabszisse einer Dirichletschen Reihe $\sum_{n=1}^{\infty} a_n n^{-s}$ als Funktion der Koeffizienten der Reihe. Archiv der Mathematik und Physik, vol. 21, pp. 326–330.

BOREL, E.
1917. Leçons sur les fonctions monogènes uniformes d'une variable complexe. Paris.

BRAY, H. E.
1919. Elementary properties of the Stieltjes integral. Annals of Mathematics, vol. 20, pp. 177–186.

# BIBLIOGRAPHY

CARLEMAN, T.
1926. Les fonctions quasi analytiques. Paris.
CARLSON, F.
1921. Über ganzwertige Funktionen. Mathematische Zeitschrift, vol. 11, pp. 1-23.
DOETSCH, G.
1937a. Theorie und Anwendung der Laplace-Transformation. Berlin.
1937b. Bedingungen für die Darstellbarkeit einer Funktion als Laplace-Integral und eine Umkehrformel für die Laplace-Transformation. Mathematische Zeitschrift, vol. 42, pp. 263-268.
DUBOURDIEU, M. J.
1939. Sur un théorème de M. S. Bernstein relatif à la transformation de Laplace-Stieltjes. Compositio Mathematica, vol. 7, pp. 96-111.
EVANS, G. C.
1927. The logarithmic potential. American Mathematical Society Colloquium Publications 6. New York.
FABER, G.
1903. Über die Fortsetzbarkeit gewisser Taylorschen Reihen. Mathematische Annalen, vol. 57, pp. 369-388.
FELLER, W.
1939. Completely monotone functions and sequences. Duke Mathematical Journal, vol. 5, pp. 662-663.
FORT, T.
1930. Infinite series. Oxford.
GRÜSS, G.
1935. Bemerkungen zur Theorie der voll-bzw. mehrfachmonotonen funktionen. Mathematische Zeitschrift, vol. 39, pp. 732-741.
HAMBURGER, H.
1920a. Bemerkungen zu einer Fragestellung des Herrn Pólya. Mathematische Zeitschrift, vol. 7, pp. 302-322.
1920b. Über eine Erweiterung der Stieltjesschen Momentproblems. Mathematische Annalen, vol. 81, pp. 235-319.
1921. Über die Riemannsche Funktionalgleichung der $\zeta$-Funktion. Mathematische Zeitschrift, vol. 10, pp. 240-254.
HANKEL, H.
1861. Über eine besondere Klasse der symmetrischen Determinanten. Thesis, University of Leipzig. Göttingen.
HARDY, G. H.
1918. Notes on some points in the integral calculus, XLVIII. Messenger of Mathematics, vol. 47, pp. 145-150.
1921. Notes on some points in the integral calculus, LIV. Messenger of Mathematics, vol. 50, pp. 165-171.
HARDY, G. H., AND LITTLEWOOD, J. E.
1912. Contributions to the arithmetic theory of series. Proceedings of the London Mathematical Society (2), vol. 11, pp. 411-478.
1928. Some properties of fractional integrals. Mathematische Zeitschrift, vol. 27, pp. 565-606.
1930. Notes on the theory of series (XI): On Tauberian theorems. Proceedings of the London Mathematical Society (2), vol. 30, pp. 23-37.
HARDY, G. H., LITTLEWOOD, J. E., AND PÓLYA, G.
1934. Inequalities. Cambridge.

# BIBLIOGRAPHY

HARDY, G. H., AND TITCHMARSH, E. C.
1929. Solution of an integral equation. Journal of the London Mathematical Society, vol. 4, pp. 300–304.

HAUSDORFF, F.
1921a. Summationsmethoden und Momentfolgen. I. Mathematische Zeitschrift, vol. 9, pp. 74–109.
1921b. Summationsmethoden und Momentfolgen. II. Mathematische Zeitschrift, vol. 9, pp. 280–299.

HELLY, E.
1921. Über lineare Funktionaloperationen. Sitzungsberichte der Naturwissenschaftlichen Klasse der Kaiserlichen Akademie der Wissenschaften, vol. 121 (part IIa, numbers I to X), pp. 265–297.

HILDEBRANDT, T. H.
1938. Stieltjes integrals of the Riemann type. American Mathematical Monthly, vol. 45, pp. 265–277.

HILDEBRANDT, T. H., AND SCHOENBERG, I. J.
1933. On linear functional operations and the moment problem for a finite interval in one or several dimensions. Annals of Mathematics, vol. 34, pp. 317–328.

HILLE, E., AND TAMARKIN, J. D.
1933a. Questions of relative inclusion in the domain of Hausdorff means. Proceedings of the National Academy of Sciences, vol. 19, pp. 573–577.
1933b. On moment functions. Proceedings of the National Academy of Sciences, vol. 19, pp. 902–908.
1933c. On the theory of Laplace integrals. I. Proceedings of the National Academy of Sciences, vol. 19, pp. 908–914.
1934. On the theory of Laplace integrals. II. Proceedings of the National Academy of Sciences, vol. 20, pp. 140–144.

HOBSON, E. W.
1926. The theory of functions of a real variable and the theory of Fourier's series. Second edition. Cambridge.

HYSLOP, J.
1924. The integral expansions of arbitrary functions connected with integral equations. Proceedings of the Cambridge Philosophical Society, vol. 22, pp. 169–185.

IKEHARA, S.
1931. An extension of Landau's theorem in the analytic theory of numbers. Journal of Mathematics and Physics, Massachusetts Institute of Technology, vol. 10, pp. 1–12.

KARAMATA, J.
1931. Neuer Beweis und Verallgemeinerung der Tauberschen Sätze, welche die Laplacesche und Stieltjessche Transformation betreffen. Journal für die reine und angewandte Mathematik, vol. 164, pp. 27–39.

KNOPP, K.
1928. Theory and application of infinite series. London and Glasgow.

LANDAU, E.
1906. Über die Grundlagen der Theorie der Fakultätenreihen. Sitzungsberichte der mathematisch-physikalischen Klasse der Kgl. Bayerischen Akademie der Wissenschaften zu München, vol. 36, pp. 151–218.
1909. Handbuch der Lehre von der Verteilung der Primzahlen. Vol. 2. Leipzig, Berlin.

1929. Darstellung und Begründung einiger neuerer Ergebnisse der Funktionentheorie. Berlin.
LAPLACE, P. S.
  1820. Théorie analytique des probabilités. Vol. 1, Part 2, Chapter 1. Paris.
LERCH, M.
  1903. Sur un point de la théorie des fonctions génératices d'Abel. Acta Mathematica, vol. 27, pp. 339–351.
LITTLEWOOD, J. E.
  1910. The converse of Abel's theorem on power series. Proceedings of the London Mathematical Society (2), vol. 9, pp. 434–448.
  See also HARDY, G. H., AND LITTLEWOOD, J. E., and HARDY, G. H., LITTLEWOOD, J. E., AND PÓLYA, G.
MATHIAS, M.
  1923. Über positive Fourier-Integrale. Mathematische Zeitschrift, vol. 16, pp. 103–125.
MELLIN, H.
  1902. Über den Zusammenhang zwischen den linearen Differential- und Differenzengleichungen. Acta Mathematica, vol. 25, pp. 139–164.
MERCER, J.
  1909. Functions of positive and negative type, and their connection with the theory of integral equations. Philosophical Transactions of the Royal Society of London, vol. 209A, pp. 415–446.
MERTENS, F.
  1874. Über die Multiplicationsregel für zwei unendliche Reihen. Journal für die reine und ungewandte Mathematik, vol. 79, pp. 182–184.
NÖRLUND, N. E.
  1926. Leçons sur les séries d'interpolation. Paris.
OSGOOD, W. F.
  1923. Lehrbuch der Funktionentheorie. Fourth edition. Leipzig, Berlin.
PALEY, R. E. A. C. See WIENER, N.
PINCHERLE, S.
  1905. Sur les fonctions déterminantes. Annales Scientifiques de l'École Normale Supérieure, vol. 22, pp. 9–68.
PITT, H. R.
  1938a. General Tauberian theorems. Proceedings of the London Mathematical Society, vol. 44, pp. 243–288.
  1938b. A remark on Wiener's general Tauberian theorem. Duke Mathematical Journal, vol. 4, pp. 436–440.
POLLARD, H.
  1940. Note on the inversion of the Laplace integral. Duke Mathematical Journal, vol. 6, pp. 420–424.
PÓLYA, G.
  1938. Sur l'indétermination d'un problème voisin du problème des moments. Comptes Rendus des Séances de l'Académie des Sciences, vol. 207, pp. 708–711.
  See also HARDY G. H., LITTLEWOOD, J. E., AND PÓLYA, G.
PÓLYA, G., AND SZEGÖ, G.
  1925. Aufgaben und Lehrsätze. Berlin.
POST, E. L.
  1930. Generalized differentiation. Transactions of the American Mathematical Society, vol. 32, pp. 723–781.

RIEMANN, B.
 1876. Über die Anzahl der Primzahlen unter einer gegebener Grösse. Gesammelte Mathematische Werke. Leipzig.
RIESZ, F.
 1909. Sur les opérations fonctionnelles linéaires. Comptes Rendus des Séances de l'Académie des Sciences, vol. 149, pp. 974–977.
RIESZ, M.
 1922. Sur le problème des moments. Arkiv För Matematik, Astronomi Och Fysik, vol. 17, no. 16, pp. 1–52.
SAKS, S.
 1933. Théorie de l'integrale. Warsaw.
 1937. Theorie of the integral. New York, Warsaw.
SHOHAT, J.
 1938. Sur les polynomes orthogonaux généralisés. Comptes Rendus des Séances de l'Académie des Sciences, vol. 207, pp. 556–558.
STIELTJES, T. J.
 1894. Recherches sur les fractions continues. Annales de la faculté des sciences de Toulouse, vol. 8, pp. 1–122.
SZEGÖ, S.
 1918. Ein Beitrag zur Theorie der Polynome von Laguerre und Jacobi. Mathematische Zeitschrift, vol. 1, pp. 341–356.
 1939. Orthogonal polynomials. American Mathematical Society Colloquium Publications 23. New York.
 *See also* PÓLYA, G., AND SZEGÖ, G.
SCHOENBERG, I. J. *See* HILDEBRANDT, T. H., AND SCHOENBERG, I. J.
TAMARKIN, J. D.
 1926. On Laplace's integral equation. Transactions of the American Mathematical Society, vol. 28, pp. 417–425.
 *See also* HILLE, E., AND TAMARKIN, J. D.
TAUBER, A.
 1897. Ein Satz aus der Theorie der unendlichen Reihen. Monatshefte für Mathematik und Physik, vol. 8, pp. 273–277.
TITCHMARSH, E. C.
 1932. The theory of functions. Oxford.
 1937. Introduction to the theory of Fourier integrals. Oxford.
 *See also* HARDY, G. H., AND TITCHMARSH, E. C.
TOEPLITZ, O.
 1911. Über allgemeine lineare Mittelbildungen. Prace Matematyczano-Fizyczne, vol. 22, pp. 113–119.
WIDDER, D. V.
 1929. A generalization of Dirichlet's series and of Laplace's integrals by means of a Stieltjes integral. Transactions of the American Mathematical Society, vol. 31, pp. 694–743.
 1931. Necessary and sufficient conditions for the representation of a function as a Laplace integral. Transactions of the American Mathematical Society, vol. 33, pp. 851–892.
 1934a. The inversion of the Laplace integral and the related moment problem. Transactions of the American Mathematical Society, vol. 36, pp. 107–200.
 1934b. Necessary and sufficient conditions for the representation of a function by a doubly infinite Laplace integral. Bulletin of the American Mathematical Society, vol. 40, pp. 321–326.

1935. An application of Laguerre polynomials. Duke Mathematical Journal, vol. 1, pp. 126–136.
1936. A classification of generating functions. Transactions of the American Mathematical Society, vol. 39, pp. 244–298.
1937. The successive iterates of the Stieltjes kernel expressed in terms of the elementary functions. Bulletin of the American Mathematical Society, vol. 43, pp. 813–817.
1938. The Stieltjes transform. Transactions of the American Mathematical Society, vol. 43, pp. 7–60.
1940. Functions whose even derivatives have a prescribed sign. Proceedings of the National Academy of Sciences, vol. 26, pp. 657–659.

See also BOAS, R. P., JR., AND WIDDER, D. V., and WIDDER, D. V., AND Wiener, N.

WIDDER, D. V., AND WIENER, N.
1938. Remarks on the classical inversion formula for the Laplace integral. Bulletin of the American Mathematical Society, vol. 44, pp. 573–575.

WIENER, N.
1933. The Fourier integral and certain of its applications. Cambridge.
1934. (With R. E. A. C. PALEY). Fourier transforms in the complex domain. American Mathematical Society Colloquium Publications 19. New York.

WINTNER, A.
1929. Spektraltheorie der unendlichen Matrizen. Leipzig.

# INDEX

The numbers refer to pages

ABEL, 89, 116.
Abelian theorems, 181, 185.
  for the Laplace transform, 180.
  for the Stieltjes transform, 183.
Abscissa of convergence of the bilateral Laplace transform, 238.
  Dirichlet series, 45.
  Laplace transform, 37, 38, 42.
  uniform convergence for the Laplace transform, 51.
Abscissas of convergence of the bilateral Laplace transform, 240.
  absolute convergence of the bilateral Laplace transform, 241.
Absolute convergence of the Laplace transform, 46.
Absolutely convergent improper Stieltjes integrals, 15.
  monotonic $(D)$, 147.
  $(\Delta)$, 148.
  $(G)$, 154.
  function, 144.
  functions, additional properties of, 156.
    analytic extension of, 146.
    analyticity of, 146.
    derivatives of, 167.
    equivalence of Bernstein and Grüss definitions, 155.
      Bernstein's two definitions, 151.
    existence of one-sided derivatives of, 149.
    · Grüss's definition of, 154.
    higher differences of, 150.
    series of, 151.
    with prescribed derivatives, 165.
Additional properties of absolutely monotonic functions, 156.
Analytic character of the generating function, 57.
  of the Stieltjes transform, 328.

Analytic extension of absolutely monotonic functions, 146.
Analyticity of absolutely monotonic functions, 146.
Application of Wiener's theorem, 221, 222.
Applications of the Laplace method, 280.
Arithmetic means, 76.
Asymptotic evaluation of an integral, Laplace's, 277, 296.
  properties of Stieltjes transforms, 329.
Axis of convergence of the bilateral Laplace transform, 238.
  of the Laplace transform, 37.
  of uniform convergence for the Laplace transform, 50.

BANACH, 33.
BERNSTEIN, 144, 147, 165, 168.
Bernstein polynomials, 101, 152.
Bernstein's theorem, 160, 162, 175, 312.
Bernoulli numbers, 264.
Bessel's function, 169.
BIEBERBACH, 95.
Bilateral Laplace integrals, product of, 257.
  transform, 237.
    abscissa of convergence of, 238.
    axis of convergence of, 238.
    abscissas of absolute convergence of, 241.
    convergence of, 240.
    integration by parts of, 239.
    necessary and sufficient conditions for representation as, 272.
    region of convergence of, 238.
    summability of, 244.
  transforms, product of, 252.
    representation of functions as, 265.
  -Lebesgue transform, inversion formulas for, 241.
  -Lebesgue transforms, 258.

# INDEX

Bilateral Laplace-Stieltjes transform, inversion formulas for, 242.
Stieltjes resultant, 248, 249, 250.
BLUMENTHAL, 135.
BOAS, 18, 138, 177, 275, 363, 383, 384, 386.
BOCHNER, 252, 273.
BOHR, 53.
Bonnet's second law of the mean, 17.
BOREL, 337.
Bound, greatest lower, 4.
  least upper, 4.
Bounded functions, 111.
  variation, determining function of, 306.
  function $\alpha(t)$ of, 361.
  functions of, 6.
  normalization of functions of, 13.
  sequence of functions of uniformly, 31.
BRAY, 25, 31.

CAHEN, 246.
CARLEMAN, 177.
CARLSON, 161.
Cauchy's integral theorem, 265.
  singular integral, 338.
  theorem, 86, 88.
Cauchy-value, 15, 35.
Cesàro's means, 76.
  method of order $p$, $(C,p)$, 121.
  summability, 114, 118, 119.
Change of variable of a Stieltjes integral, 19.
Class $A$, 204.
  norm of a function of, 205.
  $B$, 209.
  $L$, 317, 374.
  $L^p$, 312, 368.
  $M$, 213.
  $S$, 209.
  $S^*$, 211.
  $W$, 210.
Classical resultant, 91.
Completely convex functions, 177.
  monotonic functions, 144, 161, 310.
    interpolation by, 163.
    sequence, 158, 269.
      minimal, 163.
    sequences, 108.

Complex inversion formula for the Laplace transform, 63.
  Stieltjes transform, 338.
  variable Tauberian theorem, 233.
Condition $A$, 101.
  $A$ for a moment sequence, 101.
  $B$ for a moment sequence, 109.
  $C$ for a moment sequence, 111.
Conditions $A$ for the Laplace transform, 306.
  $A'$ for the Laplace transform, 308.
  $A$ for the Stieltjes transform, 361.
  $B$ for the Laplace transform, 312.
  $B$ for the Stieltjes transform, 369.
  $C$ for the Laplace transform, 315.
  $C$ for the Stieltjes transform, 372.
  $D$ for the Laplace transform, 317.
  $D$ for the Stieltjes transform, 374.
  of integrability for the Stieltjes integral, 24.
Consistent, 115.
  methods of summability, 115.
Convergence properties of the Laplace integral, 38.
Convergent improper Stieltjes integrals, 15.
Convex functions, 148, 167.
  completely, 177.
Convexity, 148.
Convolution, 84.
$(C,p)$ equivalent to $(H,p)$, 123.

Definite quadratic forms, 136.
Derivatives of absolutely monotonic functions, 167.
Determinant, Hadamard's maximum value of a, 141.
  criteria for definite sequences, 134.
Determinants, Hankel, 135.
Determining function, 38.
  belonging to $L^2$, 80, 245.
  fractional integral of, 73.
  integrals of the, 70.
  non-decreasing, 310.
  normalized, 237.
  of bounded variation, 306.
  periodic, 96.
  the integral of a bounded function, 315.
  uniqueness of the, 59.

# INDEX

Determining function, variation of the, 299.
Diagonal matrix, 114.
Difference matrix, 114.
Dini's condition, 67.
Dirichlet integral, 64.
  series, 44, 53.
  abscissa of convergence of, 45.
  $\sigma_a - \sigma_c$ for, 49.
Divergent improper Stieltjes integrals, 15.
  integrals, summability of, 75.
DOETSCH, 168.
$d_t \beta(t, \tau)$, 50.
DUBOURDIEU, 295.

Elementary properties of the Stieltjes transform, 325.
Entire function, order of an, 94, 179.
  type of an, 94, 179.
  functions, 177.
  generating function, 58.
Equivalence of Bernstein and Grüss definitions of absolutely monotonic functions, 155.
  of Bernstein's two definitions of absolutely monotonic functions, 151.
Equivalent methods of summability, 121.
Euler's constant, 229.
  theorem for homogeneous functions, 291, 303, 357, 385.
EVANS, 31.
Existence of one-sided derivatives of absolutely monotonic functions, 149.
  of Stieltjes integrals, 7.
Extension of moment operator, 127.

FABER, 96.
Factorial series, 97.
Faltung, 84.
Fatou's lemma, 314.
FELLER, 295.
FORT, 99.
Fourier development, 96.
  integrals, 204.
  -Stieltjes transforms, product of, 252.
  transform, uniqueness of, 82, 204.
  transforms, 202.
    of functions of $L$, 204.
    product of, 203.
    quotient of, 207.

Fractional integral, 71.
  of determining function, 73.
Fredholm, function of, 157.
Fubini's theorem, 26, 41, 71, 73, 74, 82, 200, 201, 203, 207, 211, 256, 259, 266, 292, 318.
Function, absolutely monotonic, 144.
  $\alpha(t)$ non-decreasing and bounded, 363.
  non-decreasing and unbounded, 365.
  of bounded variation, 361, 377.
  completely monotonic, 144, 310.
  normalized determining, 237.
  of $L^p$, 109.
  order of an entire, 179.
  $\varphi(t)$ bounded, 372.
  type of an entire, 179.
Functional, linear, 171.
Functions analytic at infinity, 93.
  bounded, 111.
  completely convex, 177.
  monotonic, 161.
  convex, 148, 167.
  entire, 177.
  logarithmically convex, 167.
  number theoretic, 224.
  of bounded variation, 6.
    moments of, 138.
    normalization of, 13.
  of $L$, Fourier transforms of, 204.
  of $L^p$, 11.
  Riemann-integrable, 190.
  slowly decreasing, 209.
  slowly oscillating, 209.

GAUSS, 224.
General Laplace-Stieltjes transform, 320.
  representation theorem for the Laplace kernel, 302.
    Stieltjes kernel, 355.
  theorems for the Stieltjes transform, 360.
Generating function, 37.
  analytic at infinity, 93.
  analytic character of the, 57.
  as factorial series, 97.
  entire, 58.
  singularities of, 58.
GOURSAT, 337.
Greatest lower bound, 4.
GRÜSS, 154.

# INDEX

Grüss's definition of absolutely monotonic functions, 154.
HADAMARD, 224.
Hadamard's maximum value of a determinant, 141.
Half-plane of absolute convergence for the Laplace transform, 46.
HAMBURGER, 58, 125, 265, 266.
Hamburger moment problem, 125, 129.
  not unique, 125.
HANKEL, 135.
Hankel determinants, 135, 167.
HARDY, 71, 77, 78, 161, 169, 193, 198, 199, 370, 387.
HAUSDORFF, 100, 113, 144, 163.
Hausdorff matrix, 118, 119.
  moment problem, 100, 273.
  summability, 113, 115.
  transformation, 118.
HELLY, 26, 31.
Helly-Bray theorem, 31, 33, 253.
Helly's theorem, 104, 176, 307, 367.
Hilbert, inequality of, 370.
Higher differences of absolutely monotonic functions, 150.
HILDEBRANDT, 3, 105.
HILLE, 113.
HOBSON, 76.
Hölder summability, 114, 118, 119.
Hölder's inequality, 110, 313, 314, 370.
  method of summability of order $p$, $(H,p)$, 120.
Homogeneous functions, Euler's theorem on, 291, 303, 357, 385.
  Laplace integral equation, 390.
l'Hospital's rule, 79.
$(H,p)$, $(C,p)$ equivalent to, 123.
HYSLOP, 387.

Ikehara's theorem, 233.
Improper Stieltjes integrals, 15.
Indefinite integrals, Stieltjes resultant of, 256.
Indeterminacy of solution of moment problem, 142.
Inequalities for Stieltjes integrals, 9.
Integral equation, 390.
  solution of an, 387.
  with Laplace kernel, 390.

Integral equation, with Stieltjes kernel, 387.
  of a bounded function, the determining function, 315.
Integrals of the determining function, 70.
Integration by parts of the bilateral Laplace transform, 239.
  of the Laplace integral, 41.
  for the Stieltjes integral, 8.
  with respect to an indefinite Stieltjes integral, 12.
Interpolation by completely monotonic functions, 163.
Inversion formula for the Laplace-Stieltjes integral, 69.
  Laplace transform, 386.
  formulas for the bilateral Laplace-Lebesgue transform, 241.
  Laplace-Stieltjes transform, 242.
  of the Laplace transform, 276.
  operator for the Laplace-Lebesgue transform, 288.
  Laplace-Stieltjes transform, 290.
  moment sequences, 107.
  the Stieltjes transform, 345, 347.
  when the determining function belongs to $L^2$, 80.
Iterated Stieltjes integrals, 25.
  transform, 383.
Iterates of the Stieltjes kernel, 259, 262.
Iteration of the Laplace transform, 325.

Jump operator for the Laplace transform, 298.
  Stieltjes transform, 351.

KARAMATA, 189, 191.
Karamata's theorem, 189.
Kernels, positive, 270.
  definite, 270.
  semidefinite, 271.
  of positive type, 270.
KNOPP, 264.

Laguerre polynomials, 168.
  Laplace transform of, 170.
Lambert series, 231.
LANDAU, 49, 56, 64, 99, 193, 195.
LAPLACE, 277.

# INDEX

Grüss's definition of absolutely monotonic functions, 154.

HADAMARD, 224.
Hadamard's maximum value of a determinant, 141.
Half-plane of absolute convergence for the Laplace transform, 46.
HAMBURGER, 58, 125, 265, 266.
Hamburger moment problem, 125, 129.
    not unique, 125.
HANKEL, 135.
Hankel determinants, 135, 167.
HARDY, 71, 77, 78, 161, 169, 193, 198, 199, 370, 387.
HAUSDORFF, 100, 113, 144, 163.
Hausdorff matrix, 118, 119.
    moment problem, 100, 273.
    summability, 113, 115.
    transformation, 118.
HELLY, 26, 31.
Helly-Bray theorem, 31, 33, 253.
Helly's theorem, 104, 176, 307, 367.
Hilbert, inequality of, 370.
Higher differences of absolutely monotonic functions, 150.
HILDEBRANDT, 3, 105.
HILLE, 113.
HOBSON, 76.
Hölder summability, 114, 118, 119.
Hölder's inequality, 110, 313, 314, 370.
    method of summability of order $p$, $(H,p)$, 120.
Homogeneous functions, Euler's theorem on, 291, 303, 357, 385.
Laplace integral equation, 390.
l'Hospital's rule, 79.
$(H,p)$, $(C,p)$ equivalent to, 123.
HYSLOP, 387.

Ikehara's theorem, 233.
Improper Stieltjes integrals, 15.
Indefinite integrals, Stieltjes resultant of, 256.
Indeterminacy of solution of moment problem, 142.
Inequalities for Stieltjes integrals, 9.
Integral equation, 390.
    solution of an, 387.
    with Laplace kernel, 390.

Integral equation, with Stieltjes kernel, 387.
    of a bounded function, the determining function, 315.
Integrals of the determining function, 70.
Integration by parts of the bilateral Laplace transform, 239.
    of the Laplace integral, 41.
    for the Stieltjes integral, 8.
    with respect to an indefinite Stieltjes integral, 12.
Interpolation by completely monotonic functions, 163.
Inversion formula for the Laplace-Stieltjes integral, 69.
    Laplace transform, 386.
    formulas for the bilateral Laplace-Lebesgue transform, 241.
    Laplace-Stieltjes transform, 242.
    of the Laplace transform, 276.
    operator for the Laplace-Lebesgue transform, 288.
    Laplace-Stieltjes transform, 290.
    moment sequences, 107.
    the Stieltjes transform, 345, 347.
    when the determining function belongs to $L^2$, 80.
Iterated Stieltjes integrals, 25.
    transform, 383.
Iterates of the Stieltjes kernel, 259, 262.
Iteration of the Laplace transform, 325.

Jump operator for the Laplace transform, 298.
    Stieltjes transform, 351.

KARAMATA, 189, 191.
Karamata's theorem, 189.
Kernels, positive, 270.
    definite, 270.
    semidefinite, 271.
    of positive type, 270.
KNOPP, 264.

Laguerre polynomials, 168.
    Laplace transform of, 170.
Lambert series, 231.
LANDAU, 49, 56, 64, 99, 193, 195.
LAPLACE, 277.

# INDEX

Determining function, variation of the, 299.
Diagonal matrix, 114.
Difference matrix, 114.
Dini's condition, 67.
Dirichlet integral, 64.
  series, 44, 53.
    abscissa of convergence of, 45.
    $\sigma_a - \sigma_c$ for, 49.
Divergent improper Stieltjes integrals, 15.
  integrals, summability of, 75.
DOETSCH, 168.
$d_t \beta(t, \tau)$, 50.
DUBOURDIEU, 295.

Elementary properties of the Stieltjes transform, 325.
Entire function, order of an, 94, 179.
  type of an, 94, 179.
  functions, 177.
  generating function, 58.
Equivalence of Bernstein and Grüss definitions of absolutely monotonic functions, 155.
  of Bernstein's two definitions of absolutely monotonic functions, 151.
Equivalent methods of summability, 121.
Euler's constant, 229.
  theorem for homogeneous functions, 291, 303, 357, 385.
EVANS, 31.
Existence of one-sided derivatives of absolutely monotonic functions, 149.
  of Stieltjes integrals, 7.
Extension of moment operator, 127.

FABER, 96.
Factorial series, 97.
Faltung, 84.
Fatou's lemma, 314.
FELLER, 295.
FORT, 99.
Fourier development, 96.
  integrals, 204.
  -Stieltjes transforms, product of, 252.
  transform, uniqueness of, 82, 204.
  transforms, 202.
    of functions of $L$, 204.
    product of, 203.
    quotient of, 207.

Fractional integral, 71.
  of determining function, 73.
Fredholm, function of, 157.
Fubini's theorem, 26, 41, 71, 73, 74, 82, 200, 201, 203, 207, 211, 256, 259, 266, 292, 318.
Function, absolutely monotonic, 144.
  $\alpha(t)$ non-decreasing and bounded, 363.
  non-decreasing and unbounded, 365.
  of bounded variation, 361, 377.
  completely monotonic, 144, 310.
  normalized determining, 237.
  of $L^p$, 109.
  order of an entire, 179.
  $\varphi(t)$ bounded, 372.
  type of an entire, 179.
Functional, linear, 171.
Functions analytic at infinity, 93.
  bounded, 111.
  completely convex, 177.
    monotonic, 161.
  convex, 148, 167.
  entire, 177.
  logarithmically convex, 167.
  number theoretic, 224.
  of bounded variation, 6.
    moments of, 138.
    normalization of, 13.
  of $L$, Fourier transforms of, 204.
  of $L^p$, 11.
  Riemann-integrable, 190.
  slowly decreasing, 209.
  slowly oscillating, 209.

GAUSS, 224.
General Laplace-Stieltjes transform, 320.
  representation theorem for the Laplace kernel, 302.
    Stieltjes kernel, 355.
  theorems for the Stieltjes transform, 360.
Generating function, 37.
  analytic at infinity, 93.
  analytic character of the, 57.
  as factorial series, 97.
  entire, 58.
  singularities of, 58.
GOURSAT, 337.
Greatest lower bound, 4.
GRÜSS, 154.

# INDEX

Laplace integral, convergence properties of the, 38.
  integration by parts of the, 41, 239.
  uniform convergence of the, 53.
  kernel, general representation theorem for, 302.
  integral equation with, 390.
  method, applications of the, 280.
  uniform convergence, 283.
  -Lebesgue transform, inversion operator for the, 288.
  transforms, product of, 91.
Laplace's asymptotic evaluation of an integral, 277, 296.
Laplace-Stieltjes integral, inversion formula for the, 69.
  -Stieltjes transform, 37.
    general, 320.
    inversion operator for the, 290.
    transforms, product of, 88.
  transform, abscissa of convergence of, 37, 38, 42.
    of uniform convergence of the, 51.
    absolute convergence of the, 46.
    axis of convergence of the, 37.
    of uniform convergence of the, 50.
    bilateral, 237.
    complex inversion formula for the, 63.
    Condition A for, 306.
      A' for, 308.
      B for, 312.
      C for, 315.
      D for, 317.
    inversion formula for, 386.
    of, 276.
    iteration of the, 325.
    of Laguerre polynomials, 170.
    jump operator for, 298.
    region of convergence of the, 35.
    representation by, 276.
    $\sigma_a - \sigma_c$ for the, 47.
    uniform convergence of the, 50.
    unilateral, 35.
    uniqueness of the, 80.
    transforms, Abelian theorems for the, 180.
Lattice points, 228.
Law of the mean, 108.

Laws of the mean for Stieltjes integrals, 16.
Least upper bound, 4.
Lebesgue set, 78, 244, 289, 345.
Lebesgue's limit theorem, 97, 104, 260.
Lebesgue-Stieltjes integral, 4, 20.
Leibniz's rule, 295, 351.
LERCH, 61.
l.i.m., 80, 245.
Limit in the mean, 80.
  theorem, Lebesgue's, 97.
Linear functional, 105, 171.
  as Stieltjes integral, 105.
LIOUVILLE, 71.
Lipschitz condition, 72, 75.
Little moment problem, 100.
LITTLEWOOD, 71, 193, 195, 198, 199, 370.
Logarithmically convex functions, 167.
Lower bound, greatest, 4.

MATHIAS, 273.
Matrix, diagonal, 114.
  difference, 114.
  Hausdorff, 118, 119.
  theory, 113.
Maximal type, 94.
Max, true, 34.
Mean convergence, 81.
  square convergence, 245.
Measurable functions, 11.
MELLIN, 246.
Mellin transform, 246.
  -Stieltjes transform, 246.
MERCER, 270, 271.
MERTENS, 89.
Method of summability $(C,p)$, 121.
  $(H,p)$, 120.
  stronger, 121.
Methods of summability, consistent, 115.
  equivalent, 121.
  permanent, 115.
  regular, 115.
Minimal completely monotonic sequence, 163.
  type, 94.
Mittag-Leffler development, 97, 261.
Moment function, uniqueness of, 60.
  of a polynomial, 102, 127.
  operator, 102, 126.

Moment operator, extension of, 127.
  problem, 100.
    indeterminacy of solution of, 142.
    little, 100.
    of Hamburger, 125.
    of Hausdorff, 100.
    of Stieltjes, 136.
  sequence, 100, 101.
    Condition $B$ for, 109.
    $C$ for, 111.
    necessary and sufficient conditions for, 103.
  sequences, inversion operator for, 107.
Moments of functions of bounded variation, 138.

Necessary and sufficient conditions for representation as a bilateral Laplace transform, 272.
NÖRLUND, 97.
Non-decreasing determining function, 310.
  functions, 6.
Non-vanishing of the zeta-function, 229.
Norm, 4.
  of a function of class $A$, 205.
Normal orthogonal set, 168.
  type, 94.
Normalization of functions of bounded variation, 13.
Normalized determining function, 237.
Number theoretic functions, 224.

One-sided derivatives of absolutely monotonic functions, existence of, 149.
  Tauberian theorem, 215.
Operational considerations, 381.
Order conditions, 357.
  of an entire function, 94, 179.
  on vertical lines, 92.
  properties of the determining function, 38.
Orthogonal set, normal, 168.
OSGOOD, 260, 261, 389.

Parseval theorem, 81, 202.
Periodic determining function, 96.
Permanent, 115.
  methods of summability, 115.
PINCHERLE, 37.

PITT, 210, 216.
Pitt's form of Wiener's theorem, 210.
Plancherel's theorem, 81, 202.
Point of increase, 6.
  of invariability, 6, 19.
POLLARD, 289, 295.
PÓLYA, 138, 172, 229, 370.
Polynomial, Bernstein, 101.
Polynomials, positive, 132, 172.
Positive definite kernels, 270.
  quadratic forms, 135.
  sequences, 132.
  kernels, 270.
  polynomials, 132, 172.
  quadratic forms, 166.
  semidefinite kernels, 271.
  sequences, 132.
  sequences, 127, 159, 266.
  type, kernels of, 270.
POST, 276, 289.
Power series, product of, 88.
Prime-number theorem, 224, 236.
Product of bilateral Laplace integrals, 252, 257.
  of Fourier transforms, 203.
  -Stieltjes transforms, 252.
  of Laplace-Lebesgue transforms, 91.
  -Stieltjes transforms, 88.
  of power series, 88.
Properties of Stieltjes integrals, 8, 12.

Quadratic forms, 273.
  definite, 136.
  positive, 166.
    definite, 135.
  semidefinite, 136.
Quotient of Fourier transforms, 207.

Region of convergence of the bilateral Laplace transform, 238.
    Laplace transform, 35.
    Stieltjes transform, 326.
Regular, 115.
  methods of summability, 115.
Relation between the Laplace transform and the Stieltjes transform, 334.
Representation by the Laplace transform, 276.
    bilateral Laplace transform, 265.

# INDEX 405

Resultant, bilateral Stieltjes, 248.
   unilateral Stieltjes, 83.
RIEMANN, 66, 71, 246.
Riemann-integrable functions, 190.
   integral, 4.
Riemann-Lebesgue theorem, 65, 68, 77, 83, 207, 235.
Riemann-Stieltjes integral, 4.
RIESZ, 105, 127.
Rolle's theorem, 168.

SAKS, 26, 324.
Saltus, 299.
SCHOENBERG, 105.
Second law of the mean, Bonnet's, 17.
Selection principle, the, 26.
Semidefinite quadratic forms, 136.
   sequences, 132.
      positive, 132.
Sequence, completely monotonic, 158, 269.
   minimal completely monotonic, 163.
   of functions of uniformly bounded variation, 31.
   positive, 127, 159, 266.
   positive definite, 132.
      semidefinite, 132.
   semidefinite, 132.
Series of absolutely monotonic functions, 151.
SHOHAT, 138.
$\sigma_a$, 46, 47.
$\sigma_a - \sigma_c$ for Dirichlet series, 49.
$\sigma_a - \sigma_c$ for the Laplace transform, 47.
$\sigma_c$, 37, 42, 43.
$\sigma_u$, 51, 52, 53.
Singular integral, 341.
Singularities of the generating function, 58.
Singular point of the Stieltjes transform, 336.
Slowly decreasing functions, 209.
   oscillating functions, 209.
Solution of an integral equation, 387.
Special Tauberian theorem, 209.
Step-function, 6.
STIELTJES, 125, 126, 325, 338.
Stieltjes integral, absolutely convergent improper, 15.
   as a Lebesgue integral, 11.
      series, 10.

Stieltjes integral, change of variable of, 19.
   conditions of integrability for the, 24.
   equation, 387.
   integration by parts for the, 8.
   linear functional as, 105.
   Tauberian theorem for the, 213.
   variation of the indefinite, 20.
   integrals, 3.
      as infinite series, 22.
      convergent improper, 15.
      divergent improper, 15.
      existence of, 7.
      improper, 15.
      inequalities for, 9.
      integration with respect to an indefinite, 12.
      iterated, 25.
      laws of the mean for, 16.
      properties of, 8, 12.
   integration of series, 10.
   kernel, general representation theorem for, 355.
   integral equation with, 387.
   iterates of the, 259, 262.
   moment problem, 136.
      not unique, 125.
      sufficient condition for the solubility of the, 140.
   resultant, 83.
      at infinity, 86, 247.
      bilateral, 249, 250.
      of indefinite integrals, 256.
      variation of, 85.
   transform, 325.
      Abelian theorems for the, 183.
      analytic character of, 328.
      complex inversion formula, 338.
      conditions $A$ for, 361.
      conditions $B$ for, 369.
      conditions $C$ for, 372.
      conditions $D$ for, 374.
      elementary properties of the, 325.
      general representation theorems for the, 360.
      inversion operator for the, 345, 347.
      jump operator for, 351.
      region of convergence of, 326.

Stieltjes transform, relation to the Laplace transform, 334.
  singular at the origin, 336.
  singular point of, 336.
  Tauberian theorems for the, 198.
  uniqueness of, 336, 349.
  variation of $\alpha(t)$ for, 353.
  transforms, asymptotic properties of, 329.
Stirling's formula, 95, 227, 231, 281, 352.
Stolz region, 240.
Stronger method of summability, 121.
Subdivision, 3.
Sufficient condition for the solubility of the Stieltjes moment problem, 140.
Summability by a matrix, 114.
  $(C,p)$, method of, 121.
  equivalent methods of, 121.
  of bilateral Laplace transform, 244.
  of divergent integrals, 75.
  stronger method of, 121.
Summable $(C,1)$, 79, 244.
SZEGÖ, 168, 169, 172, 229.

TAMARKIN, 21, 63, 113, 163.
TAUBER, 186, 188.
Tauberian theorem, 321.
  complex variable, 233.
  for the Stieltjes integral, 213.
  one-sided, 215.
  special, 209.
  Wiener's general, 212.
  theorems, 180, 185.
  for the Stieltjes transform, 198.
Taylor's formula, 146, 179, 278.
  series, 267.
  theorem, 177, 193.
TITCHMARSH, 6, 81, 97, 314, 338, 341, 372, 387, 389.
TOEPLITZ, 115.
Toeplitz's theorem, 115, 119.
Total variation, 6.
Transformation, Hausdorff, 118.
True max, 34, 316.
Type, maximal, 94.
  minimal, 94.
  normal, 94.
  of an entire function, 94, 179.

Uniform convergence, Laplace method, 283.
  of the Laplace integral, 50, 53.
Uniformly bounded variation, 31.
Unilateral Laplace transform, 35.
Unique, Hamburger moment problem not, 125.
Stieltjes moment problem not, 125.
Uniqueness of the bilateral Laplace transform, 243.
  determining function, 59.
  Fourier transform, 82, 204.
  Laplace transform, 80.
  moment function, 60.
  Stieltjes transform, 336, 349.
Upper bound, least, 4.

de la VALLÉE-POUSSIN, 224.
Variation of $\alpha(t)$ for the Stieltjes transform, 353.
  of the determining function, 299.
  indefinite Stieltjes integral, 20.
  Stieltjes resultant, 85.
  total, 6.

Wallis's product, 383.
WARSCHAWSKI, 21.
Weak compactness, 33.
WEIERSTRASS, 17, 57, 60.
Weierstrass approximation theorem, 153, 189.
Weierstrass's theorem, 106.
WIDDER, 18, 36, 82, 96, 144, 163, 168, 177, 259, 273, 275, 277, 293, 308, 325, 383, 384, 386.
WIENER, 25, 82, 84, 207, 212, 233, 318, 383.
Wiener's general Tauberian theorem, 212.
  theorem, 229.
  application of, 221, 222, 224.
  Pitt's form of, 210.
Wigert, theorem of, 96.
WINTNER, 26, 84.

Zeta-function, non-vanishing of the, 229.